Foundations of Engineering Mechanics

Series Editors: V.I. Babitsky, J. Wittenburg

Foundations of Engineering Mechanics

Series Editors: Vladimir I. Babitsky, Loughborough University, UK
Jens Wittenburg, Karlsruhe University, Germany

Further volumes of this series can be found on our homepage: springer.com

Slivker, V.I.
Mechanics of Structural Elements, 2007
ISBN 978-3-540-44718-4

Elsoufiev, S.A.
Strength Analysis in Geomechanics, 2007
ISBN 978-3-540-37052-9

Awrejcewicz, J., Krysko, V.A., Krysko, A.V.
Thermo-Dynamics of Plates and Shells, 2007
ISBN 978-3-540-37261-8

Wittbrodt, E., Adamiec-Wojcik, I., Wojciech, S.
Dynamics of Flexible Multibody Systems, 2006
ISBN 3-540-32351-1

Aleynikov, S.M.
Spatial Contact Problems in Geotechnics, 2007
ISBN 3-540-25138-3

Skubov, D.Y., Khodzhaev, K.S.
Non-Linear Electromechanics, 2007
ISBN 3-540-25139-1

Feodosiev, V.I.,
Advanced Stress and Stability Analysis Worked Examples, 2005
ISBN 3-540-23935-9

Lurie, A.I.
Theory of Elasticity, 2005
ISBN 3-540-24556-1

Sosnovskiy, L.A.,
TRIBO-FATIGUE · Wear-Fatigue Damage and its Prediction, 2005
ISBN 3-540-23153-6

Andrianov, I.V., Awrejcewicz, J., Manevitch, L.I.
Asymptotical Mechanics of Thin-Walled Structures, 2004
ISBN 3-540-40876-2

Ginevsky, A.S., Vlasov, Y.V., Karavosov, R.K.
Acoustic Control of Turbulent Jets, 2004
ISBN 3-540-20143-2,

Kolpakov, A.G.
Stressed Composite Structures, 2004
ISBN 3-540-40790-1

Shorr, B.F.
The Wave Finite Element Method, 2004
ISBN 3-540-41638-2

Svetlitsky, V.A.
Engineering Vibration Analysis - Worked Problems 1, 2004
ISBN 3-540-20658-2

Babitsky, V.I., Shipilov, A.
Resonant Robotic Systems, 2003
ISBN 3-540-00334-7

Le xuan Anh,
Dynamics of Mechanical Systems with Coulomb Friction, 2003
ISBN 3-540-00654-0

Nagaev, R.F.
Dynamics of Synchronising Systems, 2003
ISBN 3-540-44195-6

Neimark, J.I.
Mathematical Models in Natural Science and Engineering, 2003
ISBN 3-540-43680-4

(Continued after index)

V.K. Astashev · V.I. Babitsky

Ultrasonic Processes and Machines

Dynamics, Control and Applications

Translated by K. Khusnutdinova

With 183 Figures and 5 Tables

Springer

Series Editors:

V.I. Babitsky
Department of Mechanical Engineering
Loughborough University
Loughborough LE11 3TU, Leicestershire
United Kingdom

J. Wittenburg
Institut für Technische Mechanik
Universität Karlsruhe (TH)
Kaiserstrasse 12
76128 Karlsruhe
Germany

Authors:

Vladimir K. Astashev
Kirovogradskaya ul 8-5-116
117587 Moscow
Russia

Vladimir I. Babitsky
Department of Mechanical Engineering
Loughborough University
LE11 3TU Loughborough, Leicestershire
United Kingdom
v.i.babitsky@lboro.ac.uk

Translator:

Karima Khusnutdinova
Outwoods Drive 24
LE11 3LT Loughborough
United Kingdom

ISSN print edition: 1612-1384

ISBN 978-3-540-72060-7 Springer Berlin Heidelberg New York

Library of Congress Control Number: 2007926121

Springer is a part of Springer Science+Business Media

springer.com

Typesetting: by the author and Integra using Springer LaTeX package
Cover Design: deblik, Berlin

Printed on acid-free paper SPIN: 10818552 42/3100/Integra 5 4 3 2 1 0

To our lifelong collaboration and friendship

Preface

I don't mind your thinking slowly, I mind your publishing faster than you think.

Wolfgang Pauli (1900-1958)

Technologies that use high frequency (ultrasonic) vibration to intensify processes are gaining wide recognition in scientific and industrial environments. By superimposing high frequency vibration, the basic mechanical behaviour of many processes and materials is seen to be transformed. This leads to the development of new machines and processes with advanced characteristics. Despite the fact that ultrasonic technology has been employed for many years, there is no generalised understanding of ultrasonic machines and processes. Their design and development has mainly been achieved using heuristic methods based on linear acoustical considerations.

This book is intended to bridge the gap between the theory and practical use of ultrasonic technology. It presents generalised foundations for the dynamics and control of ultrasonic processing systems. The main concept presented is to consider ultrasonic systems as special kinds of vibratory machines that function by exploiting nonlinear dynamic processes. This assumes coupled considerations between the ultrasonic vibration's influence on the processes, and the consequence of the transformed processing loads on the excitation and control of the working tools' vibration.

Analysis is conducted in a unified manner and is based on structural and frequency methods that have become well established in engineering practice. These methods are adjusted by the authors for the application to nonlinear ultrasonic systems.

As a processing machine, the ultrasonic system exemplifies an electro-acoustical device with an ill-defined loading. For this reason, analytical expressions that accurately reflect an ultrasonic process' qualitative features are built up step by step into final solutions that permit a clear physical interpretation of a machine's vibrating action. In addition, experimental data and simulations are presented that confirms the basic effects and conclusions.

Recommendations for the design and tuning of ultrasonic systems are an important element for all of the considerations in this book. They are supported by useful mathematical relationships between structural design parameters and mechanical characteristics of the materials being treated.

The main applications of the ultrasonic processing systems being considered by this book are to the processing of solids. These processes include: ultrasonic erosion machining, ultrasonic hardening and welding, ultrasonically assisted plastic deformation and ultrasonically assisted turning and drilling. As a result of the recent increased industrial interest in ultrasonically assisted machining, analysis and experiments conducted on ultrasonically assisted turning and drilling are separated into a special chapter facilitating independent study.

The book is intended for engineers, researchers and advanced students who are involved in the study, design and application of ultrasonic processing systems. It is also proposed that lecturers and students should find here useful examples of practical analysis and application to nonlinear dynamical systems. In order to study this book, a reasonable knowledge of engineering based general mathematics and theory of mechanical vibration will be adequate.

This book is the outcome of many years of collaboration between the authors who have been involved with the research and development of ultrasonic processing systems. It originated in Russia, and with the support of Loughborough University and British enterprises, was extended to an international scale. Within the text, and the up to date bibliography, results from different research groups from all over the world are reflected.

Research associates and PhD students in Russia and Britain contributed to the investigation and development of ultrasonic processing systems at different stages. Their results have enriched the contents of this book. Contributors include M. Hertz, A. Kalashnikov, N. Makarov, A. Meadows and H. Wijesundara. A. Mitrofanov prepared additionally section 5.5 of Chapter 5. N. Halliwell supported greatly ultrasonic research in Loughborough and V. Silberschmidt joined this research activity.

The book's translator, K. Khusnutdinova, P. Thomas, who proof read the English translation, M. Dimenkov and I. Sokolov, who formed the final electronic version of the manuscript, were extremely co-operative during all stages of the book preparation.

The authors would like to thank all of the individuals mentioned above, and the staff of Springer-Verlag.

Vladimir Astashev and Vladimir Babitsky
Moscow-Loughborough

Contents

Introduction

Take care of the sense and the sounds will take care of themselves.

Lewis Carroll (1832-1898): Alice's Adventures in Wonderland

The recent development of modern hi-tech industries has given rise to the creation of a whole range of new materials. These include high-strength, stainless and heat resistant steels and alloys, titanium, ceramics, composites and other nonmetallic materials. These materials may not be suitable for traditional methods of machining due to the chipping or fracturing of the surface layer, or even the whole component, and results in a poor product quality.

Similarly, the creation of new materials often highlights some problems unsolvable in a framework of traditional technologies. In certain cases these problems are caused by the construction of the object and the requirements particular to it. As an example, in microelectronics, it is often necessary to connect some components without heating them or adding any intermediate layers. This forbids the use of traditional methods such as soldering or welding.

Many of these and similar problems can be successfully solved using ultrasonic technologies. The working process of an ultrasonic machine is performed by subjecting its tool to a combination of two motions. A driving motion program is required to shape the workpiece. A high-frequency (ultrasonic) vibration of specific direction, frequency and intensity is then superimposed. The construction of the machine and its elements depends critically on the process being performed by the tool. A transducer generates vibration which is then transmitted to the tool via a vibration system, often with a change of direction and amplitude.

Ultrasonic machines belong to the general class of vibration machines, but they form a special group for the following reasons.

The first reason is determined by peculiarities in the behaviour of materials and media in an ultrasonic field. This has been demonstrated in numerous experiments. Among these peculiarities is the drastic change of elastic-plastic characteristics and rheological properties observed in experiments, these include fragility, plasticity and viscosity. For example, dry friction in the region of contact between two surfaces transforms, under the influence of ultrasonic vibration, to viscous friction and a substantial decrease in the yield limit can be seen during the deformation of a specimen in the presence of ultrasonic vibration. These effects have been repeatedly observed in experiments and are now widely used in machines for plastic deformation, the drawing of wires and pipes, ultrasonic machining of brittle materials, ultrasonically assisted cutting and the welding of metals and plastics amongst others. There is not yet a proper theoretical explanation of these trends and one of the aims of this book is to fill this gap. The explanation of the observed phenomena is given on the basis of proposed nonlinear models of ultrasonic processes.

The second reason is due to the peculiarities in the construction of major parts of the machine. The main components are usually formed using vibrating bar systems consisting of heterogeneous sections and using waveguides. For this reason, to describe vibration, the main elements of the machine are modeled by systems with distributed parameters and are described by partial differential equations. The tool-workpiece interaction leads to a nonlinearity in the vibration system in its operating conditions.

Due to their construction, vibration systems are characterized by a high Q-factor. Therefore, ultrasonic machines can only work effectively in resonant regimes, supplying the tool with a sufficiently large amplitude of oscillation. In practice, ultrasonic machines are adjusted to resonance mostly under idle running, i.e. in the absence of operating loads, assuming that the influence of the loads in operating conditions is negligible. However, as it was established by the authors of this book and their colleagues, this assumption is incorrect. It is shown that loads lead, not only to a change in the resonant frequences of the vibrating system, but cause specific nonlinear distortions of its amplitude-frequency characteristics. Estimating the influence of the loads on the dynamic characteristics of the vibration systems used in ultrasonic machines is the second aim of this book.

New dynamic models of ultrasonic processes are suggested in order to achieve the aims of this book. These models allow us not only to give clear physical explanations of experimental phenomena, but also, being a part of the design model of the ultrasonic machine, to determine its main dynamic properties under operating loads. As a result, the main parameters of a process, such as its average speed (characterizing the output capacity), and its efficiency are expressed through the main constructional parameters of the machine.

The third and final aim of this book is to discover the methods of excitation and automatic maintenance of the most effective resonant regimes of ultrasonic machines under hugely variable loads. This problem is effectively solved using the autoresonant excitation of vibration which are maintained with the help of a special feedback system.

The book consists of 5 Chapters and is arranged as follows.

The physical foundations of ultrasonic processes are considered in Chapter 1. The arrangements of a number of processes are described and the literature on their experimental characteristics is discussed. Among these processes is the ultrasonic machining of brittle materials, surface hardening of metals, plastic deformation, drawing of tubes, ultrasonic cutting, welding of metals and plastics, etc. We discuss the effects revealed in experimental studies of these processes and point out that superimposing ultrasonic vibration leads to a decrease in the static forces, required for plastic deformation, fracture and to overcome friction.

The ultrasonic processes are studied using rheological models, reflecting the materials real elastic, viscous and plastic properties. The influence of ultrasonic vibration on plastic deformation and dry friction is considered. Possible regimes of interaction between the vibrating tool and the workpiece are described and the efficiency of vibro-impact interaction is discussed. A dependence of the average speed of the process on applied static forces and parameters of vibration is revealed as a result of the modeling of the processes. These theoretical results are then compared with experimental data. Having validated the theoretical models, a clear understanding is reached of the effect that ultrasonic vibration has on the process of plastic deformation and friction.

Chapter 2 deals with the dynamic characteristics of different ultrasonic machines. It begins with a description of different machine arrangements and suitable drivers such as magnetostrictive and piezostrictive transducers. Concentrators, i.e. waveguides are also considered. These are used to transmit ultrasonic vibration to the tool with an increase in amplitude. The generalized models of the machines considered in this book consist of an electromechanical chain of coupled elements: a vibration exciter, a mechanism suitable for the transmission of this vibration and a load. It is shown, that the dynamics of the machine can be analyzed using transfer functions for the exciter and operators of dynamical compliance and stiffness for the waveguides and loads. The dynamical compliances (receptances) are found for bar systems performing longitudinal and bending vibrations. With their help, resonant and antiresonant frequencies and amplitudes of the system, and parameters of different concentrators are found. The dynamic characteristics of magnetostrictive and piezostrictive transducers are considered. The possibility of using simplified dynamical models is justified by analysis.

Chapter 3 is devoted to the nonlinear processes in ultrasonic systems. Nonlinear dynamic characteristics of different processes are constructed. Elastic and dissipative components of a nonlinear load are found using the method of harmonic linearization. Methods for their simplification are suggested.

For systems performing longitudinal and bending vibrations and interacting with a concentrated nonlinear load, the periodic motions are found. Amplitude-frequency characteristics of vibrating systems are constructed, special nonlinear dynamical effects are revealed and the stability of the constructed solutions is investigated.

Periodic vibro-impact processes in a system of two coupled waveguides connected with a gap, tension, or compressed by a constant force are investigated. The amplitude-frequency characteristics are studied. It is shown, that nonlinear loads cause specific nonlinear distortions of resonance curves leading to different nonlinear effects. These include the appearance of zones of ambiguity in amplitude-frequency characteristics, the pulling of vibration into resonant zones with respect to frequency and amplitude, jumps of amplitude caused by change of frequency and the hard excitation of resonant vibration.

The vibrational displacement of a viscoelastic bar, in a medium with dry friction, under a constant force, is also considered. The bar is excited under its resonant conditions by a force acting in one of its cross-sections. The force due to dry friction is uniformly distributed along the bar. The average speed of penetration of the bar into the medium, and its dependence on the constant force and parameters of vibration are established. An estimation of the mutual influence of fast (vibrational) and slow (using an average speed) motion in the bar is given.

Chapter 4 is devoted to the problems associated with the resonant tuning of ultrasonic machines, under load, with the aim of increasing the efficiency of ultrasonic tools. Using a rheological model of ultrasonic cutting, the dependence of the productivity of the tool on its dynamic characteristics and the main constructional parameters of the machine is established. The results are compared with experimental data. The limiting capabilities of different tuning methods in vibration systems are studied. The results of experiments are discussed.

Nonlinear theory of ultrasonic concentrators working under nonlinear elastic-dissipative loads is presented. For an ultrasonic machining process, the dependence of the cutting speed and the elastic and dissipative components of the load on the concentrator vibration amplitude are determined. The nonlinear load is accounted for.

It is demonstrated that an increase in the output capacity, together with a decrease in energy consumption, can be achieved using vibrating systems with higher Q-factors and more precise resonant tuning under load. The creation of a special feedback system for the autoresonant excitation of vibration allows us, theoretically, to use these dynamic properties on a machine designed for full scale production. The possibility of creating a feedback system is discussed, dynamics of a system for the autoresonant excitation of vibration in ultrasonic machines is investigated, conditions for self-excitation of vibrations are analyzed and the existence and stability of periodic regimes are studied. It is shown, that an autoresonant system automatically supports the most effective resonant regimes under significantly varying machine parameters and operating loads.

The book finishes with Chapter 5 considering ultrasonic assisted cutting. A rheological model of the process is suggested. An experimentally established feature of ultrasonic vibration cutting is a decrease in cutting force. A theoretical explanation for this phenomenon is given. The dynamics of the tool used in ultrasonic cutting is investigated, and its amplitude-frequency characteristics are constructed. A system for the autoresonant excitation of cutter vibration is described. The Chapter is completed with materials on experimental investigation of ultrasonically assisted machining and its finite element analysis.

In the bibliography list we tried to collect the main publications on the subject obtained in different countries and often published in different languages.

The content of this book is divided into chapters, sections and subsections. The formulae, figures and tables are marked by double numbers, the first being the number of the chapter, the second being the number of the formula, figure or table within that chapter.

1

Models of ultrasonic processes

All art is but imitation of nature.

Mareus Annaeus Seneca (55BC-40AD)

1.1 Ultrasonic processes

1. The working process of an ultrasonic machine is performed when its tool interacts with the workpiece or medium to be treated. The tool is subjected to vibration with a specific direction, frequency and intensity. The vibration is generated by a transducer and is transmitted to the tool using a vibration system, often with a change in direction and amplitude.

The construction of the machine and its components is dependant on the process being performed by its tool. Let us consider some of these processes.

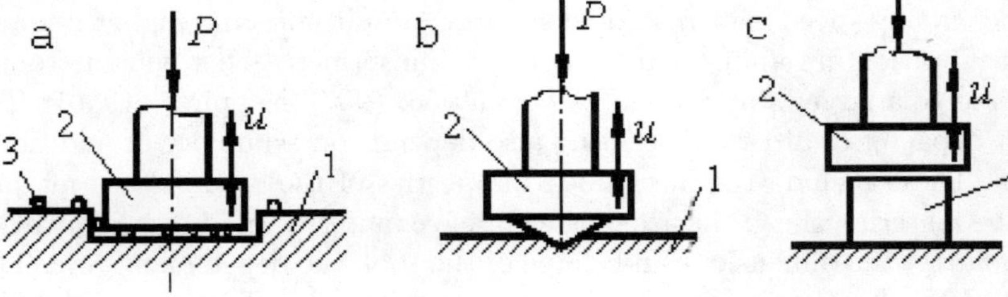

Fig. 1.1.

Fig.1.1,a shows the ultrasonic erosion process (Markov (1962, 1966, 1980), Rosenberg et al. (1964)) used to machine hard, brittle materials. The workpiece 1 is placed under the face of the tool 2 which is subjected to high-frequency vibration perpendicular to the surface being machined. Abrasive slurry is conveyed to the working zone between the face of the tool and the surface being machined. The tool moves towards the workpiece and is subjected to a static driving force P. Repetitive impacts of the tool on the grains of the abrasive material, falling from the slurry onto the treated surface, lead to the fracture of the workpiece material and to the creation of a cavity with the shape mirror-formed of the tool.

Since its invention (Farrer (1948)), the method of ultrasonic erosion machining has won recognition and is now widely used for cutting materials which are difficult to cut using traditional methods of machining. These materials include glass, ceramics, semiconductor materials, ferrite, diamond, hard alloys, etc. The development of ultrasonic machining is largely attributable to numerous experimental investigations. These investigations were mainly based on estimations of how the cutting speed is influenced by different factors. The main results of these investigations are discussed, choosing from a huge number of papers, only those which are essential to our discussions. Among these are papers on the physical nature of the fracture of materials during ultrasonic manufacturing.

Using high-speed filming, it was established (Averianova et al. (1962), Averianova & Milovidov (1964), Rosenberg & Kazantsev (1959*a*,*b*)) that material fracture only results when the tool impacts the surface being treated via the particles of an abrasive material. An investigation of the forces showed (Astashev & Sakaian (1967), Blank (1961), Kazantsev (1963*a*), Nishimura & Shimakawa (1959)) that the motion of the tool is a periodic vibro-impact process in which workpiece impacts take place with the frequency of the tool's vibration. The amplitude of the impact force transpired to be a main factor influencing the cutting speed (Kazantsev (1963*b*), Nishimura & Shimakawa (1959)). The character of an impact interaction, for example the form of the impulse, its duration, or the maximum value of the force, is dependent on the real cutting conditions.

The cutting speed is increased when larger amplitudes and higher frequencies of vibration are applied to the tool. It is also increased when the tool is subjected to a larger static force (Goetze (1956, 1957), Neppiras (1956)). The output capacity of ultrasonic cutting also depends on a number of additional factors. These include the mechanical properties of the workpiece, tool, and abrasive material; the grain size of the abrasive material and its concentration in the slurry; the dimensions and shape of the tool; the depth of cut, etc. This is caused by the physical nature of the cutting process. Brittle materials are more suitable for ultrasonic machining, and cutting speed increases with the decrease in the materials hardness (Hartley (1956)). Not only does the material of the workpiece wear, but also that of the abrasive grains. To stabilize the cutting conditions, a frequent change of the abrasive material in the working zone is necessary (Kazantsev et al. (1966)).

An important peculiarity of the ultrasonic erosion process must be noted. For material fracture to occur, the forces acting between the tool and the workpiece should be sufficient to make the stresses in the vicinity of abrasive grains higher than the ultimate stress of the material. The cutting process takes place even when a very small static force is applied to the tool. Therefore, the insertion of ultrasonic vibration in to the working zone essentially reduces the static force necessary to overcome the material's ultimate stress.

2. The decrease in static force necessary to fulfill the process characterizes the majority of ultrasonic processes where a fracture or plastic deformation of the treated medium takes place. Fig.1.1,b shows the scheme of a workpiece 1 surface hardening process (Markov (1980)), it results from a plastic deformation in the surface layer of the metal under the high-frequency vibration of the tip 2. Here again, the process requires a relatively small static force P. As a general rule, the static force required to produce a plastic deformation in a material is reduced by an order of magnitude with the application of ultrasonic vibration. If the vibration at the workpiece-tool interaction point constitutes a periodic vibro-impact process, similar to that of ultrasonic cutting (Fig.1.1,a), the decrease in the static pressure required is more pronounced and the hardening process is greatly facilitated.

A similar diagram represents the forging, riveting or upsetting process (Fig.1.1,c), which is subjected to ultrasonic vibration (Severdenko et al. (1973, 1976)). Some results are now considered, that are obtained from the plastic deformation of a specimen 1, subject to a static load P and high-frequency tool 2 vibration. The experiment is conducted on the machine shown in Fig.1.2 and is described in Izumi et al. (1966a,b).

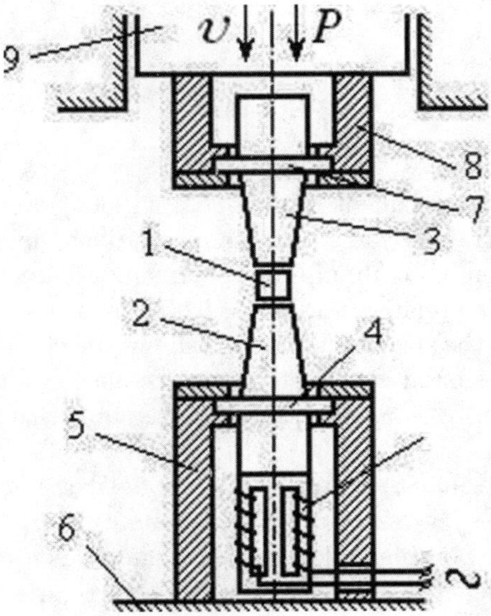

Fig. 1.2.

The specimen 1 is placed between the ends of identical bars 2 and 3. Using the shoulder 4, the bar 2 is connected with the housing 5, situated on the base 6 of a hydraulic test machine. The bar 3, using the shoulder 7 and the housing 8, is connected to the transverse beam 9 of the machine. The machine moves with a small velocity v which leads to the deformation of the specimen. Bar 2 is a waveguide, and transmits vibration to the working zone from the magnetostrictive transducer 10, which is being fed by an alternating current generator. The machine is equipped to measure the static load and strain experienced by the specimen. A control system is also used to stabilize the amplitude of vibration in bar 2 which is subjected to varying forces.

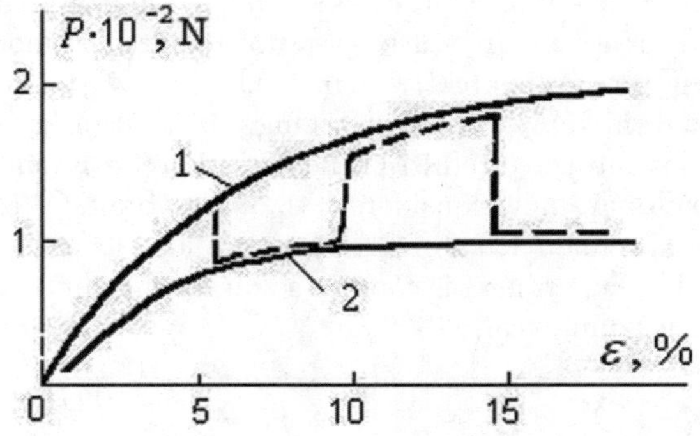

Fig. 1.3.

The results of the experiment (Izumi et al. (1966*b*)) are shown in Fig.1.3. Curve 1 shows a "strain – force" diagram, obtained for a static (with no vibration) compression of an aluminium specimen of diameter $d = 14mm$ and height $H = 20mm$. The relative strain $\varepsilon = h/H$ is shown on the x-axes, where h is the shortening of the specimen. Curve 2 shows the change in static force P whilst compressing a similar specimen in the presence of ultrasonic vibration. In both cases the compression process was carried out with the same constant speed $v = 0.5mm/min$.

Note that in experiments Izumi et al. (1966*a*,*b*) the vibration employed was at a frequency of $f \approx 22kHz$, which is usual for an ultrasonic process. The amplitude at the working end of the waveguide was $a \approx 10\mu m$. Vibration at such a high frequency and small amplitude are indistinguishable by an unaided eye. The observer has the impression that the softening of the material and the change of its elastic-plastic properties take place under the influence of ultrasound. It is important that such a softening can only be observed in the presence of vibration. The results of an experiment investigating the shortening of specimens under periodically imposed ultrasonic vibration are shown in Fig.1.3 by a dashed line and confirm the previous observation.

These effects are confirmed in numerous experiments and are generalized in monographs Severdenko et al. (1973, 1976). Some typical properties of material behavior during a deformation processes with an ultrasonic influence are now considered. Every material has a threshold for the static loading or level of ultrasonic vibration that it can withstand. Exceeding this threshold with either static loading or with ultrasonic vibration will lead to the occurrence of residual plastic deformation. The application of ultrasonic vibration to the deformation zone always results in the decrease of the static load required for plastic deformation to occur. This static force P depends on the loading speed v. The compliance of the material with respect to static force becomes greater with an increase in vibration intensity. In some experiments, the static forces were decreased by more than 85 %.

The abnormal behavior of materials when subjected to a static load in the presence of an ultrasonic field gave rise to a number of hypotheses on the influence of ultrasound on materials (Severdenko et al. (1976)). Some hypotheses state that such a behavior is related to the influence of ultrasonic vibration on the distribution and mobility of dislocations within the internal structure of the material. Others state that the increase in the plasticity of materials in an ultrasonic field is linked to thermal effects caused by a dissipation of energy during the ultrasonic vibration of the specimen. However, a hypothesis of that kind does not explain experimental phenomena with sufficient accuracy. Although the proposed mechanisms can not be rejected completely, they only appear to play a secondary role in the processes under consideration. This statement is based on the fact that the frequencies of the order $10^4 - 10^5 Hz$, which are used in ultrasonic processes are far below the $10^8 - 10^{10} Hz$ range where vibration can significantly influence crystal lattices and dislocation lines (Nozdrev & Fedorischenko (1978)). Therefore, it is possible that the change in the structure of the material (Langenecker (1966)), and of the density of dislocations observed in individual crystallographic investigations of specimens treated using ultrasonic vibration, is in fact not a reason for the abnormal behavior of materials in the ultrasonic field, but a consequence of the plastic deformation.

Considering the influence of thermal effects, it must be noted that the decrease in static force required in the presence of ultrasonics takes place even when the environment is maintained at constant temperature (Langenecker (1966)). Moreover, experimental comparison of the influence of ultrasound and temperature showed that to achieve an equal ductility, the required thermal energy should be several times higher than in the ultrasonic case. Therefore, it is stated in Izumi et al. (1966*b*), Langenecker (1966) that the thermal effect can not be considered a main factor in the decrease of a material's yield point under the influence of ultrasonics.

Taking these statements into account, a different approach is suggested by us for the description of ultrasonic processes. It is based on the use of rheological models of materials (Reiner (1958)) and reflects their real elastic, viscous and plastic properties. Different aspects of such an approach will be considered in detail.

3. Practical application of ultrasonics to a number of processes using plastic deformation or the fracture of materials is advocated by the decrease in static load required as described above. Fig.1.4,a shows a sheet metal pressing process (Severdenko et al. (1973)) acting on a workpiece 1. The punch 2, vibrates in the direction of the static force P. This arrangement is similar to those shown in Fig.1.1,b,c.

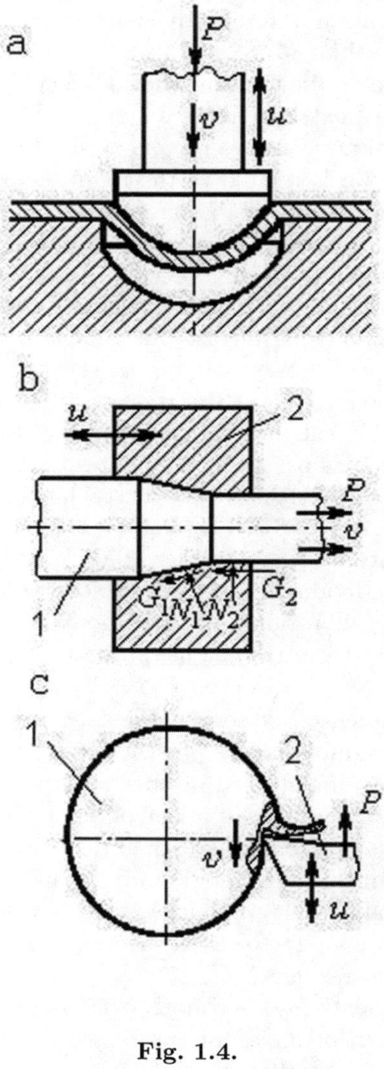

Fig. 1.4.

During the pressing process, the conditions of plastic deformation change as the geometry of the workpiece changes. The movement of the punch causes this, and it is an important feature of the pressing process. The punch is subjected to both deformation and friction forces. The frictional forces act in the direction tangential to the surface of the punch, opposite to the relative velocity between the punch and the workpiece. The friction forces can be essential to the general balance of forces acting on the punch, especially during the pressing of materials that resist deformation. As shown in experiments, excitation of vibration within a punch leads to a decrease in the static loads required to overcome frictional forces.

Frictional forces are essential in the bar drawing process which is used to manufacture wires and pipes. An arrangement of this type of process is shown in Fig.1.4,b.

The workpiece 1 is pulled through the hole of the drawing die 2. The workpiece is squeezed and is subject to a plastic deformation in the conical section of the orifice. The final sizing of the workpiece takes place in the cylindrical part. The forces acting within the deformation zone are shown in Fig.1.4,b, for a drawing process, in which a blank is pulled through a fixed die. N_1, N_2 are the forces of normal pressure determined by the elastic-plastic properties of the material, G_1, G_2 are dry friction forces,

$$G_1 = \vartheta N_1, \quad G_2 = \vartheta N_2,$$

where, ϑ is the coefficient of friction.

The required drawing force is given by,

$$P = N_1(\sin\theta + \vartheta\cos\theta) + \vartheta N_2, \tag{1.1}$$

where, θ is the angle of the conical section in the die.

As seen from equation (1.1), force P depends on the properties of the material being machined, the coefficient of friction, and the geometry of the die.

This statement also holds true for an ultrasonic drawing process (Nosal & Rimsha (1966), Robinson (1964), Severdenko et al. (1973, 1976)), in which the die is vibrated in the direction that the workpiece is pulled in. As in all of the processes considered earlier, superimposing ultrasonics in to the operating zone leads to a decrease in the static force required. Even in initial experiments (Robinson (1964)) of steel and copper wire drawing, the static force was reduced by up to 50 %.

Experiments to determine the main cause of the decrease in drawing force required are described in papers Nosal & Rimsha (1966), Robinson (1964). In Robinson (1964) a decrease in the metal's resistance to deformation is considered to be the main factor while in Nosal & Rimsha (1966) it is considered to be a decrease in external friction. In section 1.3 we shall show that each factor in the relation (1.1) may be significant and is dependant on the real conditions of the deformation process.

Due to the relative simplicity of realization, regulation and registration of the main parameters, ultrasonic drawing remains one of the most experimentally studied processes of ultrasonically assisted deformation. Experiments have revealed (Nosal & Rimsha (1966), Panov & Piskunov (1966)) that ultrasonic exposure only leads to decrease in static force when the drawing speed v is smaller than the amplitude of the die's vibrational speed

$$v < a\omega, \tag{1.2}$$

where a is the amplitude, and ω is the angular frequency of vibration.

The application of ultrasonics to drawing processes has transpired to be most effective when applied to high-strength materials. Normally, such materials are almost considered as unsuitable for plastic deformation due to the appearance of cracks and breaks at the exit of the drawing die (Goliamina (1979)). Conversely, the decrease in static force during ultrasonically assisted drawing allows an increase in the grade and the reduction or prevention of wire breakages when processing very plastic metals such as lead.

The properties described above are typical for all processes in which the tool is excited with ultrasonic vibration. They are also present in cutting processes when ultrasonic vibration is superimposed on to the cutting tool (Isaev & Anokhin (1961), Kumabe (1979), Markov (1962, 1966, 1980)). Fig.1.4,c shows a turning process in which a workpiece 1 is cut by a lathe tool 2. The cutting forces are mainly generated by the plastic deformation of the layer being cut, and on overcoming the external friction forces present on the working surfaces of the tool. When ultrasonic vibration is applied to the cutter, both of the main cutting force components are reduced. The result is most pronounced if the direction of vibration coincides with the direction of cutting velocity, as shown in Fig.1.4,c. Similar results are obtained for other types of metal machining, for example, in gear shaping.

The dependence of cutting force P on the cutting speed v is obtained in experiments (Isaev & Anokhin (1961)) and is shown in Fig.1.5. Curve 1 relates to traditional turning, while curve 2 is encountered when ultrasonic vibration is superimposed on to the cutting tool during turning. An important observation of these experiments is a "disappearance" of the cutting force for values of speed v close to zero. Here, of course, a constant component of cutting force is considered. This is either measured from the torque experienced by the workpiece, or by the strain experienced by the cutter during turning.

As the cutting speed is increased to the value $v = a\omega$, the cutting force monotonically increases up to the value that it would take in the absence of vibration. Such a relation between a constant force and the relative speed of displacement is typical for systems with dry friction, under the influence of vibration, within the frictional contact zone. Experimental data is described in Arkhangelsky (1963), Pohlman & Lehfeldt (1966). It must be noted however, that frictional forces are not the most important forces in the process under consideration. It is more likely that the forces due to plastic deformation dominate. We see, from experimental results, that the static force is

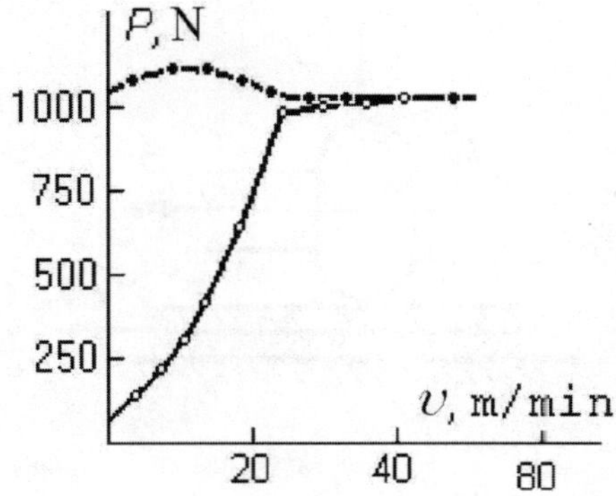

Fig. 1.5.

almost completely eliminated when speeds v are low in the processes of plastic deformation with the application of ultrasonics.

A physical mechanism for this phenomenon will now be modelled, and is based on real elastic-plastic properties of materials.

4. During the processes described above, friction usually deteriorates the normal working process and, therefore, represents an unfavourable factor. When accompanying the main processing loads in the ultrasonic welding of metals (Kuznetsov & Loschilov (1972), Nikolaev (1962), Nikolaev & Olshansky (1966)), however, friction is important for the formation of good quality welded joints (Mitzkevich & Pugachev (1979), Silin et al. (1962)).

Fig.1.6,a shows a process for welding metals ultrasonically. A welding tip, 1, is pressed on to the components to be welded, by a constant force N, and experiences vibrations in a plane perpendicular to this pressing force. Due to the vibration, contact surfaces become smoother and oxide films are removed as they are pressed together by the normal force. Work done to overcome friction results in the local heating of parts within the contact zone by up to 0.4-0.6 of the fusion temperature (Holopov (1972)). This all assists the formation of stable metallic bonds between the welded components (Holopov (1972), Weare et al. (1960)).

The mechanism of joint formation in ultrasonic welding is similar to the metal adhesion mechanism found in friction welding.

During the formation of a welded joint, the tip may be moved in the direction of vibration (Fig.1.6,a) or perpendicular to it. In both cases, the required static force P is less than that in the absence of vibration ($P < \vartheta N$). Again, it is seen that the static force required to overcome friction is decreased.

Ultrasonic welding can be used to connect components of the same metal or different metals, including those unsuitable for traditional types of welding. Combinations of materials suitable for ultrasonic welding are described

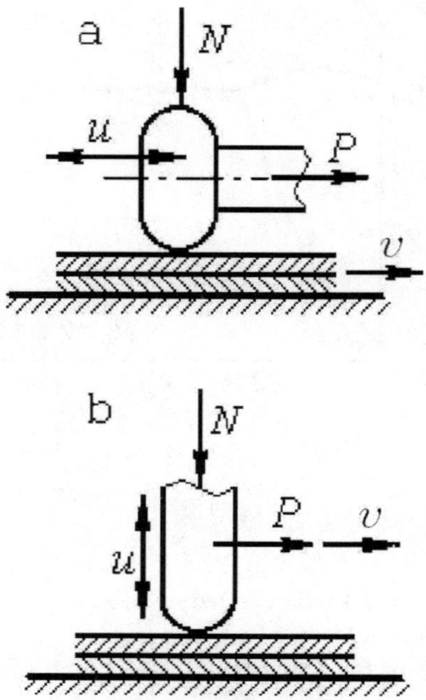

Fig. 1.6.

in Daniels (1965), Drews (1970), Holopov (1972), Agranat (1974). In a joint formed using this method, a metal remains almost unchanged, i.e., almost no structural changes, or changes of its physical and mechanical properties take place. The welded joint demonstrates high strength and plasticity, and low level of residual stresses. All these factors promote the extensive use of ultrasonic welding in the instruments industry, electronics, and microelectronics.

An alternative arrangement (Fig.1.6,b) for the welding of plastics and synthetic fabrics (Kuznetsov & Loschilov (1972), Nikolaev (1962), Nikolaev & Olshansky (1966)) differs from that used for metals. In this case, a force P presses the welding tip on to the surface of the components to be connected. The direction of tip vibration coincides with the direction of action of the pressing force. This is similar to the processes of hardening and plastic deformation considered previously. Localised heating takes place in the materials being connected (caused by periodic deformation), when the welding tip is vibrated in the frequency range 20 - 100 kHz. The heating is caused by internal friction energy losses and is localized because of the low heat conductivity of the polymers being welded. Finally, a softening of the materials occur in the welding zone and their plastic deformation results in a strong joint. During the formation of the welded joint, the material is conveyed to the welding tip to produce a continuous joint. Again, the drawing force is smaller than the frictional force present in the absence of vibration.

We have now considered a few examples of when ultrasonic vibration is used in industrial processes. These examples do not describe all of the useful applications of ultrasonic energy (Graf (1975), Goliamina (1979)). Alternative applications include ultrasonic crystallization and dispersion, ultrasonic cleaning, degassing and flotation (Abramov (1972), Gershgal & Freedman (1976), Agranat (1974)), ultrasonic echolocation, non-destructive flaw detection and microscopy. This is an incomplete list of the applications of ultrasonics in metallurgy, engineering, the food industry, the instruments industry, medicine and biology.

Such a diversity of applications occurs because of the specific physical properties of ultrasonic emission. These include the simplicity of ultrasonic vibration excitation, the transmission of the vibration to the medium being treated (using waveguides), the possibility of focusing and concentrating the ultrasonic energy in a narrow working zone, and the high directionality and penetrability of ultrasonic radiation.

Mechanisms for the influence of ultrasonics on solids, fluids, gases and their mixtures are different as well. The processes considered above are characterized by a strong mutual influence between the source of ultrasonic vibration and the medium being treated. Numerous aspects of such interaction are studied in this book.

1.2 The influence of ultrasonics on the processes of plastic deformation

1. An approach, based on rheological models (Reiner (1958)) of materials, is suggested in this book. This reflects their real elastic, viscous and plastic properties, and aids the description of ultrasonic processes. Such an approach allows us to explain the physical mechanisms of the effects of ultrasonic influence on the processes of plastic deformation as described above (Astashev (1983)).

In order to do this, a "strain-force" diagram is considered (Fig.1.7) for the static compression of a specimen. There are regions of elastic and plastic deformation (Rabotnov (1979)). In an elastic zone, if the load P does not exceed an elastic limit D, the force P and the strain h are related by a one-to-one relation $P = P(h)$. We shall assume, that in this zone, the material behaviour obeys Hook's law

$$P = k_0 h = \varepsilon E S, \tag{1.3}$$

where h is the deformation of the specimen, $k_0 = ES/H$ is its static stiffness; S and H are the cross section and height of the specimen respectively; E is the elastic modulus of the material and $\varepsilon = h/H$ is the relative deformation.

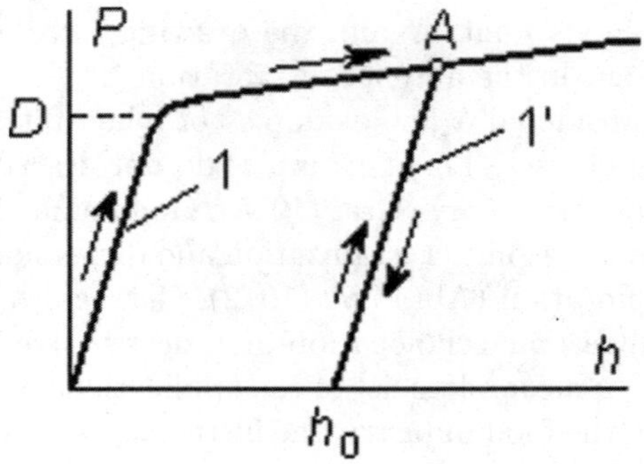

Fig. 1.7.

In a material's plastic zone in which the material experiences both stress and strain, the materials reaction to the process depends on whether the deformation increases or decreases: curve 1 corresponds to an increase and curve 1' corresponds to a decrease in deformation starting from point A. Curve 1' is practically parallel to the initial part of curve 1 as shown in Fig.1.7 by arrows. After a complete unloading there is a residual plastic deformation h_0 and during reloading, the material is in an elastic state, determined by a curve 1', until the force P reaches the value D_1, corresponding to the point A, i.e. $P = D_1$. Further loading causes a plastic deformation along curve 1.

The diagram of loading shown in Fig.1.7 is typical for an elastic-plastic material that experiences work hardening, i.e. the force in the zone of plasticity rises as the deformation increases. During a cyclic loading of such a material, its elastic limit rises from one cycle to the next and depends on the material's complete history of loading. (The process of vibrational hardening (Fig.1.1,b) is based precisely on this effect). In the case of an ideal elastic-plastic material, a plastic deformation takes place when the material is subjected to a constant force $P = D$ (dashed line in Fig.1.7), and there is no hardening.

It is convenient to describe the mechanical behaviour of a real material using idealized models consisting of elementary parts (Mase (1970)). The model depicted in Fig.1.8,a consists of an elastic element k_0 and a pair of dry friction components D connected in series. Under a static loading, this gives the characteristic of an ideal elastic-plastic deformation (Fig.1.8,b). Another model (Fig.1.8,c), consists of two elastic elements k_1, k_2 and a pair of frictional components D. This leads to the characteristic of an elastic-plastic deformation with hardening (Fig.1.8,d), $k_0 = k_1 + k_2$. It must be noted that similar models are successfully used for the investigation of vibration in elastic-plastic solids (Palmov (1998)) and also in impact phenomena.

2. Let us consider the process of material deformation under vibrational loading using the models described above. First we will consider the behaviour of an ideal elastic-plastic material.

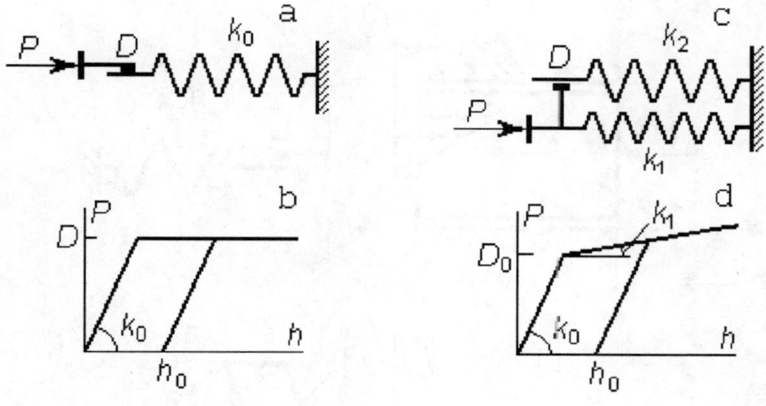

Fig. 1.8.

Let the deformation of the material (Fig.1.9,a) result from its interaction with tool 1. The vibrational motion of tool 1 with respect to the carriage 2 obeys the law

$$u^0 = a \sin \omega t,$$

where a is the amplitude and ω is the angular frequency. Carriage 2 is subject to a constant static force P and, depending on the conditions of deformation, either moves with a constant speed v, or remains in equilibrium ($v = 0$). We shall consider a general case in which the full motion of the tool is described as follows

$$u(t) = vt + u^0(t) = vt + a \sin \omega t. \tag{1.4}$$

Similar to that of static loading, the following experiment is carried out: the force P is slowly increased and the movement of the carriage is tracked. (This is how the characteristics shown in Fig.1.3 were obtained). It is assumed that, for every given value of P we have a steady-state periodic regime of deformation. This assumption allows the interaction between the tool and the material to be considered during one period of tool vibration alone. The following notation (Fig.1.9) is now introduced: Δ is the distance between the centreline of the tool's vibration and the height of the unstrained specimen; $h = u_m - \Delta$ is the displacement of the carriage with respect to the specimen and is equal to the maximum deformation over a period; f is the force of interaction between the tool and the specimen. Values of $\Delta > 0$ correspond to the adjustment of the tool in which an initial clearance is present, while $\Delta < 0$ corresponds to an initial interference between the material and the tool.

It is assumed that the specimen being deformed is short. An elastic wave will therefore propagate along it in a period of time smaller than the period of loading, i.e. $H/c \ll T = 2\pi\omega$, where c is the speed of sound in the

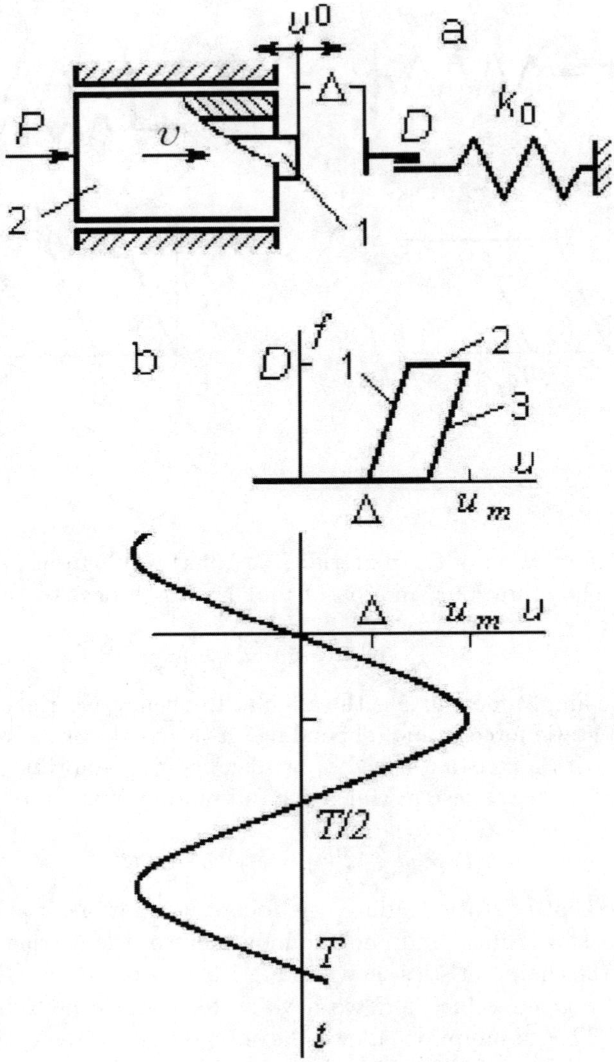

Fig. 1.9.

material. Due to this assumption, the dynamical characteristic $f = f(u, \dot{u})$ determines the relation between the force f, displacement u and speed $\dot{u}$ of the tool (Fig.1.9,b) and coincides with the diagram of static loading (Fig.1.8,b) up to the notations and the constant shift Δ. The dynamic characteristic is represented as follows

$$f = f(u,\dot{u}) = \begin{cases} 0 & u \le \Delta, \dot{u} > 0, \\ k_0(u-\Delta) & \Delta \le u \le \Delta + \frac{D}{k_0}, \dot{u} > 0, \\ D & \Delta + \frac{D}{k_0} \le u \le u_m, \dot{u} > 0, \\ D + k_0(u - u_m) & u_m - \frac{D}{k_0} \le u \le u_m, \dot{u} < 0, \\ 0 & u \le u_m - \frac{D}{k_0}, \dot{u} < 0, \end{cases} \tag{1.5}$$

where

$$u_m = a[\sqrt{1 - (\frac{v}{a\omega})^2} + \frac{v}{a\omega}\arccos(-\frac{v}{a\omega})] \tag{1.6}$$

is the maximum of the function (1.4) over a period.

Due to the periodic nature of the processes under consideration, the force of interaction

$$F(t) = f[u(t), \dot{u}(t)] \tag{1.7}$$

is a periodic function of time with a period $T = 2\pi/\omega$. Taking this into consideration, and using the theorem of momentum, the relation between the constant force P and the parameters of motion of the tool is obtained in the following form:

$$P = \frac{1}{T}\int_{t_1}^{t_1+T} F(t)dt = \frac{1}{T}\int_{t_1}^{t_1+T} f[u(t), \dot{u}(t)]dt. \tag{1.8}$$

Fig.1.10 illustrates the possible occurrences when specimen deformation takes place using a vibrating tool. The deformation in this case does not exceed the limits of the initial linear part of the dynamic characteristic ($0 < h \le D/k_0$). Since the carriage stays in its periodic regime, it remains in dynamic equilibrium with speed $v = 0$.

Fig.1.10 shows a graph of the dynamic characteristic $f(u, \dot{u})$, a motion graph (1.4) below it, and a force-time dependence(1.7) on the right-hand side.

If $h < 2a$, then the influence of the tool on the specimen takes the form of a periodic sequence of impulses (Fig.1.10,a)

$$f(t) = k_0(a \sin \omega t - \Delta), \quad t_1 \le t \le t_2, \tag{1.9}$$

where $t_1 = \frac{1}{\omega}\arcsin\frac{\Delta}{a}, t_2 = \frac{\pi}{\omega} - t_1$.

Substituting (1.9) into (1.8) and taking into account that $\Delta = a - h$, we obtain

$$P = \frac{ak_0}{\pi}[\sqrt{1 - (1 - \frac{h}{a})^2} - (1 - \frac{h}{a})\arccos(1 - \frac{h}{a})]. \tag{1.10}$$

If the vibration amplitude $a > D/2k_0$, we have a vibro-impact regime of loading which exists in the region of elastic deformation.

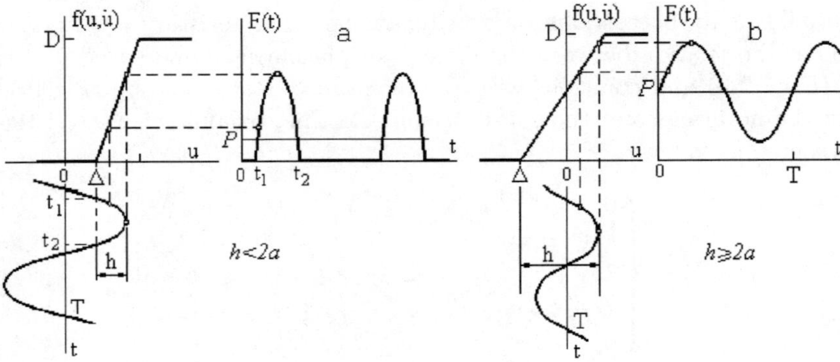

Fig. 1.10.

When the vibration amplitude $a < D/2k_0$ and carriage displacement $h \geq 2a$, the regime of loading is continuous (Fig.1.10,b). The force of interaction in the region of elastic deformation is described by relation (1.9), where $t_1 = -\frac{\pi}{2\omega}$. According to (1.9), the constant force is equal to

$$P = k_0(h - a). \tag{1.11}$$

The dependence of the static force P, on displacement h, is shown in Fig.1.11 for different values of D/k_0a. The dashed line divides the diagram into two parts. The lower part corresponds to regimes of vibro-impact deformation, while the upper part leads to continuous impact-free deformation. The equation for the separating line $P = k_0h/2$ is obtained from (1.10) and (1.11) by substituting $a = \frac{h}{2}$. For comparison, a static loading curve (1.11), when $a = 0$, is shown on the graph. Horizontal lines correspond to static force values equal to $P = D_a$, where

$$D_a = D \begin{cases} 1 - \frac{k_0 a}{D}, & a \leq \frac{D}{2k_0}, \\ \frac{1}{\pi}[\sqrt{\frac{2k_0 a}{D}} - 1 - (\frac{k_0 a}{D} - 1)\arccos(1 - \frac{D}{k_0 a})], & a \geq \frac{D}{2k_0}. \end{cases} \tag{1.12}$$

Irreversible plastic deformation appears if these values are exceeded. To find these expressions, (1.10) and (1.11) should be used, substituting $h = D/k_0$.

Formula (1.12) determines the materials observed yield point $\sigma_a = D_a/S$ (where S is the cross-sectional area of the specimen) under the influence of ultrasonics. Its dependence on the amplitude of vibration is shown in Fig.1.12. $\sigma_y = D/S$ and $E = k_0 l/S$ is the yield point and elastic modulus of the material respectively; l is the length of the specimen tested.

3. Let us consider regimes that cause plastic deformation ($h > D/k_0$). Assuming that during the process of deformation, the carriage 2 (Fig.1.9,a)

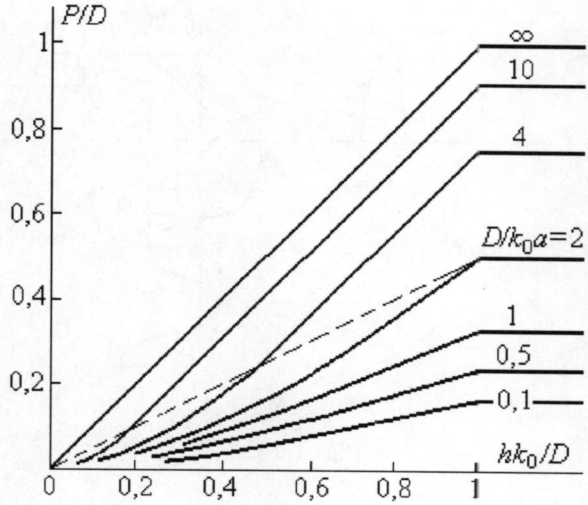

Fig. 1.11.

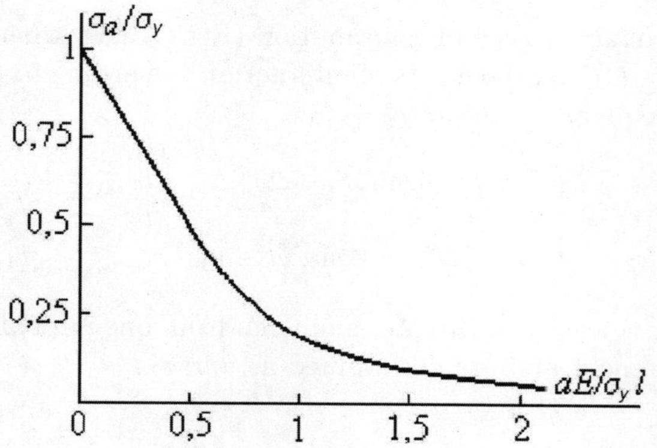

Fig. 1.12.

moves with a constant speed v, the full motion of the tool 1 is described by equation (1.4). Fig.1.13 demonstrates the situations possible for this type of deformation.

It is clear from Fig.1.13 that the residual deformation for a period of vibration is

$$h_0 = u_m - \Delta - \frac{D}{k_0}.$$

The average speed of deformation is therefore

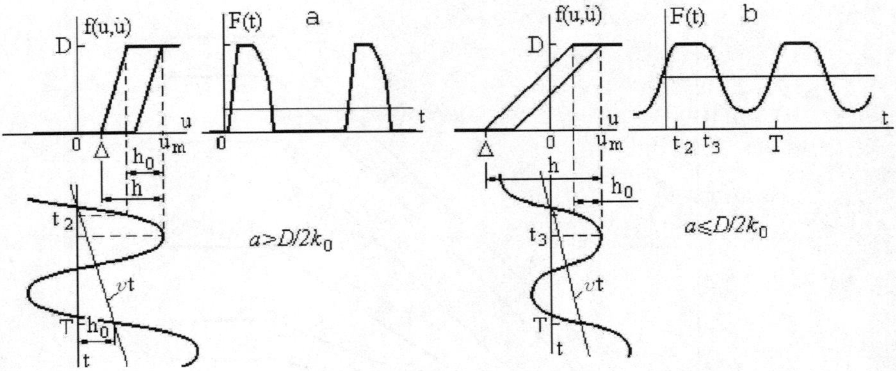

Fig. 1.13.

$$v = \frac{h_0}{T} = \frac{u_m \omega}{2\pi}\left(1 - \frac{\Delta}{u_m} - \frac{D}{k_0 u_m}\right). \tag{1.13}$$

In a vibro-impact regime of deformation (Fig.1.13,a), when the vibration amplitude is $a > D/2k_0$, the forces of interaction, according to (1.5) and (1.4), are a periodic sequence of impulses

$$F(t) = \begin{cases} k_0(vt + a\sin\omega t - \Delta), & t_1 \leq t \leq t_2, \\ D & t_2 \leq t \leq t_3, \\ D + k_0(vt + a\sin\omega t - u_m), & t_3 \leq t \leq t_4. \end{cases} \tag{1.14}$$

$t_i, (i = 1, ..., 4)$ are the switching moments from one part of the characteristic (1.5) to the next and are determined as follows:

$$\sin\tau_1 = \tfrac{1}{a}(\Delta - \tfrac{v}{\omega}\tau_1), \;\; \sin\tau_2 = \frac{1}{a}(\Delta + \frac{D}{k_0} - \frac{v}{\omega}\tau_2),$$
$$\tau_3 = \arccos(-\tfrac{v}{a\omega}), \;\; \sin\tau_4 = \frac{1}{a}(u_m - \frac{D}{k_0} - \frac{v}{\omega}\tau_4), \tag{1.15}$$

where $\tau_i = \omega t_i$, and $\tau_{1,2} \in [-\tau_2, \tau_3]$, $\tau_4 \in [\tau_3, 2\pi + \tau_1]$; u_m is given by (1.6).

Substituting (1.14) into (1.8),

$$P = \frac{k_0}{2\pi}\begin{bmatrix} \Delta\tau_1 - (\Delta + \frac{D}{k_0})\tau_2 + u_m\tau_3 - (u_m - \frac{D}{k_0})\tau_4 + \\ +a(\cos\tau_1 - \cos\tau_2 + \cos\tau_3 - \cos\tau_4) + \\ +\frac{v}{2\omega}(\tau_2^2 - \tau_1^2 + \tau_4^2 - \tau_3^2) \end{bmatrix}. \tag{1.16}$$

From (1.13),

$$\Delta = u_m - \frac{D}{k_0} - \frac{2\pi v}{\omega}. \tag{1.17}$$

Using (1.17) the solutions to the equations (1.15) may be found. Substituting these solutions in to (1.16), the speed v of plastic deformation is found to be dependant on the static force P. A similar dependence can be found for the regimes of continuous deformation (Fig.1.13,b), for which the amplitude is $a \leq \frac{D}{2k_0}$ for the whole range of the carriage's speed v. The impulse regimes will transform into continuous regimes for relatively high speeds v, when the limiting values $\tau_1 = -\tau_3$ and $\tau_4 = 2\pi + \tau_1$ are reached in (1.15). For continuous regimes of deformation, the same relations (1.14) - (1.16) are valid, if the substitutions

$$\tau_1 = -\tau_3 = -\arccos(-\frac{v}{a\omega}), \ \tau_4 = 2\pi + \tau_1. \tag{1.18}$$

are used.

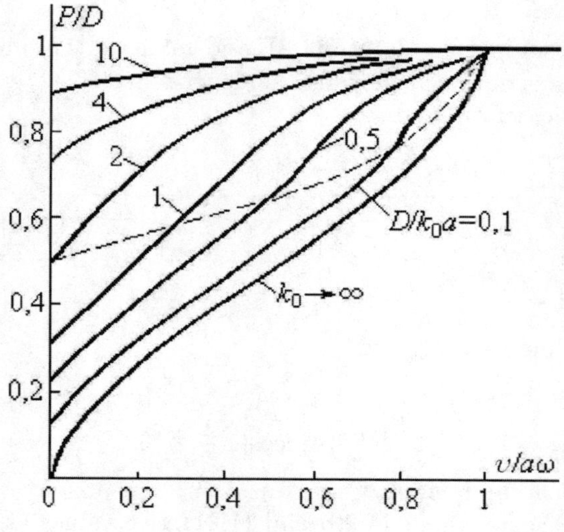

Fig. 1.14.

The results of these calculations are shown in Fig.1.14, where the dependence of plastic deformation speed on the static force P is shown for different values of D/k_0a. The dashed line separates the regions of vibro-impact (lower region) and impact-free (higher region) deformation as used in Fig.1.11. When $v = 0$, equations (1.14) to (1.18) give the value $P = D_a$, as determined by the expressions (1.12).

According to (1.6), (1.15), and (1.18), the relations found have real solutions for deformation speeds $v \leq a\omega$. If $v = a\omega$, (1.18) and (1.15) show that $\tau_1 = \tau_2 = -\tau_3 = -\tau_4 = -\pi$. Substituting these values into (1.16) we get $P = D$. A further increase in speed v, does not change the static force P required for plastic deformation. This is mentioned in a number of experiments (Isaev & Anokhin (1961), Markov (1962, 1966, 1980), Panov &

Piskunov (1966), Severdenko et al. (1976)), which use ultrasonics in various processes (refer to the section 1.3 and formula (1.2)). It must be noted that both continuous and impulsive regimes were observed in a number of experiments (Astashev & Sakaian (1967), Kazantsev (1963*a*), Rosenberg & Kazantsev (1959*b*)). In each case it was found that impulsive regimes are always more efficient.

The speeds used in ultrasonic processes are often small $v \ll a\omega$. In these cases it can be assumed that the small speed v does not influence the interaction force (1.14) over a period of vibration. For this reason, the moments t_i of switching are not changed within the characteristic (1.5). Under these assumptions, from (1.15)

$$\tau_1 = \arcsin\frac{\Delta}{a},\ \tau_2 = \arcsin(\frac{\Delta}{a} + \frac{D}{k_0 a}), \tau_3 = \frac{\pi}{2},\ \tau_4 = \arcsin(1 - \frac{D}{k_0 a}). \quad (1.19)$$

Substituting (1.14) and $v = 0$ into (1.8), and taking (1.17) into account, the explicit formulae is obtained showing the dependence of static force P on the speed v of plastic deformation:

$$P = \frac{k_0 a}{2\pi}\left[\begin{array}{l} \frac{\pi}{2}(\frac{2D}{k_0 a} - 1) + \sqrt{1 - (1 - \frac{D}{k_0 a})^2} - \\ -\sqrt{1 - (1 - \frac{2\pi v}{a\omega})^2} + \sqrt{1 - (1 - \frac{D}{k_0 a} - \frac{2\pi v}{a\omega})^2} + \\ +(1 - \frac{D}{k_0 a})\arcsin(1 - \frac{D}{k_0 a}) - (1 - \frac{2\pi v}{a\omega})\arcsin(1 - \frac{2\pi v}{a\omega}) + \\ +(1 - \frac{D}{k_0 a} - \frac{2\pi v}{a\omega})\arcsin(1 - \frac{D}{k_0 a} - \frac{2\pi v}{a\omega}) \end{array}\right] \quad (1.20)$$

for impulse loading regimes, and

$$P = \frac{k_0 a}{2\pi}\left[\begin{array}{l} 2\pi(\frac{D}{k_0 a} - 1 + \frac{\pi v}{a\omega}) - \sqrt{1 - (1 - \frac{2\pi v}{a\omega})^2} + \\ +(1 - \frac{2\pi v}{a\omega})\arccos(1 - \frac{2\pi v}{a\omega}) \end{array}\right] \quad (1.21)$$

for continuous loading regimes.

When $v = 0$, expressions (1.20) and (1.21) give values (1.12). Estimates show that these approximate relations are valid when the speed of deformation does not exceed the value $v = 0.25a\omega$.

4. The consideration above allows us to explain the "unhardening" effect of material behaviour observed in experiments, i.e. the decrease in static force required for plastic deformation under the influence of vibration. Following from (1.12), any arbitrarily small levels of vibration cause a decrease in static force and leads to residual deformation. Alternatively, this force can be made arbitrarily small by the increase of vibration amplitude.

The fact that the static force required for plastic deformation decreases with an increase in vibration amplitude was observed in all ultrasonic experiments (Izumi et al. (1966*b*,*a*), Severdenko et al. (1973, 1976), Goliamina (1979)). Fig.1.13 helps us to clearly understand the nature of this phenomenon: during every deformation cycle, the material's yield limit is being overcome for a short time. A small residual deformation appears, which accumulates from one cycle to the next. If the static deformation force is held constant and the progress of the process is slow, which is typical for the majority of experiments, there is a distinct impression that the yield point of the material is decreased, i.e., the material "unhardens" under the influence of ultrasonic vibration.

It is necessary to point out again, that the observed phenomenon of material "unhardening" is only an impression. The real elastic-plastic properties of the material do not change: in every deformation cycle, the force acting on the material is enough to overcome its yield limit. This remark is important as a number of papers (Izumi et al. (1966*b*,*a*), Langenecker (1966), Goliamina (1979)) state wrongly that a change in material properties takes place under the influence of ultrasonics. Usually, such a statement is based on the incorrect interpretation of experimental data or by methodic mistakes. This is the case in (Izumi et al. (1966*b*), Langenecker (1966)) whilst making force estimates in the deformation zone. In Izumi et al. (1966*b*) it was erroneously assumed, that the maximum deformation force cannot exceed the force present at the nodal point in bar 2 of the vibration system (Fig.1.2). This is situated in the place where the membrane 4 is fixed.

The experimental data from paper Izumi et al. (1966*a*) is in good accordance with the results of the analysis carried out above. Data on the deformation of specimens with diameter $d = 14mm$ and height $H = 15mm$, made from different materials, led in Izumi et al. (1966*a*) to the following empirical relation between an observed yield limit σ and vibration amplitude

$$\sigma = \sigma_0 - Ka, \tag{1.22}$$

where σ_0 is the actual yield limit found in a static deformation process and K is a constant coefficient dependant on the material.

From the first relation (1.12), we obtain an expression for force

$$D_a = D - k_0 a, \tag{1.23}$$

Dividing this relation by S and using notation $\sigma = D_a/S$, $\sigma_0 = D/S, K = k_0/S = E/H$, where S and H are the cross-sectional area and height of the specimen respectively, we obtain a relation of the form (1.22).

Fig.1.15 shows this dependence (1.23) for different materials. The data used for these calculations is given in the Table 1.1 below. The points in Fig.1.15 represent experimental data (Izumi et al. (1966*a*)). The values of yield limit σ_0 correspond to static deformation results and are shown by points for $a = 0$; elastic modulus values E are chosen in accordance with the handbook Teumin (1959). It must be noted that these experiments were carried out, using small amplitude vibration, satisfying the condition for when continuous deformation regimes occur. These are the conditions under which relation (1.23) was derived.

5. Following on from (1.12), Fig.1.12. and Fig.1.13, various materials respond to vibration in plastic deformation process differently. Materials with a bigger elastic modulus E and specimens with a bigger static stiffness k_0 are

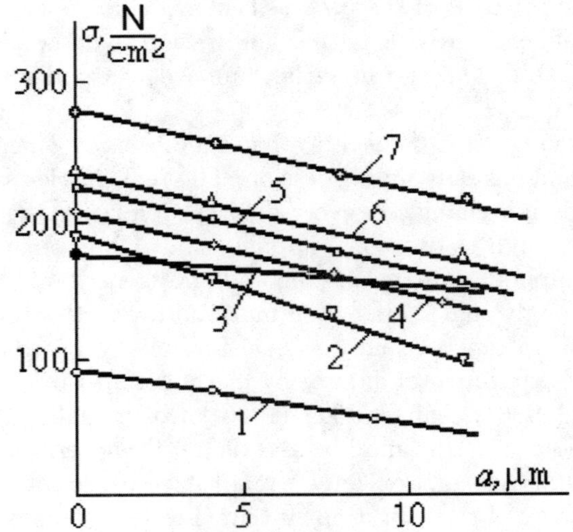

Fig. 1.15.

Table 1.1.

No.	Material	$E, \frac{N}{mm^2}$	$\sigma_0, \frac{N}{mm^2}$
1.	Aluminum	71000	98
2.	Copper	125000	192
3.	Magnesium	40000	180
4.	Brass with $Cu-Zn$ content: $7-3$ $9-1$ $8-2$ $6-4$	100500	208 228 239 282

more sensitive to the influence of ultrasonics. The achievable degree by which the static force may be decreased during plastic deformation can be estimated by considering the limiting case $k_0 \to \infty$. This leads to the characteristics of an ideal rigid-plastic material (Mase (1970), Rabotnov (1979)) and is shown in Fig.1.16. Such a material is assumed to be absolutely rigid until the force reaches a threshold value D. At this point, irreversible plastic deformation occurs. In order to investigate the properties of a rigid-plastic material with respect to static loading under vibrational influence, the limit $k_0 \to \infty$ is considered in the relations (1.12) and (1.15) to (1.17). This gives the following relations: $D_a = 0$,

$$\tau_2 = \tau_1, \sin \tau_1 = \frac{1}{a}(\Delta - \frac{v}{\omega}\tau_1),$$

$$\tau_4 = \tau_3 = \arccos(-\frac{v}{a\omega}), \quad \Delta = u_m - \frac{2\pi v}{\omega}, \tag{1.24}$$

$$P = D\frac{(\tau_3 - \tau_1)}{2\pi}. \tag{1.25}$$

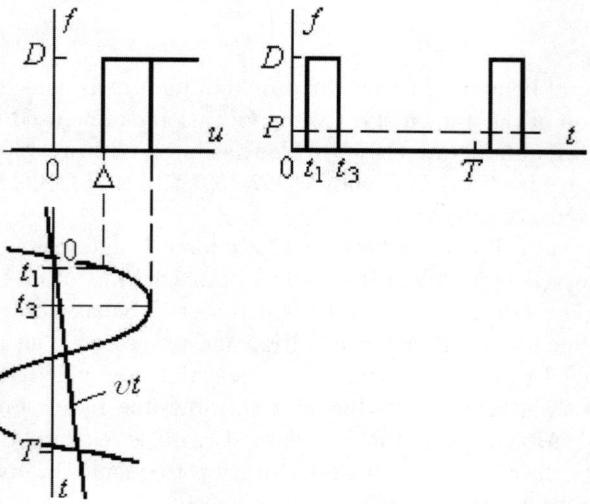

Fig. 1.16.

Using these relations, a curve of $k_0 \to \infty$ is plotted (Fig.1.14), and shows the dependence of the constant force P, on the deformation speed v of the rigid-plastic material.

Accounting for the assumptions used in the derivation of the approximate analytical expressions (1.20) and (1.21), i.e. by substituting $\sin \tau_1 = \Delta/a, \tau_3 = \pi/2, u_m = a$ into (1.24) and from (1.25)

$$P = \frac{D}{2\pi} \arccos(1 - \frac{2\pi v}{a\omega}). \tag{1.26}$$

It is convenient to rewrite (1.26) as

$$v = \frac{a\omega}{\pi} \sin^2 \frac{\pi P}{D}. \tag{1.27}$$

Relation (1.26), in addition to (1.20) and (1.21), is valid for $v < 0.25a\omega$. The equality (1.27) for forces $P < 0.35D$ also corresponds to the range of speeds $v < 0.25a\omega$.

The relations obtained above and the curves shown in Fig.1.14 can be interpreted as some material behavioural characteristics, when subjected to an ultrasonic field, with respect to applied static loads.

It is interesting that the character of the material's behaviour changes in vibrational conditions: elastic-plastic materials behave like viscous-elastic-plastic materials with a reduced yield point; rigid-plastic materials do not hold static load and yield when subjected to small static forces P in a similar way to viscous media. It is especially obvious if, assuming $P \ll D$, (1.27) is written in the form

$$v = a\omega\pi \left(\frac{P}{D}\right)^2 \tag{1.28}$$

The material behaviour under ultrasonic influence, as considered above, is a manifestation of the known effect when vibration can be used to smooth out nonlinearities (Kubota et al. (1977)). This has been extensively investigated in systems with dry friction (Andronov (1967*a*,*b*, 1970, 1975), Blekhman (1999)), with the presence of vibration.

The single-valued relation between static force P and average deformation speed v, observed experimentally in Isaev & Anokhin (1961), Izumi et al. (1966*a*), Markov (1962), is an important material behavioural feature when under the influence of vibration in ultrasonic processes. The characteristics shown in Fig.1.14 allow materials to be selected, for which the application of ultrasonics is advantageous due to a considerable reduction in the static force required. Alternatively the characteristics observed can be used to select values of static force, amplitude and vibration frequency in order to provide the required performance specified by the speed v.

6. In order to describe the decrease in static force required to cause plastic deformation, rheological models of ideal elastic and rigid-plastic materials have been used. Real metals, as used in experiments and processes, have characteristics which exhibit hardening in their plastic zone. A rheological model and characteristic of a material which exhibits hardening is shown in Fig.1.8,c,d. During a real process, the deformation conditions of the workpiece often change. As an example, in the process of sheet metal forming (Fig.1.4,a), the geometry of the workpiece changes as the die deforms it. For this reason, the dynamic characteristics of the working process change as both the force of plastic deformation, and stiffness of the specimen in its region of elastic

deformation, are dependant on the history of loading, and be will changed with an increase in residual plastic deformation of the material. This is shown in Fig.1.17,a.

To describe a process of this kind, the results obtained above can be used. The workpiece's yield limit D, and static stiffness k_0 are considered to be functions of deformation h.

$$D = D(h), \quad k_0 = k_0(h), \tag{1.29}$$

The small changes in these parameters over one cycle of loading are neglected. Thus, a change in the plasticity threshold of a material with linear hardening (Fig.1.8c,d) under the influence of vibration can be found using relations (1.12). In accordance with (1.29) and Fig.1.8,d the following substitution should be made.

$$D = D_0 + k_1(h - \frac{D_0}{k_0}). \tag{1.30}$$

A similar approach can be used when a "force-deformation" diagram is defined by a graph, as is the case after performing static experiments. It is clear, that the plastic deformation speed of a material that is subjected to a constant force (or a static force when the speed is constant) will also change with an increase in residual deformation.

Consider now the deformation process of a rigid-plastic material with hardening. The characteristics of this are shown in Fig.1.17,b.

$$D = D_0 + k_1 h. \tag{1.31}$$

(This expression can be derived from (1.30) when $k_0 \to \infty$).

Our consideration is now restricted to deformations under small static forces $P \ll D$. By substituting (1.31) into (1.28), we obtain the speed of the process.

$$v = \pi a \omega [\frac{P}{D_0 + k_1 h}]^2. \tag{1.32}$$

The deformation speed graph (1.32), as a function of residual deformation for a constant static force, is shown in Fig.1.18.

Using (1.32) the time necessary to obtain the given value h_0, of the workpiece's residual deformation may be found.

$$t = \int_0^{h_0} \frac{dh}{v(h)} = \frac{h_0}{\pi a \omega} \left(\frac{D_0}{P}\right)^2 \left[1 + \frac{k_1 h_0}{D_0} \left(1 + \frac{3k_1 h_0}{D_0}\right)\right]. \tag{1.33}$$

When the parameters D and k_0 are given functions (1.29) of deformation h, the behaviour of the material in its plastic flow region can be estimated by using the results represented in Fig.1.14. For a given level of deformation h, the values of functions (1.29) may be evaluated, and the parameter $D/k_0 a$ may be calculated. The curve corresponding to this parameter (Fig.1.14) may be used to find the point with the ordinate equal to P/D. The abscissa of this

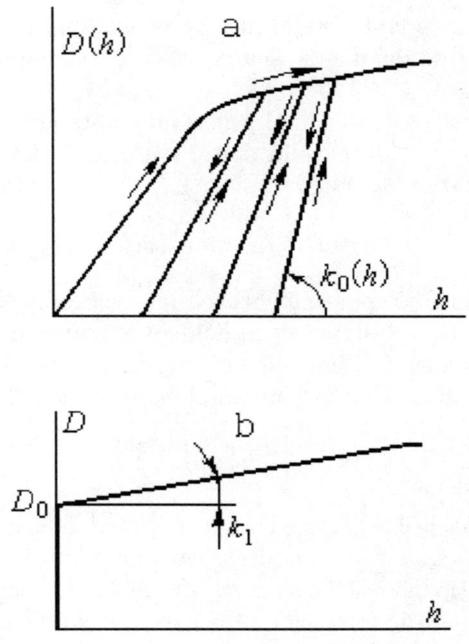

Fig. 1.17.

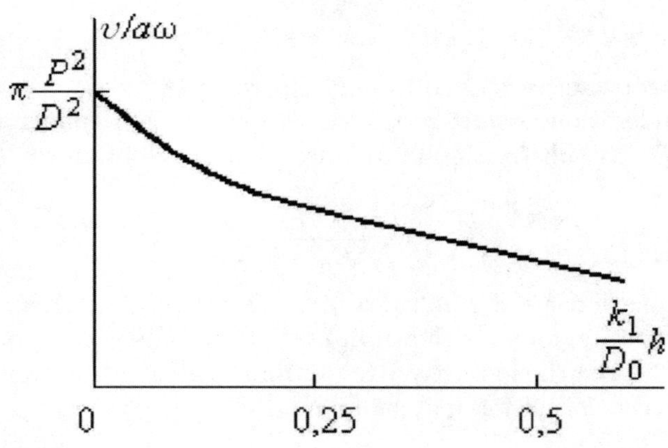

Fig. 1.18.

point determines the current speed of plastic deformation. This allows the dependence $v = v(h)$ of the deformation speed on the value of deformation to be found for a given static force P. The integral in (1.33) also determines the time and therefore the law $h = h(t)$ of deformation.

For a given deformation speed v, such a procedure allows us to determine the dependence of static force P on the deformation h.

The suggested approach allows different ultrasonic processes to be described, in which the deformation of materials takes place. As shown, the effects predicted by theoretical considerations are confirmed by numerous experiments. It should be noted however, that a lot of experimental data has poor accuracy and can hardly be compared with the results obtained in this section. This may be caused by an indistinct description of the experimental methods and test conditions. In a number of cases, poor data accuracy is caused by neglecting the effects that the nonlinear working process have on the vibration condition of the tool. In these cases, interpretation of experimental data requires some additional analysis of the dynamics of the ultrasonic vibration system. The nonlinear load, caused by the workpiece being treated, must be accounted for. Using such an approach, the main effects observed during the ultrasonic machining of brittle materials will be explained. This can be found in section 5.1.

1.3 The influence of ultrasonics on the processes with dry friction

1. In ultrasonic processes there are different cases in which friction appears during the interaction of the tool with the medium. They differ in the relative positions of the vibration direction, and the slow sliding of the bodies that are in contact.

A model of a system containing dry friction, similar to the one demonstrated earlier in Fig.1.8, is shown in Fig.1.19. The carriage 2 moves, subject to a static force P, with an average speed v along the guide bar 3. A bush 1, connected with the carriage and able to perform relative vibration with respect to it, is in contact with the bar. A dry friction force Q is developed in the contact zone. It acts in the direction opposite to that of a uniform motion velocity in the absence of vibrations. A static force $P = Q$ is needed to maintain the uniform motion. The influence of vibration in the contact zone, on the static force necessary to overcome friction is now estimated.

The case in which the vibration direction coincides with the velocity v of the carriage is a typical occurrence in ultrasonic processes. As an example, in ultrasonic machining (Fig.1.1,a), the friction forces appear between the side surfaces of the tool and the walls of the hole. A similar situation appears in the process of drawing (Fig.1.4,b), turning (Fig.1.4,c) and the ultrasonic welding of metals when using a longitudinal vibration motion on the welding tip (Fig.1.6,a).

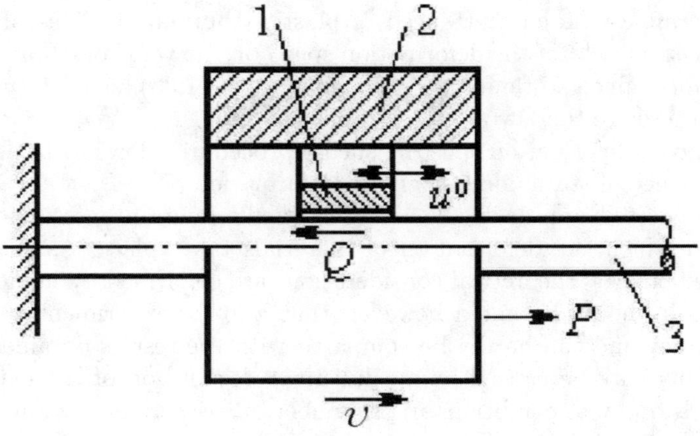

Fig. 1.19.

As in subsection 1.2.1, consider the steady motion of the bush 1 (Fig.1.19), as

$$u(t) = vt + u^0(t) = vt + a\sin\omega t. \tag{1.34}$$

The nonlinear dynamic characteristic of the dry friction force is presented as follows:

$$f(\dot{u}) = Q \text{ sgn } \dot{u}.$$

Due to the periodicity of the process under consideration (1.34), the frictional force developed in the contact zone

$$F(t) = f(\dot{u}(t)) = Q \text{ sgn } (v + a\omega\cos\omega t) \tag{1.35}$$

is a periodic function with the period $T = 2\pi/\omega$.

The relation between the static force P and the parameters of the system's motion can be written in a similar way to (1.8)

$$P = \frac{1}{T}\int_0^T F(t)dt = \frac{Q}{T}\int_0^T \text{sgn } (v + a\omega\cos\omega t)dt. \tag{1.36}$$

The characteristics of the friction present in the contact zone $f(\dot{u})$ are shown in Fig.1.20 together with a graph of the speed $\dot{u}(t)$ and a plot of the friction force $F(t)$ as a function of time. Evaluation of (1.36) gives

$$P = \frac{Q}{\pi}[\pi - (\tau_2 - \tau_1)], \tag{1.37}$$

where

$$\tau_1 = \omega t_1 = \arccos(-\frac{v}{a\omega}) = \frac{\pi}{2} + \arcsin va\omega,$$
$$\tau_2 = \omega t_2 = 2\pi - \tau_1, \tag{1.38}$$

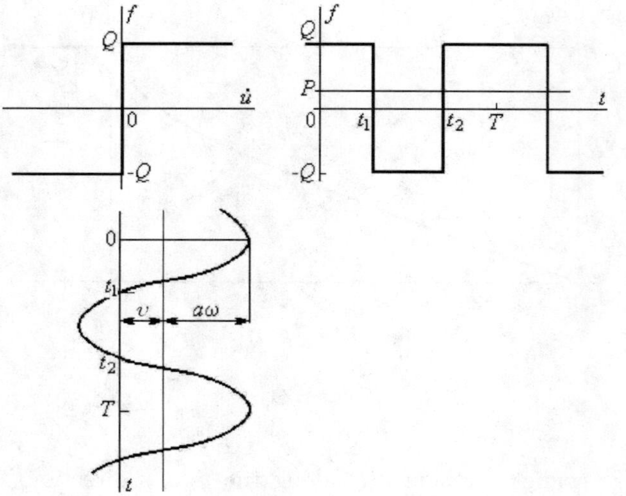

Fig. 1.20.

t_1, t_2 are the moments, when the speed $\dot{u}(t)$ (Fig.1.20) changes its sign.

Substituting (1.38) into (1.37), reveals

$$P = \frac{2Q}{\pi} \arcsin \frac{v}{a\omega}. \tag{1.39}$$

The dependence of the static force P on the ratio of speeds $\frac{v}{a\omega}$, in accordance with (1.39), is shown in Fig.1.21. It can easily be seen, that the vibration in the contact zone leads to a decrease in the static force required to overcome friction for sliding velocities $v < a\omega$. For very small sliding velocities $v \ll a\omega$, (1.39) gives

$$P = \frac{2Q}{\pi a\omega} v, \tag{1.40}$$

Therefore, for small speeds v, the behaviour of a system with respect to static force in vibrating conditions is analogous to its motion in a linear viscous medium. This phenomenon is known as an effect of vibrational smoothing of nonlinearities, previously mentioned in subsection 1.2.5.

2. In practice, certain ultrasonic processes exhibit instances where vibration is performed in a sliding plane perpendicular to the velocity of the slow motion. This happens, for example, in the ultrasonic welding of metals, with a transverse motion of the welding tip, as described in subsection 1.1.4. In the model (Fig.1.19), this corresponds to a torsional vibration acting on the bush 1 around the guide bar 3. Such an arrangement is sometimes used in the process of ultrasonic wire and rod drawing (Severdenko et al. (1976)), mostly for calibration. The layout in Fig.1.22 shows an element's friction force formation. The element moves in the plane Π, along the x-axes, with a speed v and

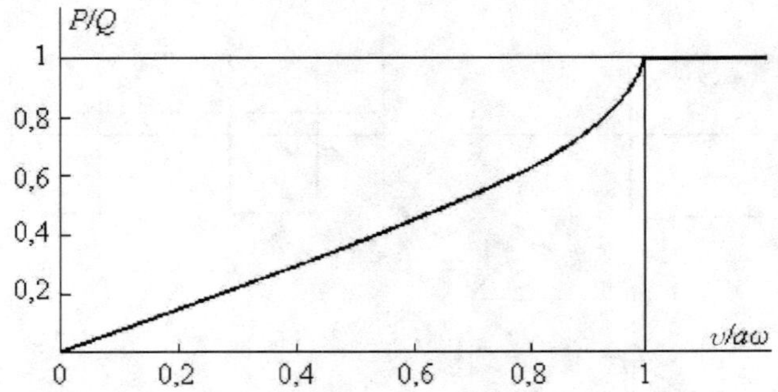

Fig. 1.21.

performs T-periodic vibration $\dot{u}(t)$ along the y-axes direction. The friction force Q has a constant value and direction opposite to that of the velocity vector $v_0(t)$ along which its slides in the plane.

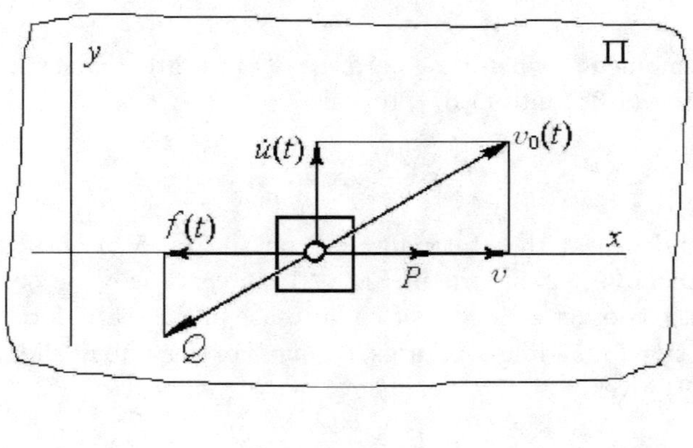

Fig. 1.22.

The x-component of the friction force

$$F(t) = Q\frac{v}{\sqrt{v^2 + [\dot{u}(t)]^2}} \tag{1.41}$$

is a periodic function of time with the period T.

To overcome friction in the direction of constant velocity v, a constant force is required

$$P = \frac{1}{T}\int_0^T F(t)dt = \frac{Q}{T}\int_0^T \frac{vdt}{\sqrt{v^2 + [\dot{u}(t)]^2}} \tag{1.42}$$

Substituting $\dot{u}(t) = a\omega \sin \omega t$ into (1.42) and introducing the notation $\tau = \omega t$, we obtain

$$P = \frac{Q}{2\pi} \int_0^{2\pi} \frac{d\tau}{\sqrt{1 + (\frac{a\omega}{v})^2 \sin^2 \tau}} = \frac{2Q}{\pi} \int_0^{\pi/2} \frac{d\tau}{\sqrt{1 + (\frac{a\omega}{v})^2 \sin^2 \tau}}. \quad (1.43)$$

Integral (1.43) can not be evaluated as an elementary function, however, the following estimation can be made:

$$F(\tau) = [1 + (\frac{a\omega}{v} \sin \tau)^2]^{-1/2} \leq 1$$

and, therefore,

$$\int_0^{\pi/2} F(\tau) d\tau < \frac{\pi}{2}. \quad (1.44)$$

Therefore, as it follows from (1.43) and (1.44), for any value of $v/a\omega$ the static force necessary to overcome friction satisfies the inequality $P < Q$.

To find the upper and lower estimates of the force P, $\sin \tau$, $(\tau \in [0, \frac{\pi}{2}])$ are replaced in (1.43) by the functions $\Phi_1(\tau)$ and $\Phi_2(\tau)$ such that

$$\Phi_1(\tau) \leq \sin \tau \leq \Phi_2(\tau), \ (\tau \in [0, \frac{\pi}{2}]).$$

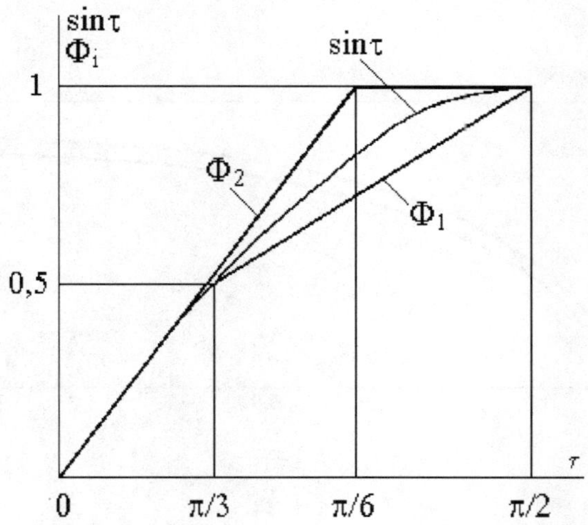

Fig. 1.23.

A real value of the force P can now be determined by (1.43) and will be in the interval $P_2 \leq P \leq P_1$ where,

$$P_i = \frac{2Q}{\pi} \int_0^{\pi/2} \frac{d\tau}{\sqrt{1 + [\frac{a\omega}{v} \Phi_i(\tau)]^2}}, \ (i = 1, 2). \tag{1.45}$$

Functions $\Phi_1(\tau)$ and $\Phi_2(\tau)$ are chosen as shown in Fig.1.23, i.e. the first curve is inscribed and the second is circumscribed with respect to $\sin\tau$. The following functions shall be used:

$$\begin{aligned} \Phi_1(\tau) &= \frac{3}{\pi}\tau - (\frac{3}{2\pi}\tau - \frac{1}{4})\eta(\tau - \frac{\pi}{6}), \\ \Phi_2(\tau) &= \tau - (\tau - 1)\eta(\tau - 1), \end{aligned} \tag{1.46}$$

where $\eta(\tau)$ is a unit step function.

Segments of the piecewise linear function $\Phi_1(\tau)$ pass through the points $(0,0), (\frac{\pi}{6}, \frac{1}{2}), (\frac{\pi}{2}, 1)$ belonging to $\sin\tau$. Segments of the function $\Phi_2(\tau)$ are tangential to $\sin\tau$ at points $(0,0)$ and $(\frac{\pi}{2}, 1)$. Substituting (1.46) into (1.45) and integrating gives

$$P_1 = \frac{2}{3} Q \frac{v}{a\omega} \ln \left[\frac{2(1 + \sqrt{(\frac{v}{a\omega})^2 + 1})^2}{\frac{v}{a\omega}(1 + \sqrt{(\frac{2v}{a\omega})^2 + 1})} \right], \tag{1.47}$$

$$P_2 = \frac{2}{\pi} Q \frac{v}{a\omega} \left\{ \ln \left[\frac{1 + \sqrt{(\frac{v}{a\omega})^2 + 1}}{\frac{v}{a\omega}} \right] + \frac{\frac{\pi}{2} - 1}{\sqrt{(\frac{v}{a\omega})^2 + 1}} \right\}. \tag{1.48}$$

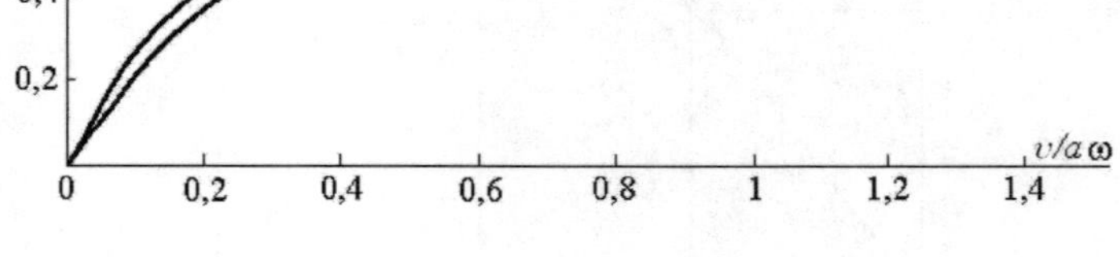

Fig. 1.24.

The upper P_1 and lower P_2 estimates of the static force P, as functions of the speed ratios $\frac{v}{a\omega}$, are shown in Fig.1.24 in accordance with (1.47) and (1.48).

Considering the appropriate limits in (1.47) and (1.48), it can be shown that $P_i \to 0$ when $(\frac{v}{a\omega}) \to 0$ and $P_i \to Q$ when $(\frac{v}{a\omega}) \to \infty$. Due to the close proximities of P_1 and P_2, there is no need to calculate more accurate values of the static force P necessary to overcome friction. Unlike in the previous case (Fig.1.21), the insertion of transverse vibrations into the friction zone leads to a decrease in static force for any sliding speed v.

Comparing Fig.1.21 and Fig.1.24, it can be seen that, for the vibration speed $a\omega$ and for a small sliding speed v, the most pronounced decrease in friction is achieved when the direction of vibration and sliding coincide. This is confirmed in experiments. The results from experiments on the estimation of the influence of ultrasonics on the decrease in static force necessary to overcome contact friction can be found in Pohlman & Lehfeldt (1966). The arrangement of the experiment is shown in Fig.1.25. A spherical tip, connected to the vibration system 1, is pressed on to the plane of the disk 2 which rotates with a constant speed. The vibration system is fixed in such a way that the direction of vibration either coincides (position I) with the direction of the disk's linear velocity, or is perpendicular to it (position II). The change in static force is estimated by the driving torque applied to the disk. In this experiment tip vibration was at a frequency $f = 22kHz$ and an amplitude $a = 12 - 18\mu m$.

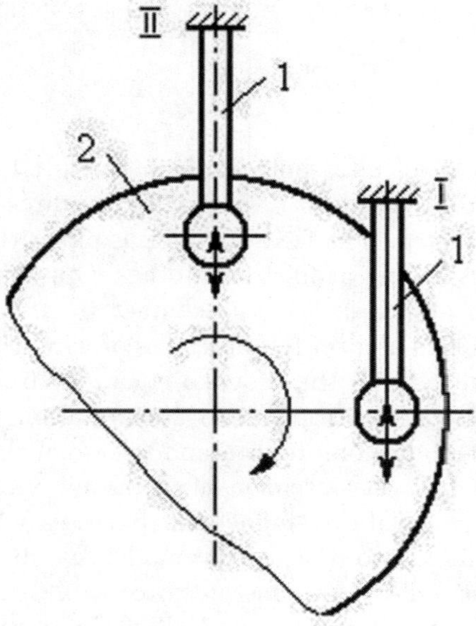

Fig. 1.25.

Experimental data on the dependence of static force P on the linear speed v of the disk is presented in Fig.1.26. Curve 1 corresponds to the case when the tip vibration direction coincides with the velocity of the disk (position I in Fig.1.25), while curve 2 corresponds to the case when they are perpendicular (position II). Comparison of the experimental results with the ones shown in Fig.1.21 and Fig.1.24 demonstrates good qualitative and quantitative agreement.

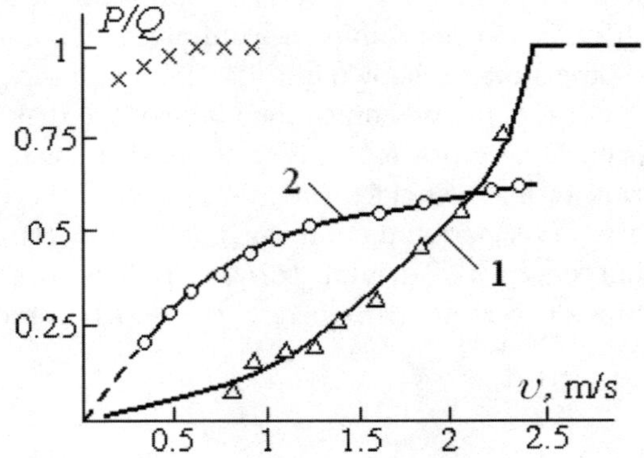

Fig. 1.26.

3. During the process of ultrasonic hardening (Fig.1.1,b) and welding of polymers (Fig.1.6,c), tool vibration is excited in the direction perpendicular to the surface being treated. The feed direction is along the surface, leading to the contact surfaces sliding against each other. Opposing the sliding motion, forces of friction are present in the contact zone. A similar layout of vibration excitation is also used in the processes of wire flattening and workpiece drawing using ultrasonics (Severdenko et al. (1976), Vagapov (1987)). Fig.1.19 shows that this case corresponds to the excitation of radial vibrations in a bush, causing a periodic compression and release of the guide bar.

In order to clarify how the insertion of ultrasonic vibration, as used in the arrangement described above, influences processes with dry friction in the contact zone, a simple two dimensional model may be used as shown in Fig.1.27. The plate 1 is pulled by a constant force P under the tip 2, which is pressed on to the plate by the static force G. The tip is subjected to periodic vibrations $u(t)$ in the direction perpendicular to the contact plane. Due to the interaction of the tip with the plate, a normal reaction force f_n appears. Its value depends on the rheological properties of the contact and, as it was shown in the previous section, is determined by a nonlinear dynamic characteristic $f_n = f_n[u(t), \dot{u}(t)]$. Regardless of the particular form of this characteristic, the following relation between f_n and G is valid for steady-state vibration:

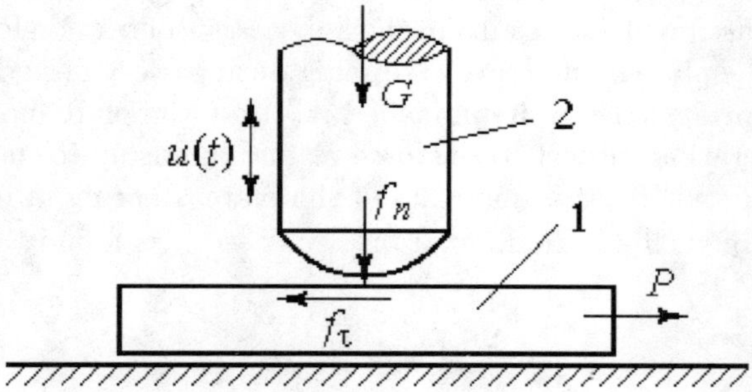

Fig. 1.27.

$$\int_0^T f_n[u(t), \dot{u}(t)]dt = GT. \tag{1.49}$$

At the same time, a dry friction force appears in the contact zone,

$$f_\tau \leq \vartheta f_n, \tag{1.50}$$

acting in the direction opposite to the direction that the contact surfaces are sliding in relative to each other.

The equality in (1.50) corresponds to the situation when the plate does not stop during its motion due to friction. A steady motion may now be considered in which no stops occur. The following expression may be written for the impulsive force, when the plate is subjected to a constant force P, and a periodic friction force $f_\tau = \vartheta f_n[u(t), \dot{u}(t)]$,

$$\vartheta \int_0^T f_n[u(t), \dot{u}(t)]dt = PT. \tag{1.51}$$

Comparing (1.49) and (1.51), it is found that a steady motion (with no stops) takes place under the same conditions that relate the forces acting in the system with the absence of tip vibration where,

$$P = \vartheta G \tag{1.52}$$

The previous statement is confirmed by the experimental results described in Pohlman & Lehfeldt (1966), and shown in Fig.1.26 by crosses.

If the static force $P < \vartheta G$, then the motion of the plate relative to the tip is periodically stopped by friction. Certain simplifying assumptions may be applied to the tip's contact characteristics with the plate, and allows some estimates to be made for the plates motion relative to the tip. Let it be assumed that the stiffness of the contact region is so high that the time of interaction is negligible compared to the period T of vibration. The existence of a periodic vibro-impact process (Babitsky (1998), Babitsky & Krupenin

(2001)) is now assumed and results in the appearance of friction force impulses. This leads to the plate momentarily stopping during every period of vibration. After the tip springs back to its unloaded position, the plate moves on with a constant acceleration subject to the force P, and builds up to the next impact speed $v_0 = \frac{P}{M}T$, where M is the mass of the plate. Bearing in mind that the average speed of such a periodic motion is $v = \frac{v_0}{2}$, it is found that

$$v = \frac{\pi}{\omega M} P, \tag{1.53}$$

or, for a static force,

$$P = \frac{\omega M}{\pi} v. \tag{1.54}$$

Thus, if the speed v satisfies the inequality

$$v < v^* = \frac{\pi}{\omega M} \vartheta G, \tag{1.55}$$

the vibrational influence leads to a decrease in the static force required to overcome friction.

It is interesting to note that, unlike the cases considered in sections 1 and 2, the value of the static force does not depend on the vibration intensity. Due to the possibility of using small amplitudes of vibration, this arrangement is appropriate for decreasing friction in kinematic pairs. It may be incorporated in to the supports of devices with light parts where the insensitivity due to threshold values of dry friction forces can be eradicated. Dry friction may be decreased using high-frequency vibration.

Returning to the consideration of ultrasonic processes, the influence of the contacting part's rheological properties on the value of static force is estimated. It is assumed that the reaction force f_n is described by the characteristic of a rigid-plastic material (Fig.1.16), with a yield limit D. This means that the normal force appearing in the contact zone equals $f_n = D$, and the duration of contact is $t_c = \frac{TG}{D}$. During the time $T - t_c$, the plate accelerates subject to the force P and develops a speed $v_0 = \frac{(T-t_c)P}{M}$. The distance covered in this time is equal to $s_1 = \frac{1}{2}v_0(T - t_c)$.

When in contact, the plate is retarded by the force $\vartheta D - P$ until it stops. The distance covered during this time is equal to $s_2 = \frac{1}{2}v_0^2 M/(\vartheta D - P)$.

By calculating the average speed of motion $v = (s_1 + s_2)/T$, we obtain

$$v = \frac{\pi}{M\omega} P \frac{\vartheta(D - G)}{(\vartheta D - P)}. \tag{1.56}$$

As $P < \vartheta G$, it is easily shown that for a constant static force P, the speed (1.56) is smaller than (1.53); the speed's boundary value, v^*, is given in both cases by (1.55). (The parameter v^* is obtained from (1.53) and (1.56) for $P = \vartheta G$).

The dependence of static force P on the speed of the plate's motion is shown in Fig.1.28. Curve 1 corresponds to the relation (1.54). Curves 2 and

3 represent relation (1.56) for $D/G = 2$ and $D/G = 5$ respectively. As a general rule, ultrasonic processes are advisable for situations when $G \ll D$. Therefore, estimates of the decrease in static force, in the presence of vibration perpendicular to the sliding surfaces, can be made using the simple relation (1.54), following from (1.56) when $D \to \infty$.

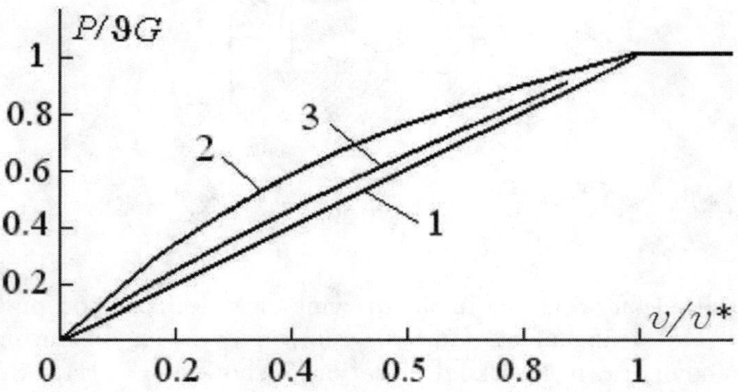

Fig. 1.28.

Here, it is seen again, that ultrasonic vibration in the contact zone completely removes insensitivity to static loading which is caused by dry friction. This effect was experimentally observed in Fridman & Levensque (1959). A qualitative change of the friction's character for small sliding speeds in the presence of ultrasonic vibration perpendicular to the friction surface was noticed in experiments and is reported in Arkhangelsky (1969).

Vibration regimes, with a periodic loss in contact between the rubbing surfaces have now been considered. These regimes are typical for ultrasonic systems because of the large acceleration experienced by the vibrating parts. (For a frequency $f = 20kHz$ and amplitude $a = 10\mu m$. the speed of vibration is equal to $a\omega = 1.2m/s$ and the acceleration is equal to $a\omega^2 = 1.510^5 m/s^2$).

To conclude, it must be pointed out that the effects revealed above are essential in brittle material die forming. The compression of the workpiece 1 (Fig.1.29) inserted into a die cavity 2 results in a flow of material inside the cavity. At the same time friction forces Q appear on the contact surfaces hindering deformation. As a result, the total stress in the surface layer of the material can be high enough to cause crack formation. The cleaving phenomenon on a material's surface is typical for brittle materials with a high yield limit. Ultrasonic influence prevents the accumulation of shear stresses, making the influence of friction forces on the contact surfaces of the workpiece and die almost negligible. This effect, together with the decrease in static force required for deformation, leads to a qualitative improvement in the plastic properties of brittle materials under the influence of ultrasonics. This has been observed experimentally in Goliamina (1979).

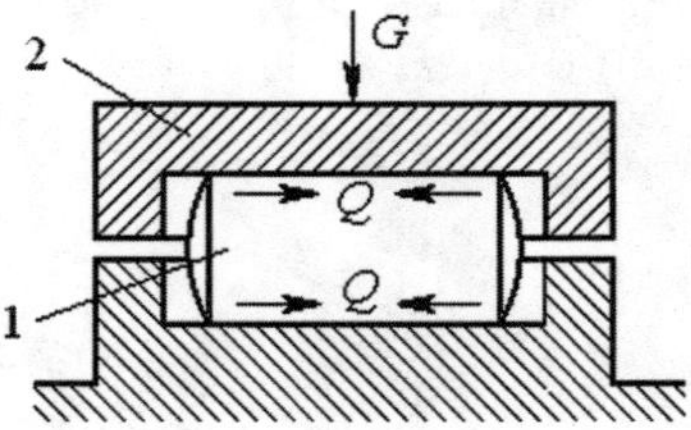

Fig. 1.29.

4. The situation occurring in the drawing of wires, rods and pipes is now considered. The arrangement of forces, acting on a drawing die during a traditional method of drawing (without vibration) is shown in Fig.1.4,b. Subsection 1.3.1 contains an estimate of the influence of vibration on the static force required to overcome friction on the calibration surface of a drawing die. In this section, the influence of vibration on the resistance, appearing in the conical part of a drawing die, is estimated.

A model of a rod 3 drawing process, through a drawing die 1, is shown in Fig.1.30. The drawing die 1 performs harmonic vibration with respect to the carriage 2 and moves with it at a constant speed v subject to the static force P. As before, it is assumed that the motion of the drawing die is determined by (1.34).

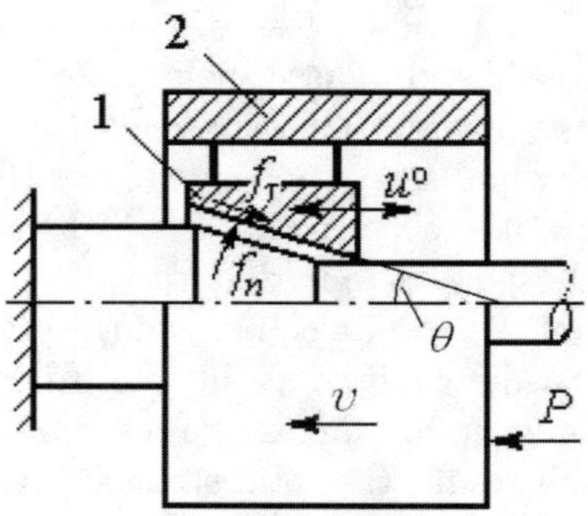

Fig. 1.30.

The displacement of the drawing die forces the conical surfaces to meet along the normal to the cone

$$u_n(t) = u(t)\sin\theta = (vt + a\sin\omega t)\sin\theta, \tag{1.57}$$

and causes their relative tangential displacement.

$$u_\tau(t) = u(t)\cos\theta = (vt + a\sin\omega t)\cos\theta, \tag{1.58}$$

This generates a normal reaction force f_n, and a dry friction force f_τ, during the contact of these surfaces.

Similar to in (1.5), we assume that the normal force f_n, is related to the displacement (1.57) by the characteristic of an elastic-plastic material

$$f = \begin{cases} 0 & u_n \le \Delta_n, \dot{u}_n > 0, \\ k_n(u_n - \Delta_n) & \Delta_n \le u_n \le \Delta_n + \frac{D_n}{k_n}, \dot{u}_n > 0, \\ D_n & \Delta_n + \frac{D_n}{k_n} \le u_n \le u_{nm}, \dot{u}_n > 0, \\ D_n + k_n(u_n - u_{nm}) & u_{nm} - \frac{D_n}{k_n} \le u_n \le u_{nm}, \dot{u}_n < 0, \\ 0 & u_n \le u_{nm} - \frac{D_n}{k_n}, \dot{u}_n < 0, \end{cases} \tag{1.59}$$

where k_n and D_n are the stiffness of the linear part of the characteristic and its yield limit respectively; Δ_n is the coordinate of the initial contact between the two surfaces; u_{nm} is the maximum value of the function (1.57) over the period.

The characteristics of dry friction may be written as follows

$$f_\tau = \vartheta f_n(u_n, \dot{u}_n)\ \mathbf{sgn}\ \dot{u}_\tau(t) = \vartheta f_n(u_n, \dot{u}_n)\ \mathbf{sgn}\ \dot{u}(t). \tag{1.60}$$

Projecting the forces (1.59) and (1.60) on to the direction of velocity v (Fig.1.30), we find the total force acting in the direction of movement. This total force is a result of the workpiece's plastic deformation and the frictional forces on the contacting surfaces

$$f_\nu = f_n\sin\theta + f_\tau\cos\theta = f_n\sin\theta[1 + \vartheta\ \mathtt{ctg}\ \theta \mathbf{sgn}\ \dot{u}(t)]. \tag{1.61}$$

Taking (1.57) and (1.59) into account, (1.61) can be rewritten as follows

$$f_\nu(u_n, \dot{u}_n) = f(u, \dot{u})[1 + \vartheta\ \mathtt{ctg}\ \theta\ \mathbf{sgn}\ \dot{u}] = f(u, \dot{u}) + f_1(u, \dot{u}), \tag{1.62}$$

where $f(u, \dot{u})$ is determined by the expression (1.5), and the following substitution should be made,

$$k_0 = k_n\sin^2\theta, \quad D = D_n\sin\theta. \tag{1.63}$$

Parameters $\Delta = \Delta_n/\sin\theta$ and $u_m = u_{nm}/\sin\theta$ have the same meaning as in (1.5).

The equality (1.62) can be considered as the effective dynamic characteristic of the interaction between the drawing die and the workpiece. The first term in (1.62) describes the material deformation forces, while the second represents the frictional forces. These components are shown in Fig.1.31,a,b; the total combined characteristic (1.62) is depicted in Fig.1.31,c. Arrows are used to show the directions of loading and unloading over a deformation cycle. The following notation is used in Fig.1.31,c:

$$D_1 = D\vartheta \; \mathtt{ctg}\; \theta, \tag{1.64}$$

$$D_0 = D + D_1 = D(1 + \vartheta \mathtt{ctg}\; \theta). \tag{1.65}$$

Parameter (1.65) determines the drawing force required without any ultrasonic influence.

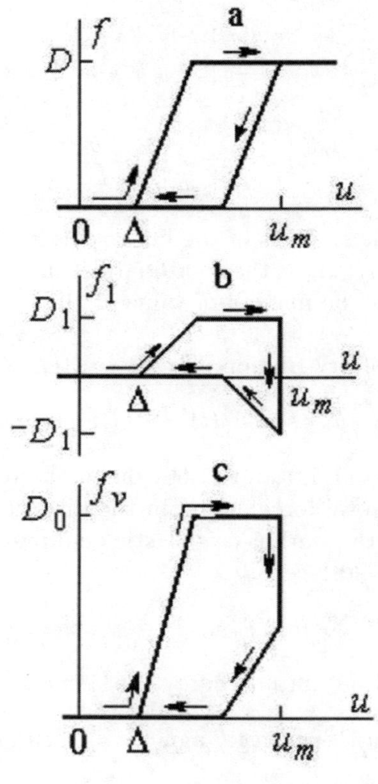

Fig. 1.31.

The static force required to maintain an ultrasonic drawing process is given by the formula

$$P = \frac{1}{T} \int_0^T f_\nu[u(t), \dot{u}(t)]dt. \tag{1.66}$$

Evaluating the integral (1.66) using (1.62), (1.5) and (1.14) gives

$$P = \frac{k_0}{2\pi}\left[\begin{array}{l}(1+\vartheta\ \mathtt{ctg}\ \theta)[\Delta\tau_1 - (\Delta + \frac{D}{k_0})\tau_2 + a(\cos\tau_1 - \cos\tau_2)+ \\ +\frac{v}{2\omega}(\tau_2^2 - \tau_1^2)] + (1-\vartheta\ \mathtt{ctg}\ \theta)[u_m\tau_3 - (u_m - \frac{D}{k_0})\tau_4 + \\ +a(\cos\tau_3 - \cos\tau_4) + \frac{v}{2\omega}(\tau_4^2 - \tau_3^2)] + \frac{2D}{k_0}\tau_3\vartheta\ \mathtt{ctg}\ \theta\end{array}\right], \quad (1.67)$$

where parameters $\tau_i (i = 1, ..., 4)$ can be found from equations (1.15), Δ is related to the speed of the process by the equality (1.17), and parameters k_0, D are given by (1.63).

The dependence of static force P on the drawing speed v for different values of the ratio $D/k_0 a$ is shown in Fig.1.32. Calculations were carried out for the following parameter values: coefficient of friction $\vartheta = 0.1$, drawing die angle $\theta = 15^0$. As for the processes of plastic deformation, considered in section 1.2, both continuous and impulse regimes of deformation can take place during ultrasonic drawing. The relation (1.67) holds in all cases, but for regimes of continuous deformation, the equalities (1.18) should be taken into account. In Fig.1.32, the dashed line separates regions of impulse (below) and continuous (above) deformation.

If the deformation speed $v \leq a\omega$, then according to (1.6), (1.15), and (1.18), the variable P, as determined by (1.67) takes real values. For $v = a\omega$, (1.18) and (1.15) yield $\tau_1 = \tau_2 = -\tau_3 = -\tau_4 = -\pi$. Substituting these parameters into (1.67) we obtain $P = D_0$, where D_0 is given by (1.65). A further increase in the drawing speed does not change the static force.

For $v = 0$, taking (1.17) in to consideration, (1.15) gives

$$\tau_1 = \arcsin\frac{\Delta}{a}, \quad \tau_2 = \tau_3 = \frac{\pi}{2}, \quad \tau_4 = \pi - \tau_1. \quad (1.68)$$

Substituting (1.68) into (1.67) we get the relations (1.12), determining the threshold values of static force. Residual deformations appear if the forces exceed these values. The threshold values do not depend on the frictional forces that appear on the contact surfaces. It can easily be shown that ultrasonic vibration eliminates the influence of friction in the process of material deformation when inside its elastic zone. In this case, the reduced dynamic characteristic (1.62) has the form as shown in Fig.1.33. The thin line shows the component $f(u)$ which is determined by elastic forces. Characteristics of this type, with triangular hysteresis loops, are typical for systems in which dry friction are present (Panovko (1960)). Describing the behaviour of such a material under vibrational loading, in a way similar to that in section 1.2.2, it is found that the static force and the deformation are related by the formulae (1.10) and (1.11). These are the same formulae, as determined by (1.63), used to relate the deformation of a purely elastic material, with its stiffness k_0. Such a deformation process is described by the characteristic shown in Fig.1.12,a. Finally, it is found that the plastic deformation in a drawing process starts with the same value of static force as in the absence of friction. Comparison of Fig.1.32 with Fig.1.12 makes this clear.

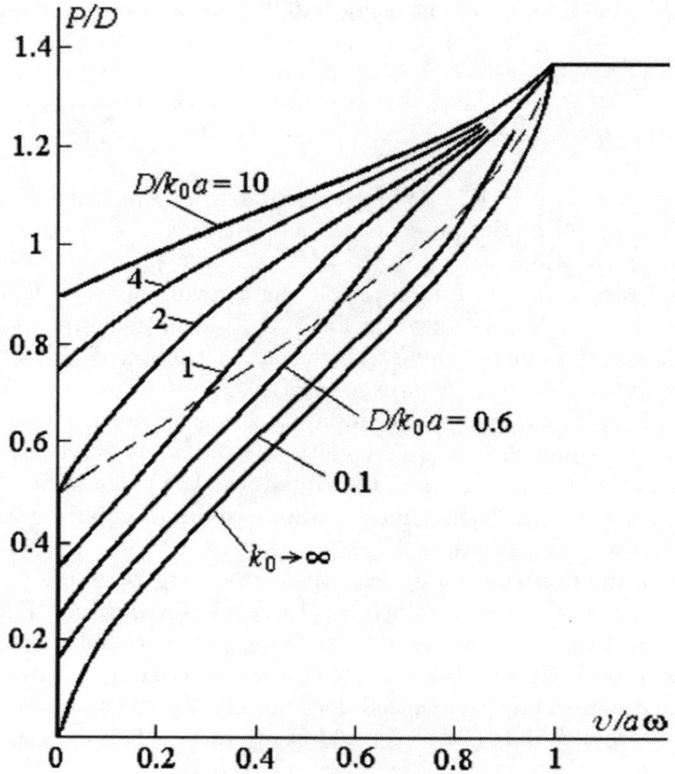

Fig. 1.32.

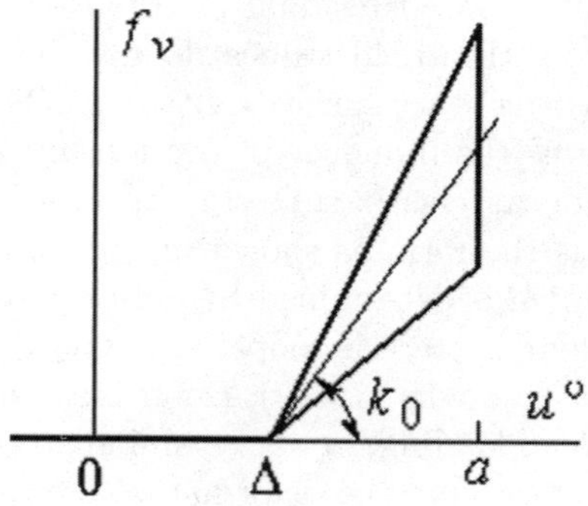

Fig. 1.33.

A decrease in the required drawing force under ultrasonic influence happens due to both a reduction in the yield limit of the material, and a decrease in the frictional force. How these components contribute to the total decrease depends on the relations between the materials elastic-plastic properties, its coefficient of friction, the parameters of vibration in the drawing die and the dies geometrical characteristics. In certain conditions the contribution of one of these components can prevail, as often happens in experiments (Nosal & Rimsha (1966), Robinson (1964)).

The decrease in the required static force whilst drawing is most pronounced when drawing rigid-plastic materials. Fig.1.32 shows a curve marked with the index $k_0 \to \infty$ corresponding to this type of material. The curve is plotted using equations (1.24) and (1.25), obtained in section 2.5, in which the yield limit D should be replaced with D_0 from (1.65).

The approach suggested in this section can be used for the description of various ultrasonic processes in which both forces of deformation, and frictional forces are pronounced. The results obtained here allow us to explain many of the effects observed experimentally. As an example, the plots in Fig.1.32 are close to the dependence of cutting force, on cutting speed in the turning of a workpiece with the imposition of ultrasonic vibration applied to the cutting tool. The original results are taken from paper Isaev & Anokhin (1961) and are shown in Fig.1.5. (See also section 5.1). The same results allow us to explain the drastic improvement in properties of brittle materials under the influence of ultrasonics in various processes (Markov (1962, 1966, 1980), Severdenko et al. (1973, 1976), Agranat (1974), Pantale et al. (2004), Perepeliatnik (1961)).

A similar approach can be applied to describe the effect of vibrational transportation (Blekhman (1999)). This can also happen under the influence of ultrasonics, if the vibration exciter is in contact with the element being displaced, and the transducer is orientated correctly (Arkhangelsky (1963), Bancevicius & Ragulskis (1981), Pohlman & Lehfeldt (1966)).

2

Dynamic characteristics of ultrasonic machines

Size determines an object, but scale determines art.

Robert Smithson (1938-1973)

2.1 Ultrasonic machine arrangements

1. As shown in the previous chapter, the output capacity of an ultrasonic process increases when the tool vibration speed $\omega a = 2\pi f a$ is increased. The lower frequencies in the ultrasonic range $f = 1.5 \cdot (10^4 - 10^5)$ Hz are the most suitable for the realization of intensive high - frequency vibration. It is for this reason that the processes and devices described above are classified as being ultrasonic.

Some typical ultrasonic machine arrangements will now be considered and some features particular to this special group of vibration machines will be discussed.

Fig.2.1, a shows the arrangement of an ultrasonic erosion machine used to cut brittle materials (Rosenberg et al. (1964)). The vibration exciter, a magnetostrictive transducer 1, is fixed in the body 2 of the acoustic head using the shoulder 3 and the thin-walled cup 4. The winding of the transducer is supplied with an alternating current, at ultrasonic frequency, by the generator 5. The alternating magnetic field induced by the current in the core of the transducer, which is made from magnetostrictive material, is transformed into a mechanical vibration in the core. The waveguide, or concentrator , 6 transmits this vibration to the tool 7. The concentrator takes the form of a bar with a variable cross section. It is specially designed to transmit vibration from the transducer, to the tool, with an increase in amplitude (Merkulov (1957), Merkulov & Kharitonov (1959), Teumin (1959)).

The workpiece 10 is placed under the tool, on a plate 8, in a tray 9, within an abrasive slurry . The body of the acoustic head is adjusted to the base's guides 11 and is subjected to a static force P which drives the tool in the direction necessary to machine the workpiece. This is the way in which the ultrasonic cutting of brittle materials, as shown in Fig.1.1, a, is implemented.

The acoustic head is the main component of any ultrasonic machine. The arrangement of a plastic deformation machine (Fig.1.2), containing the same

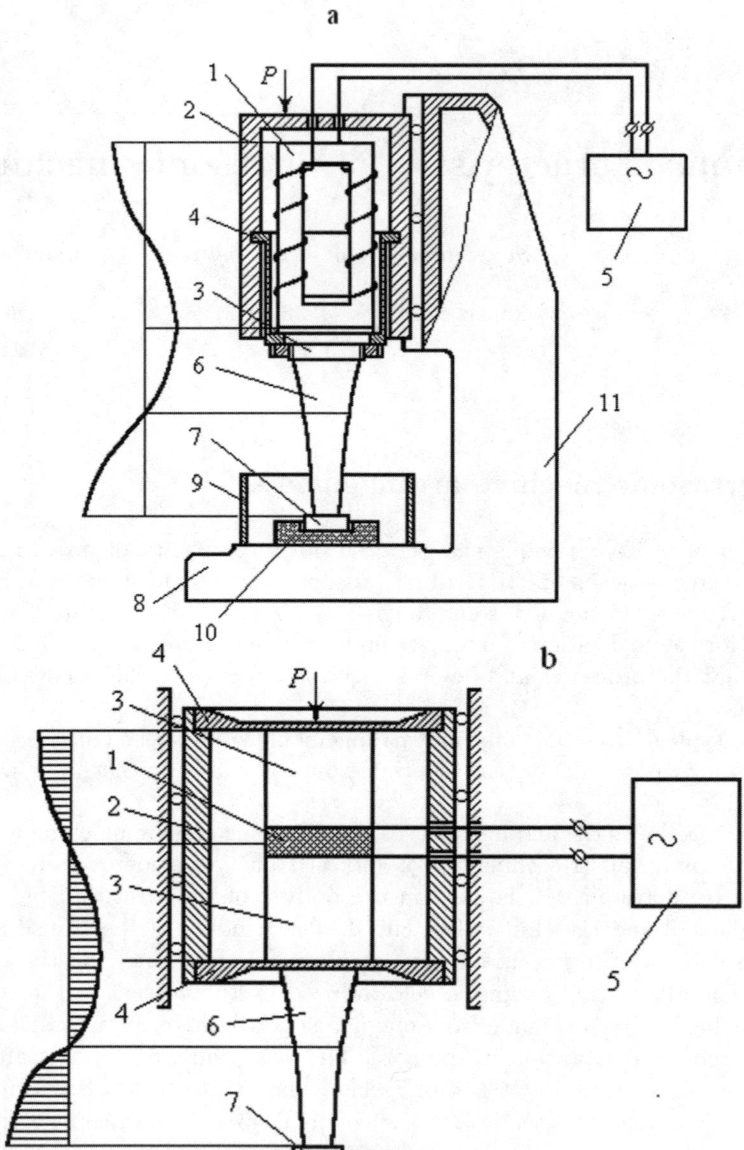

Fig. 2.1.

components, has already been shown in section 1.2. Ultrasonic surface hardening machines (Fig.1.1, b), sheet forming machines (Fig.1.4, a), and machines for the welding of polymeric materials (Fig.1.6, b) are built in a similar manner.

Depending on the type of process, the power of the vibrating system being used varies between tens of Watts and tens of Kilowatts. Piezoelectric transducers are often used to excite vibration in a system with a small power consumption. An arrangement of such a system (Donskoy et al. (1968)) is shown schematically in Fig.2.1, b. The piezoelectric element 1 is positioned between parts 3, which are connected with the body 2 of the acoustic head, by means of elastic membranes 4. The electric generator 5 feeds the piezoelectric element through the end plates. The vibration generated by the piezoelectric element is transmitted through the concentrator 6, to the tool 7.

Achieving the maximum amplitude of tool vibration is necessary for the ultrasonic process to be effective. For this reason, the vibration system is adjustment to operate within its resonant regime. The system is excited at a frequency that coincides with one of its eigenfrequencies. Adjustment of the system's parameters is usually made under idle conditions; it is assumed that the influence of the working loads on the resonant frequency of the system is negligible. This assumption is only correct if the loads applied to the tool during the working process are dissipative and cause a decrease in vibration amplitude. This takes place, for example, during the ultrasonic welding of metals (Fig.1.6, a) in which the only forces acting on the welding tip are due to dry friction.

In plastic deformation and fracture processes (Fig.1.1, 1.4), the role of the elastic force is essential. Investigations carried out in Astashev (1968, 1972), Astashev & Sakaian (1967) show that an increase in driving force leads to a decrease in vibration amplitude. This happens because, rather than dissipating the additional energy in the working zone, the system's resonant frequency is changed. This change is a consequence of the pronounced nonlinear distortions which are caused by the tool - workpiece interaction.

The diverse nonlinear effects that take place in an ultrasonic system during its working process will be discussed in detail in the Chapter 3.

2. Waveguides are used to transmit ultrasonic energy from its source to the load generated by the working process. They are one of the main components of any ultrasonic system and have very specific properties. The bar elements used in the vibration system have lengths that are comparable with the length of an elastic wave within the material. Diagrams showing the vibration amplitude within the cross sections of the vibration system are shown in Fig.2.1. The total length of the complete system is approximately equal to the vibration's wavelength. The vibration system is designed to generate standing waves with displacement nodes and antinodes located in specific cross sections of the bar system.

The acoustic nature of an ultrasonic system is favourable for the energy transport from the vibration exciter to the working zone. A machine designed to create deep holes (Petruha et al. (1970), Tchechina & Yahimovich (1974*b*,*a*)) is shown in Fig.2.2. The vibration is transmitted to the diamond tool 4 using the concentrator 2 and a bar - waveguide 3. The workpiece 5 rotates with an angular speed Ω around the axes of the hole while the cutting tool, together with the acoustic head, is driven with constant speed v along the axes. The difference in the way that the tool approaches the workpiece, compared to that shown in Fig.2.1, must be appreciated. From now on, if the tool is driven by a constant axial force, as in Fig.2.1, the system will be categorised as the force approach. If the speed of tool progression into the workpiece is specified, as in Fig.2.2, the system will be categorised as the kinematic approach. Both progression mechanisms will be considered in Chapter 3 and it will be shown that they cause different dynamic properties in a vibration system. This should be considered when adjusting a machine to its most efficient resonant regime.

An ultrasonic drilling machine is built in a similar manner. A waveguide transmits ultrasonic energy to the working zone via a pipe that is fed into the hole. The upper end of the pipe is connected with the vibration exciter and the lower end carries a turbodrill. The longitudinal vibration of the pipe leads to a decrease in the frictional forces at the side wall and improves the feed of the pipe in to the hole.

An improvement of this nature requires additional energy expenditure within the system. This additional energy is used to fulfil the working process as well as to compensate for the loses caused by the internal friction in the waveguide's material. The vibration assisted feed of a bar system, subject to a distributed dry friction force (Astashev & Hertz (1978), Astashev et al. (1981)), will be considered in section 3.4.

The high penetrability of ultrasonics predetermines additional energy loses in the compliant joints between the vibration system and the acoustic head. Loses are also present in the workpiece's fastening assembly. These loses are caused by the variable reaction components present at the fastening points. Acoustic isolation arrangements (Severdenko et al. (1976)) are used to reduce these loses. The idea behind acoustic isolation is to fasten the vibration system at its nodal points, i.e., in the cross sections with minimal vibration amplitude. The arrangement shown in Fig.2.3 shows how vibration is transmitted from the electro-acoustic transducer 1 through the concentrator 2 and the treated workpiece 3 to the reflector 4. The reflector 4 consists of a bar with a length equal to half the wavelength of the vibration within the material. As a result, a standing wave is generated within the concentrator and the reflector. A diagram showing this standing wave is included in Fig.2.3. The concentrator and reflector are fastened to the machine using the membranes 5 and 6 which are situated in the nodal positions of the standing wave. This provides acoustic isolation between the vibrating components and the main body of the machine. The main body includes the acoustic head housing 7 and the base 8 which are now only subjected to a static load P. An ultrasonic plastic deformation machine (Fig.1.2) described in section 1.2 is constructed in a similar manner.

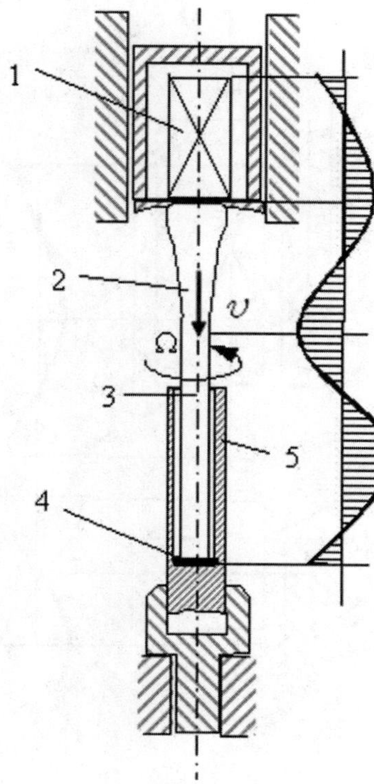

Fig. 2.2.

The main advantage of using the arrangement shown in Fig.2.3 for the processes of die forging, press work and welding of polymeric materials (Holopov (1972), Kuznetsov & Loschilov (1972), Severdenko et al. (1973, 1976)) is that any variation in the static load leaves the vibration amplitude at the working end of the concentrator almost unchanged. This is the case if the vibration system is excited at the machine's resonant frequency as found under its idling conditions. This is customary in ultrasonic machines. In this arrangement, the contact made between the half-wave reflector and the concentrator does not disturb the vibration regime established in the transducer - concentrator system. Vibration is now performed in unison with the reflector, working as a single bar system. In such a regime, the working ends of the concentrator and the reflector move in phase, and the amplitude of their relative displacements is small. The efficiency of the ultrasonic vibration's effect on the workpiece's material is insignificant.

A different vibration regime was revealed in Astashev & Babitsky (1974); the working ends of a concentrator and reflector perform vibration in opposite phase. (The corresponding diagram of the reflector tip's displacement under

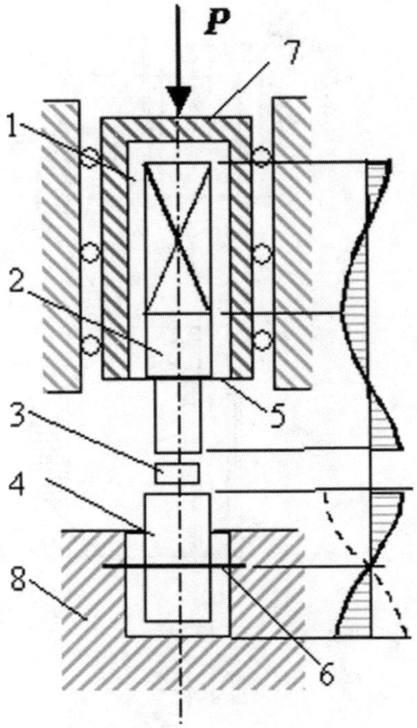

Fig. 2.3.

these circumstances is shown in Fig.2.3 by a dashed line). The workpiece is subjected to a vibro-impact action. This is most intensive at resonance, which occurs under specific conditions which are dependant on the characteristics of the vibration system, static pressing force and excitation frequency. Under these precise conditions, the ultrasonic influence is most pronounced. The vibro-impact process taking place during the interaction of the waveguides will be considered in section 3.3.

The workpiece itself constitutes a passive waveguide diverting some of the energy from the active vibration system through the working zone. The arrangement for an ultrasonic drill used to machine holes in long bars is shown in Fig.2.4. The vibration present in the tubular concentrator 2 is excited by the transducer 1. Abrasive slurry is conveyed to the cutting zone (Kazantsev (1963*a*)) via a hole in the concentrator. The periodic impacts between the concentrator and the surface being treated are transmitted through the abrasive grains, and lead to the erosion of bar 3. Longitudinal vibration is simultaneously excited in this bar. Again, adjustments to the vibration system should be made so that the vibro-impact process is maintained at resonance. This ensures that the impacts are of the highest intensity possible. It is obvious

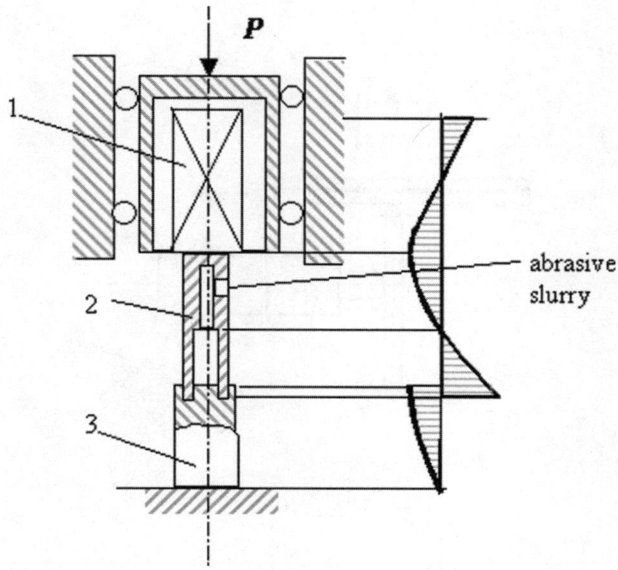

Fig. 2.4.

that the conditions in which resonance occurs will change as the concentrator progresses into bar 3.

Vibration excitation is also observed within a workpiece during an ultrasonic drawing process (Severdenko et al. (1973, 1976)). An arrangement of an ultrasonic machine used for wire drawing is shown in Fig.2.5. The drawing die 3, is attached to the output end of a concentrator 2, whose vibration is excited by an electro-acoustic transducer 1. The wire 4 is drawn through the die and moves from the feed roller 5 to the take up reel 6. If the rollers are a significant distance from the drawing die, a travelling wave may be exited in the wire. This wave gradually decays as it moves further away from its source.

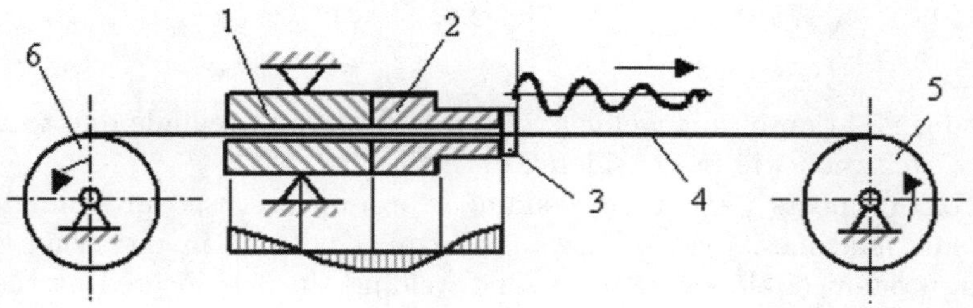

Fig. 2.5.

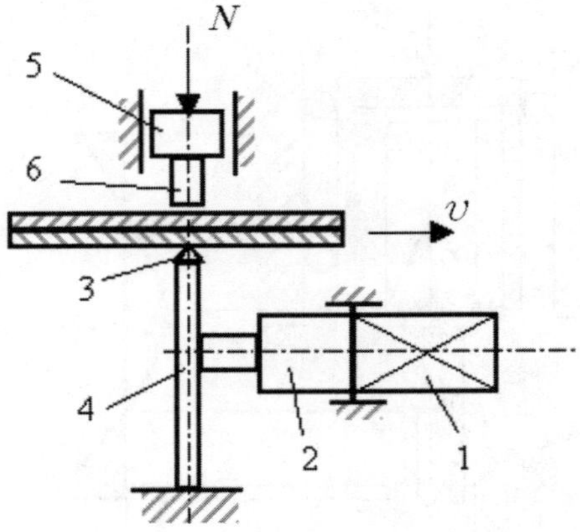

Fig. 2.6.

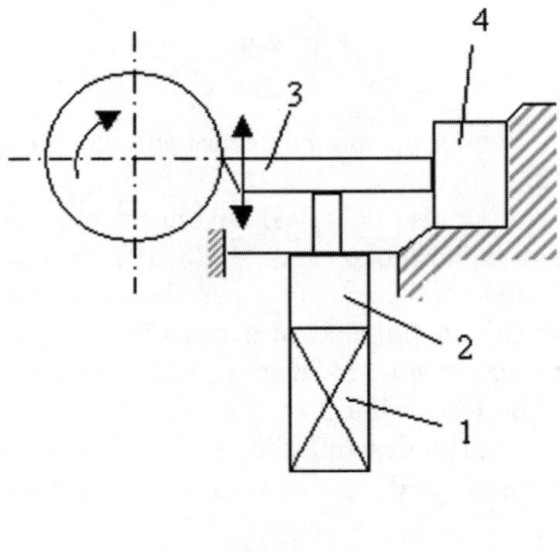

Fig. 2.7.

The generation of a travelling wave in a passive waveguide due to vibro-impact excitation will be considered in section 2.2.

3. bar elements performing bending vibration are used in a number of ultrasonic machines. Fig.2.6 shows the arrangement of an ultrasonic metal welding system (Holopov (1972)). The welding tip 3 is located at the end of bar 4. The bending vibration present in bar 4 is excited by transducer 1 and concentrator 2, which perform longitudinal vibration. The carriage 5 compresses the metal sheets 7 on to the tip by means of the intermediate bar 6 which applies the static force N.

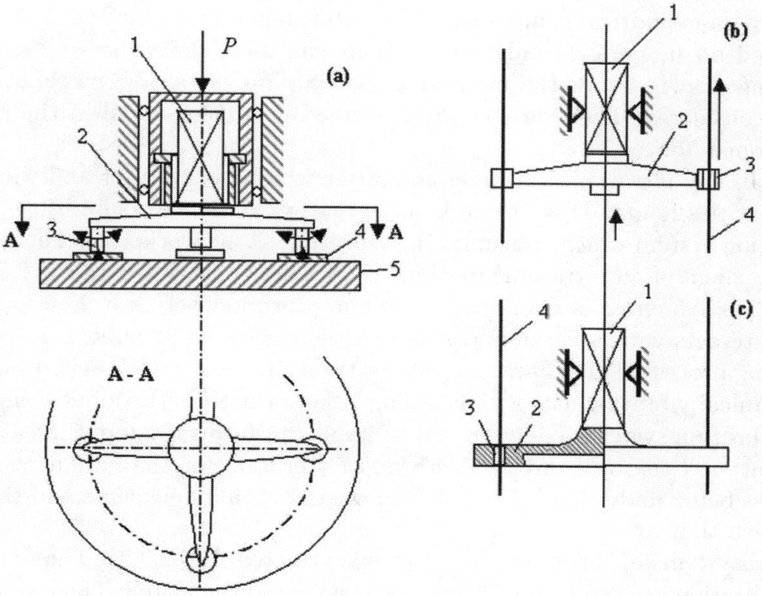

Fig. 2.8.

A similar arrangement is used for turning hard-to-machine materials (Kumabe (1979), Markov (1962, 1966, 1980)) in which ultrasonic vibration is superimposed on to the cutting tool (Fig.2.7). The cutter 3 is fastened into the tool holder 4 and is subjected to bending vibration which is transmitted from the transducer 1 using the concentrator 2.

In an ultrasonic machine used for the processing of brittle materials (Fig.2.8, a), an electro-acoustic transducer 1 excites vibration which is transmitted to the tools 3 by crisscross connecting bars 2 which perform bending vibration. The workpieces 4 are situated in four positions on the table 5. The cross sectional areas of the bars 2 are reduced as they approach the positions of the tools. In this arrangement, the bars work as waveguides, increasing the vibration amplitude applied to the tools. The tools, if excited according to this arrangement, perform both progressive vibration in the direction of cut, and angular vibration as shown by the arrows. This type of vibration system is only used in applications which do not require high accuracy machining.

A similar vibration system may also be used in other processes. As an example, in an ultrasonic drawing machine (Fig.2.8, b) the concentrators 2, which perform bending vibration, carry the drawing dies 3 through which the wire is pulled. To exclude undesirable angular vibration in the dies, the system may be transformed as shown in Fig.2.8, c. In this case, a transducer 1 excites vibration within a membrane 2, which is reinforced on its peripheral edges by a cylindrical component. The latter, together with the dies 3 evenly situated on it, perform only longitudinal vibration along the wire 4 which is being drawn. Thus, the main arrangements for ultrasonic machines have been considered. Their designs are described in further detail in the papers mentioned above.

4. In any ultrasonic machine, no matter what it is used for and whatever peculiarities it may have; there is an actuator exciting the vibration, and a vibration system which transmits the vibration to the workpiece. The general arrangement of an ultrasonic machine can be drawn as shown in Fig.2.9. The vibration system 2 is connected with the vibration actuator 1 at point 0, and interacts with a varying load 3 which is generated at point l during the working process. The vibration actuator transforms electrical oscillations into mechanical vibration at position 0. The actuator itself is the input element of the vibration system. The vibration system transforms the input, creating at the output l the most favourable form of vibration for the fulfilment of the process being undertaken. Fig.2.9, a shows the chain of elements and the way in which they are connected.

Subsystems of this chain may now be extracted (Fig.2.9,b). These include the vibration exciter 1, the vibration system 2 and the load 3. The interaction forces between the individual components may now be used to replace the components. These forces are applied at connecting points. The forces acting on the neighbouring subsystems at these connecting points have the same magnitude but opposite direction, i.e., they differ only by their sign

$$f_{21} = -f_{12}, \quad f_{23} = -f_{32}. \tag{2.1}$$

The first subscript indicates the number of the subsystem which experiences the action of the force, the second being the number of the subsystem generating the force. For the sake of simplicity, the following notation shall be used,

$$f_{21} = f_0, \quad f_{23} = f_l. \tag{2.2}$$

Consider the subsystem 2, which is subjected to the forces f_0 and f_l. The one-dimensional motion of the x-th element of the harmonically vibrating system may be described as,

$$u_x(t) = m_x + u_x^0(t) = m_x + a_x e^{j(\omega t + \phi_x)} \tag{2.3}$$

where a_x and ϕ_x are the amplitude and phase, respectively, of the periodic component of the element x motion, m_x is the constant component of element x motion and $j = \sqrt{-1}$.

It is more convenient to rewrite the periodic component in (2.3) in the form

$$u_x^0(t) = \tilde{a}_x e^{j\omega t}, \tag{2.4}$$

where

$$\tilde{a}_x = a_x e^{j\phi_x} = a_x(\cos\phi_x + j\sin\phi_x) \tag{2.5}$$

is the complex amplitude of vibration.

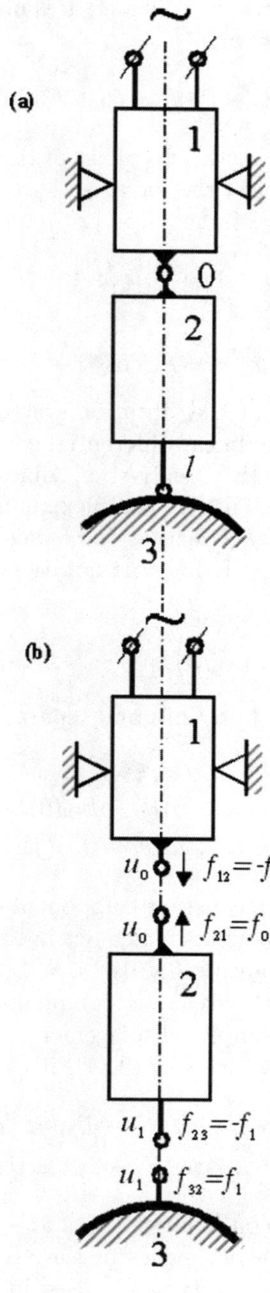

Fig. 2.9.

It is assumed that the force $f_s(t)$, which acts on element s of the system, is also a harmonic function of time,

$$f_s(t) = P_s + f_s^0(t) = P_s + F_s e^{j(\omega t+\theta_s)}, \tag{2.6}$$

where F_s and θ_s are the amplitude and phase of the periodic component, and P_s is the constant component of the force.

The periodic component of the force (2.6) may now be rewritten in a similar manner to (2.4),

$$f_x^0(t) = \tilde{F}_s e^{j\omega t}, \tag{2.7}$$

where,

$$\tilde{F}_s = F_s e^{j\theta_x} = F_s(\cos\theta_s + j\sin\theta_s). \tag{2.8}$$

5. It is now considered that the vibration system is linear and stationary, and its motion is described by linear differential equations with constant coefficients. In order to describe the steady-state vibration, dynamic compliance (receptance) $L_{sx}(j\omega)$ is used. This is a complex quantity, which establishes the relationship between the force f_s acting on element s, and the displacement u_x of the arbitrary element x. Using this notation, the motion of element x can be described as

$$u_x(t) = L_{0x}(j\omega)f_0(t) - L_{lx}(j\omega)f_l(t). \tag{2.9}$$

Substituting (2.4) - (2.7) into (2.9) and separating the constant and variable components,

$$m_x = L_{0x}(0)P_0 - L_{lx}(0)P_l, \tag{2.10}$$
$$\tilde{a}_x = L_{0x}(j\omega)\tilde{F}_0 - L_{lx}(j\omega)\tilde{F}_l. \tag{2.11}$$

Equality (2.10) relates the constant component of element x displacement with the constant force components which act in the connecting points of the system. In this case, the quantity $L_{sx}(0), (s = 0, l)$ is the static compliance and can be obtained from the dynamic compliance $L_{sx}(j\omega)$ when $j\omega = 0$. Equality (2.11) relates the complex displacement amplitude to the force. For elements $x = 0$ and $x = l$, by virtue of (2.11), it may be written that,

$$\tilde{a}_0 = L_{00}(j\omega)\tilde{F}_0 - L_{l0}(j\omega)\tilde{F}_l, \tag{2.12}$$
$$\tilde{a}_l = L_{0l}(j\omega)\tilde{F}_0 - L_{ll}(j\omega)\tilde{F}_l. \tag{2.13}$$

The values of dynamic compliances in (2.9) - (2.13) depend on the inertial, elastic and dissipative properties of the vibration system. The exact expressions for elements that are typically used in ultrasonic machine vibration systems will be constructed in the sections 2.2 and 2.3. As follows from (2.11) - (2.13), the dynamic compliance $L_{sx}(j\omega)$ is equal to the ratio between element x complex displacement amplitude $\tilde{a}_x$ and the force $\tilde{F}_s$ which acts on element s,

$$L_{sx}(j\omega) = \frac{\tilde{a}_x}{\tilde{F}_s}. \tag{2.14}$$

In the following, dynamic stiffness $W_{sx}(j\omega)$ is often used. Dynamic stiffness is the inverse of dynamic compliance,

$$W_{sx}(j\omega) = L_{sx}^{-1}(j\omega) = \frac{\tilde{F}_s}{\tilde{a}_x}. \tag{2.15}$$

It must be noted that to describe linear system vibration in acoustics (Lependin (1978)), it is customary to use a value of impedance $Z_{sx}(j\omega)$. This is equal to the ratio between complex force amplitude $\tilde{F}_s$ and vibration velocity $j\omega\tilde{a}_x$, i.e.

$$Z_{sx}(j\omega) = \frac{\tilde{F}_s}{j\omega\tilde{a}_x}. \tag{2.16}$$

Comparing (2.15) and (2.16), it can be seen that the dynamic stiffness and impedance of a system are related where,

$$W_{sx}(j\omega) = j\omega Z_{sx}(j\omega). \tag{2.17}$$

The quantities (2.14) and (2.15) used in this book not only describe the periodic components of the processes under consideration, but also provide relationships between the constant components as seen in (2.10). Chapter 3 will show that these relationships are essential in the analysis of nonlinear working processes.

Equalities (2.12) and (2.13) are four-pole circuit equations (Kharkevich (1973)). They relate the two input and two output quantities of a vibration system. Two additional conditions on $\tilde{F}_0, \tilde{F}_l, \tilde{a}_0, \tilde{a}_l$ are necessary to form the complete description of a four-pole circuit. These conditions are determined by the physical nature of the processes under consideration. In order to quantify them, the neighbouring subsystems are considered.

6. Consider subsystem 1, which is a vibration exciter. This subsystem may also be regarded as a four-pole circuit with electrical input, and mechanical output sides. Let ν, q be the instantaneous voltage and charge at the input of the exciter. Steady-state regimes are considered, in which the voltage and charge are harmonic functions,

$$\nu(t) = \tilde{\nu}_0 e^{j\omega t}, \quad q(t) = \tilde{q}_0 e^{j\omega t}. \tag{2.18}$$

where $\tilde{\nu}_0$ and $\tilde{q}_0$ are the complex amplitudes of voltage and charge at the input of the four-pole circuit.

The harmonic force, with amplitude $\tilde{F}_{12} = -\tilde{F}_0$, acts at the mechanical output side of the four-pole circuit. Vibration with an amplitude $\tilde{a}_0$ is excited. Two linear relations exist between the electrical input and the mechanical output quantities (Kikuchi (1969)),

$$\begin{aligned} \tilde{F}_{12} &= \alpha_{11}(j\omega)\tilde{a}_0 + \alpha_{12}(j\omega)\tilde{q}_0, \\ \tilde{\nu}_0 &= \alpha_{21}(j\omega)\tilde{a}_0 + \alpha_{22}(j\omega)\tilde{q}_0. \end{aligned} \tag{2.19}$$

The complex coefficients $a_{mn}(j\omega)$ $(m,n=1,2)$ are dependant on the parameters of the mechanical and electrical parts of the actuator. Their exact form for a typical electro-acoustic transducer will be constructed in section 2.4. The physical meaning of these coefficients becomes clear by letting either $\tilde{a}_0$ or $\tilde{q}_0$ equal to zero:

$\alpha_{11}(j\omega) = \tilde{F}_{12}/\tilde{a}_1|_{\tilde{q}_0=0}$ is the dynamic stiffness of the mechanical part of the actuator in the absence of current;

$\alpha_{12}(j\omega) = \tilde{F}_{12}/\tilde{q}_0|_{\tilde{a}_0=0}$, $\alpha_{21}(j\omega) = \tilde{\nu}_0/\tilde{a}_0|_{\tilde{q}_0=0}$ are the coefficients of electro-mechanical coupling and determine the mutual influence of the electrical and mechanical parts of the actuator $\alpha_{12}(j\omega) = \pm\alpha_{21}(j\omega)$ depending on the type of the transducer;

$\alpha_{22}(j\omega) = \tilde{\nu}_0/\tilde{q}_0|_{\tilde{a}_0=0}$ is the inverse of the actuator's electrical circuit's complex capacitance with no displacement. This quantity may be given in terms of the impedance $Z(j\omega)$ of the circuit,

$$\alpha_{22}(j\omega) = j\omega\tilde{\nu}_0/\tilde{i}_0 = j\omega Z(j\omega), \tag{2.20}$$

where $\tilde{i}_0 = j\omega\tilde{q}_0$ is the complex amplitude of the current.

According to (2.19), in the absence of transducer loading, i.e., when $F_{12} = -F_0 = 0$, the vibration amplitude $\tilde{a}_0^*$ is equal to,

$$\begin{aligned}\tilde{a}_0^* &= \frac{\alpha_{12}(j\omega)}{\alpha_{12}^2(j\omega) - \alpha_{11}(j\omega)\alpha_{22}(j\omega)}\tilde{\nu}_0 = \\ &= -\frac{\alpha_{12}(j\omega)}{\alpha_{11}(j\omega)}\tilde{q}_0 = -\frac{\alpha_{12}(j\omega)}{j\omega\alpha_{11}(j\omega)}\tilde{i}_0.\end{aligned} \tag{2.21}$$

The value of one of the electrical quantities is usually given. This may be the voltage $\tilde{\nu}_0$, if the transducer is fed from a voltage source, or current $\tilde{i}_0 = j\omega\tilde{q}_0$ if a current source is used. In either case, the vibration amplitude $\tilde{a}_0^*$ of the unloaded transducer characterizes the excitation of the system. Following from (2.19), by virtue of (2.1), (2.2), and (2.21), the transducer's vibration amplitude when loaded by a mechanical vibration system can be evaluated by,

$$\tilde{a}_0 = \tilde{a}_0^* - L_n(j\omega)\tilde{F}_0, \tag{2.22}$$

where $L_n(j\omega)$ is the transducer's dynamic compliance.

$$L_n(j\omega) = \alpha_{11}^{-1}(j\omega) \tag{2.23}$$

when a current source is used, and

$$L_n(j\omega) = \left[\alpha_{11}(j\omega) - \frac{\alpha_{12}(j\omega)\alpha_{21}(j\omega)}{\alpha_{22}(j\omega)}\right]^{-1} \tag{2.24}$$

when a voltage source is used.

Equality (2.22) relates the displacement amplitude $\tilde{a}_0^*$ and force $\tilde{F}_0$ at the input to the four-pole circuit (2.12), (2.13). Using this relation, it is found from (2.12) that,

$$\tilde{F}_0 = \frac{\tilde{a}_0^* - L_{l0}(j\omega)\tilde{F}_l}{L_{00}(j\omega) + L_n(j\omega)}, \tag{2.25}$$

$$\tilde{a}_0 = \frac{\tilde{a}_0^* L_{00}(j\omega) + L_n(j\omega) L_{l0}(j\omega)\tilde{F}_l}{L_{00}(j\omega) + L_n(j\omega)}. \tag{2.26}$$

Substituting (2.25) into (2.13), the tool's vibration amplitude is obtained in the form,

$$\tilde{a}_l = \tilde{a}_l^* - L_M(j\omega)\tilde{F}_l, \tag{2.27}$$

where,

$$\tilde{a}_l^* = \frac{L_{0l}(j\omega)}{L_{00}(j\omega) + L_n(j\omega)}\tilde{a}_0^* \tag{2.28}$$

is the tool's vibration amplitude under idle conditions, i.e., in the absence of load ($\tilde{F}_l = 0$);

$$L_M(j\omega) = L_{ll}(j\omega) - \frac{L_{0l}^2(j\omega)}{L_{00}(j\omega) + L_n(j\omega)} \tag{2.29}$$

is the complete dynamic compliance of the vibration system including the actuator and is adjusted to the tool. The equality

$$L_{sx}(j\omega) = L_{xs}(j\omega) \tag{2.30}$$

is taken into account in (2.29). This equality holds true for any linear system due to the generalized Rayleigh reciprocity principle (Rayleigh (1945)).

7. Equation (2.27) determines the tool's vibration amplitude when it's subjected to the load generated by the interaction between the tool and workpiece. As shown in the previous chapter, in general, the force generated by the tool-workpiece interaction is described by a nonlinear dynamic characteristic,

$$f_l = f_l[u_l(t), \dot{u}_l(t)]. \tag{2.31}$$

The characteristics of a typical process will be considered in section 3.1. With certain assumptions about the type of process under consideration, characteristics (2.31) allow the relationships between the complex force amplitude $\tilde{F}_l$ and displacement $\tilde{a}_l$ to be found,

$$\tilde{F}_l = \tilde{F}_l(\tilde{a}_l). \tag{2.32}$$

Formulae (2.27) and (2.32) form a closed system of equations for the unknown process parameters: $\tilde{a}_l$ and $\tilde{F}_l$. These problems constitute the main content of this book and will be considered in great detail in the following chapters.

In some cases, the load's characteristic may be obtained by the linear dependence of the force f_l on the tool's displacement $u_l(t)$,

$$f_l(t) = W_H(j\omega)u_l(t) = L_H^{-1}(j\omega)u_l(t), \tag{2.33}$$

where $W_H(j\omega)$ and $L_H(j\omega)$ are the dynamic stiffness and compliance of the load.

This is the case, for example, if the tool-workpiece interaction takes place during the continuous deformation of a specimen (Fig.2.4, b) within an elastic zone of the material's characteristics. Taking into account the type of processes under consideration (2.4), the following relation can be written for a linear elastic-viscous material

$$f_l(t) = k_0 u_l(t) + \beta_0 \dot{u}_l(t) = (k_0 + j\omega\beta_0) u_l(t), \tag{2.34}$$

where k_0 and β_0 are the static stiffness and coefficient of resistance to deformation for the specimen.

Comparing (2.33) and (2.34), it is found that,

$$W_H(j\omega) = k_0 + j\omega\beta_0. \tag{2.35}$$

Using (2.4) and (2.7), the complex amplitude of the force (2.33) may be expressed as,

$$\tilde{F}_l = W_H(j\omega)\tilde{a}_l. \tag{2.36}$$

Substituting (2.36) into (2.27), the tool's complex vibration amplitude is found under linear loading conditions,

$$\tilde{a}_l = \tilde{a}_l^* \left[1 + \frac{W_H(j\omega)}{W_M(j\omega)}\right]^{-1}, \tag{2.37}$$

where $W_M(j\omega) = L_M^{-1}(j\omega)$ is the dynamic stiffness of the vibration system when adjusted to the tool.

If follows from (2.37), that the loading on the vibration system changes the tools vibration amplitude from the amplitude $\tilde{a}_l^*$ observed under idle conditions. It must be noted that the loading on the tool can cause both a decrease or an increase in vibration amplitude, and is dependant on the ratios of the vibration system's dynamic stiffness and the load experienced by the system. This also holds true for a machine working under the influence of nonlinear loads. Different load characteristics will be considered in the Chapter 3. As follows from the relationships constructed above, the dynamic behaviour of ultrasonic machines is hugely determined by the dynamic behaviour of their parts, or specifically, their dynamic compliances and stiffnesses. These expressions may now be obtained for some typical ultrasonic machine elements.

2.2 Longitudinal vibration within an ultrasonic bar system

1. ultrasonic machine vibration systems contain many different parts performing longitudinal vibration. The most common components of this type are bar

waveguides which have specific cross-sections that vary along the length of the bar. These should be considered as distributed parameter systems. Strictly speaking, in any real system, the inertial, elastic and dissipative properties are continuously distributed within its whole volume. Under certain conditions, a distributed system of this nature may be simplified and the corresponding elements may be replaced by lumped masses, concentrated elasticity, and a damper, to form an idealized model. (The conditions in which this is possible are obtained below). Tools that are attached to the concentrator , membranes that fasten the vibration system to the acoustic head and some parts of the concentrator -transducer joint for example may be modelled by concentrated elements.

For this reason, an ultrasonic machine's vibration system is a combination of distributed and concentrated elements. The main aim of this section is to quantify the dynamic compliance and stiffness for typical ultrasonic bar systems which perform longitudinal vibration.

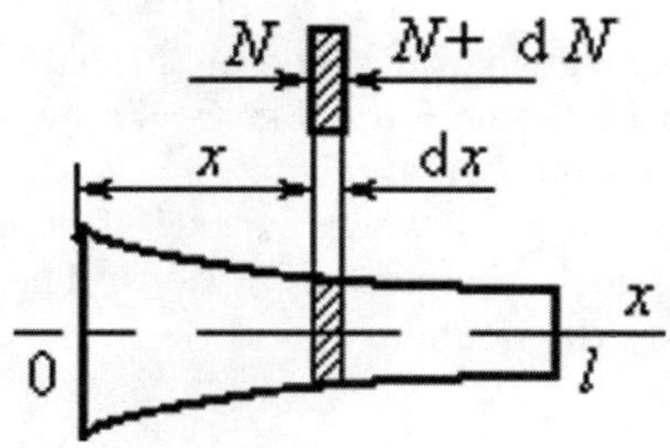

Fig. 2.10.

Consider a bar (Fig.2.10) with length l and cross-section $S = S(x)$ given by a known function. In order to describe the vibration in the bar, the cross-sections are assumed to remain flat and the particles lying in the cross-sections are assumed to move only in the x-direction. Let $u(x,t)$ describe the displacement of the cross-section x from its position within the undeformed bar. Consider harmonic vibration of the type

$$u(x,t) = \tilde{a}_x e^{j\omega t} = a_x e^{j(\omega t + \phi_x)}. \tag{2.38}$$

($\tilde{a}_x = \tilde{a}(x)$ is the complex amplitude; a_x and ϕ_x are the amplitude and phase, respectively, of the vibration at the specified cross-section x). The dependence of the stress σ on the relative strain ε is given in the form

$$\sigma = \tilde{E}\varepsilon = \tilde{E}\frac{\partial u}{\partial x}, \tag{2.39}$$

where

$$\tilde{E} = E\left(1 + j\frac{\psi}{2\pi}\right). \tag{2.40}$$

E is the elastic modulus and ψ is the coefficient of absorption which characterizes the dissipation of energy due to the internal friction within the material (Panovko (1960), Frolov (1995)).

Under these assumptions, the equation of motion for element dx can be written in the form (see Fig.2.10),

$$\rho S_x \frac{\partial^2 u(x,t)}{\partial t^2} dx = dN_x = \frac{\partial(S_x \sigma)}{\partial x} dx,$$

which, after the substitution of (2.39), yields the equation for longitudinal vibration within a bar,

$$\rho \frac{\partial^2 u(x,t)}{\partial t^2} = \tilde{E}\left[\frac{\partial^2 u(x,t)}{\partial x^2} + \frac{S'_x}{S_x} \cdot \frac{\partial u(x,t)}{\partial x}\right], \tag{2.41}$$

where, ρ is the bar material's density and $S_x = S(x)$ is the area of cross-section x; $S'_x = dS(x)/dx$.

Consider an arrangement in which one of the bar's ends is free and the other is subjected to a harmonic force. Let $x = s$ be the cross-section that is subjected to the excitation $f_s(t) = F_s e^{j\omega t}$. $s = l$ when the excitation is at the right end of the bar and $s = 0$ when the excitation is at the left end. The coordinate of the free end is therefore $x = l - s$. The solution of equation (2.41) should now satisfy the boundary conditions:

$$\frac{\partial u(x,t)}{\partial x}|_{x=l-s} = 0,$$

$$\tilde{E} S_x \frac{\partial u(x,t)}{\partial x}|_{x=s} = (-1)^{1-s/l} F_s e^{j\omega t}. \tag{2.42}$$

The factor $(-1)^{1-s/l}$ in the second condition expresses the well-known sign rule (Feodosiev (1974)) for forces acting at different ends of a bar. Substituting (2.38) into (2.41) and (2.42) gives,

$$\tilde{E}\left[\tilde{a}''_x + \frac{S'_x}{S_x}\tilde{a}'_x\right] + \rho\omega^2 \tilde{a}_x = 0, \tag{2.43}$$

$$\tilde{a}'_x|_{x=l-s} = 0, \ \tilde{E} S_x \tilde{a}'_x|_{x=s} = (-1)^{1-s/l} F_s. \tag{2.44}$$

The final solution depends on the waveguide's form and is determined by the function S_x which determines the change in cross-section. Once the complex amplitude $\tilde{a}_x$ is determined, formulae (2.14) and (2.15) allow the dynamic compliance $L_{sx}(j\omega)$ and dynamic stiffness $W_{sx}(j\omega)$ to be found.

Solutions for the types of waveguides which are most commonly used in ultrasonic machines are now constructed.

2. Considering a bar with a constant cross section $S_x = S = const$. Equation (2.41) takes the form

$$\tilde{E}\tilde{a}_x'' + \rho\omega^2\tilde{a}_x = 0.$$

Solutions are sought which satisfy the boundary conditions (2.44), (2.14) is taken into account and the dynamic compliance is subsequently obtained,

$$L_{sx}(j\omega) = (-1)^{s/l}\frac{\lambda}{\omega^2\rho S}\frac{e^{\lambda x} + e^{-\lambda x}e^{2\lambda(l-s)}}{e^{\lambda s} - e^{-\lambda s}e^{2\lambda(l-s)}}, \tag{2.45}$$

where

$$\lambda = j\omega\sqrt{\frac{\rho}{\tilde{E}}} = \frac{j\omega}{c\sqrt{1 + j\psi/2\pi}}. \tag{2.46}$$

$c = \sqrt{E/\rho}$ is the speed of sound in the bar material.

If the exciting force acts on either the right $s = l$, or the left $s = 0$ of the bar, (2.45) gives,

$$L_{lx}(j\omega) = -\frac{\lambda}{\omega^2\rho S}\frac{e^{\lambda x} + e^{-\lambda x}}{e^{\lambda l} - e^{-\lambda l}} = -\frac{\lambda}{\omega^2\rho S}\frac{\cosh\lambda x}{\sinh\lambda l}, \tag{2.47}$$

or

$$L_{0x}(j\omega) = -\frac{\lambda}{\omega^2\rho S}\frac{e^{\lambda(l-x)} + e^{-\lambda(l-x)}}{e^{\lambda l} - e^{-\lambda l}} = -\frac{\lambda}{\omega^2\rho S}\frac{\cosh\lambda(l-x)}{\sinh\lambda l}, \tag{2.48}$$

respectively.

In the following, it is assumed that the dissipative component of the force in (2.39) is substantially smaller than the elastic component, i.e., the absorption coefficient ψ is small ($\psi \ll 1$). Expanding the expressions containing ψ with a small parameter power series and retaining only the linear terms, equality (2.46) becomes,

$$\lambda = \left(j + \frac{\psi}{4\pi}\right)\frac{\omega}{c} \tag{2.49}$$

Two particular cases will first be considered.

Let the bar length be so small, that $|\lambda|l \ll 1$. By virtue of (2.49) this condition is equivalent to

$$l \ll \frac{c}{\omega} = \frac{\Lambda}{2\pi}, \tag{2.50}$$

where, Λ is the wavelength at the frequency ω.

As $0 \le x \le l$, $|\lambda|x \ll 1$. Expanding the exponents in (2.47) and (2.48) in a small parameter power series and retaining the fist order terms gives,

$$L_{lx}(j\omega) = L_{0x}(j\omega) = L(j\omega) = (-M\omega^2)^{-1}, \tag{2.51}$$

where $M = \rho l S$ is the mass of the bar.

Equality (2.51) represents the dynamic compliance of the localized mass M. A short bar i.e. with a length l much shorter than the wavelength of vibration within it, can be idealized as a rigid body.

Secondly, a semi-infinite bar is considered. It is excited in cross section $x = 0$, the limit $l \to \infty$ is applied to (2.48) and (2.49) is accounted for. Within acceptable accuracy,

$$L_{0x}(j\omega) = \frac{1}{j\omega w} \exp\left[-\left(j + \frac{\psi}{4\pi}\right)\frac{\omega x}{c}\right], \tag{2.52}$$

where, $w = S\sqrt{E\rho}$ is the bar's wave resistance.

The dynamic compliance at the point of excitation, $x = 0$, may be written,

$$L_{00}(j\omega) = (j\omega w)^{-1}, \tag{2.53}$$

as such, the semi-infinite bar is seen to behave as a localized linear damper in which the damping coefficient is equal to the wave resistance.

If the semi-infinite bar is excited by a harmonic force $f(t) = F_0 e^{j\omega t}$, the longitudinal vibration that exists within the bar may be described by the function

$$u(x,t) = L_{0x}(j\omega)F_0 e^{j\omega t},$$

or, taking (2.52) into account,

$$u(x,t) = \frac{F_0}{j\omega w} \exp\left(-\frac{\psi\omega}{4\pi c}x\right) \exp\left[j\omega\left(t - \frac{x}{c}\right)\right]. \tag{2.54}$$

This expression represents a travelling harmonic wave. The waves amplitude is seen to decrease exponentially as it propagates along the bar, and travels at a speed c.

Fig.2.11 shows the form of a travelling wave in a semi-infinite bar; the dashed line corresponds to the amplitude envelope,

$$a_x = \frac{F_0}{\omega w} \exp\left(-\frac{\psi\omega}{4\pi c}x\right). \tag{2.55}$$

It must be noted that, according to (2.55), a finite bar of sufficient length may be approximated as being a semi-infinite bar,

$$l > \frac{4\pi c}{\psi\omega} = \frac{2\Lambda}{\psi}. \tag{2.56}$$

With such a length, the wave reflected from the free end of the bar, i.e. returning to the source, is damped by a factor of at least e^2.

3. The lengths, l, of the waveguides used in ultrasonic machines are usually comparable to the wavelength Λ of the elastic wave. From (2.45) and with analogy to (2.54), it may be shown that the displacement of cross sections of a bar of finite length which is subjected to a harmonic excitation $f_s(t)$, are formed

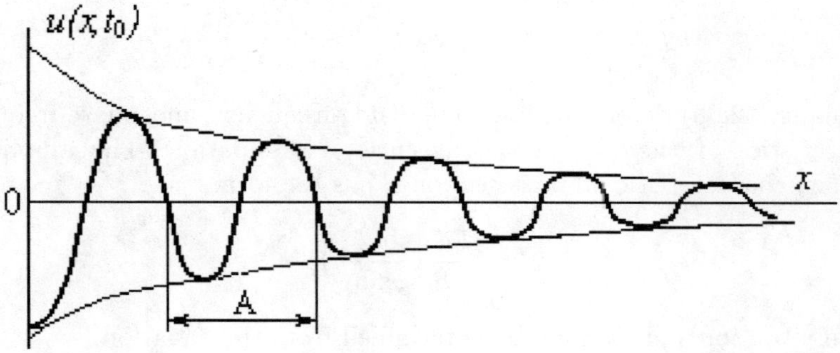

Fig. 2.11.

by the superposition of waves travelling forwards, $u_{1x}(x,t) = \tilde{a}_{1x}e^{j\omega(t-x/c)}$ and backwards, $u_{2x}(x,t) = \tilde{a}_{2x}e^{j\omega(t+x/c)}$ along the bar.

For the convenience of future consideration, the expressions for dynamic compliance, (2.47) and (2.48) may be rewritten by substituting (2.49). Only the linear terms of the small parameter expansion, ψ, are retained. As a result, (2.47) gives,

$$L_{lx}(j\omega) = -\frac{l}{ES\xi}\frac{\cos\xi\frac{x}{l} - j\frac{\psi}{4\pi}\left(\cos\xi\frac{x}{l} - \xi\frac{x}{l}\sin\xi\frac{x}{l}\right)}{\sin\xi - j\frac{\psi}{4\pi}\xi\cos\xi}, \tag{2.57}$$

where $\xi = \omega l/c$.

Comparing (2.47) and (2.48), it can be seen that the expression for $L_{0x}(j\omega)$ can be obtained from (2.57) by substituting $l - x$ for x. This expression is not given here. In the following, the values of dynamic compliance for the bar's end cross sections are most predominantly required. Following from (2.47) and (2.48),

$$L_{00}(j\omega) = L_{ll}(j\omega), \quad L_{l0}(j\omega) = L_{0l}(j\omega). \tag{2.58}$$

The first of these equalities is the consequence of the homogeneous bar's symmetry; the second reflects the displacement reciprocity principle (Rayleigh (1945)), which holds for all linear systems.

If the force acting on the bar is $f_l(t) = F_l e^{j\omega t}$, the complex amplitude of the bar's displacement is equal to,

$$\tilde{a}_x = F_l L_{lx}(j\omega). \tag{2.59}$$

In the absence of dissipation ($\psi = 0$), and taking (2.57) into account,

$$\tilde{a}_x = a_x e^{-j\phi_x} = \tilde{a}_0 \cos \xi \frac{x}{l} = -\frac{F_l l \cos \xi \frac{x}{l}}{ES\xi \sin \xi}. \tag{2.60}$$

Equality (2.60) determines the amplitude-frequency and phase-frequency characteristics of the bar in the absence of dissipation. The amplitude-frequency characteristic at cross section x has the form

$$a_x = \frac{F_l l}{ES} \left| \frac{\cos \xi \frac{x}{l}}{\xi \sin \xi} \right|. \tag{2.61}$$

The vibration's phase may be determined from the equation

$$\cos \phi_x = -\mathbf{sgn} \left(\frac{\cos \xi \frac{x}{l}}{\sin \xi} \right). \tag{2.62}$$

These characteristics are shown in Fig.2.12, a, b for a bar's end cross section $x = l$ (dashed lines).

At the bar's resonant frequencies,

$$\xi = \xi_n = \pi n, \quad (n = 0, 1, 2, ...), \tag{2.63}$$

the amplitude of vibration, $a_x \to \infty$. These peaks correspond to normal modes of vibration (Fig.2.13, a). It follows from (2.60) that,

$$a_x = a_0 \cos \pi n \frac{x}{l}. \tag{2.64}$$

According to (2.61), "antiresonant" frequencies also exist,

$$\xi = \xi_n^* = (2n + 1)\pi/2, \tag{2.65}$$

when vibration is excited at point $a_l^* = 0$.

The vibration amplitude at the bar's free end is given by

$$a_0^* = \frac{2F_l l}{\pi ES(n + 1)}. \tag{2.66}$$

The antiresonant vibration form is shown in Fig.2.13, b. At the frequencies given by (2.65), the bar's node appears at the point of excitation. As the excitation frequency is increased, the node is shifted along the bar. At antiresonant frequencies, the vibration is seen to exhibit a change in phase. It must be noted, that "antiresonance" represents the well known effect of dynamic vibration damping (Kolovsky (1999), Frolov (1995)). The amplitude-frequency and phase-frequency characteristics, in the presence of energy dissipation ($\psi \neq 0$), are shown in Fig.2.12, a, b by continuous lines. The energy dissipation restricts the resonant vibration's amplitude. According to (2.59), (2.57), and (2.63), the end cross section's resonant amplitudes are given by,

$$a_0^0 = a_l^0 = \frac{4\pi F_l l}{\psi \xi_n^2 ES} = \frac{4F_l l}{\pi \psi ES n^2} \tag{2.67}$$

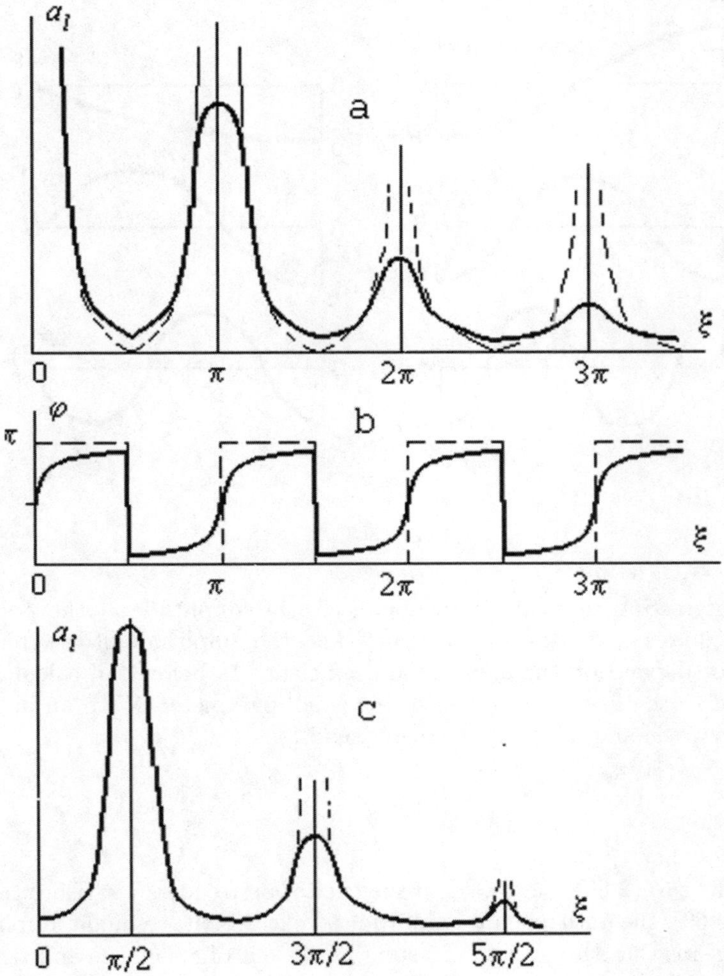

Fig. 2.12.

and decrease proportionally to the square of the mode number n. The influence of dissipation on the mode of vibration is insignificant (Fig.2.13, a).

The amplitudes of vibration corresponding to "antiresonant" frequencies are given by,

$$a_l^* = \frac{\psi F_l l}{4\pi ES}, \quad a_0^* = \frac{\delta_F}{\xi_n^*} = \frac{2F_l l}{\pi(n+1)ES}. \tag{2.68}$$

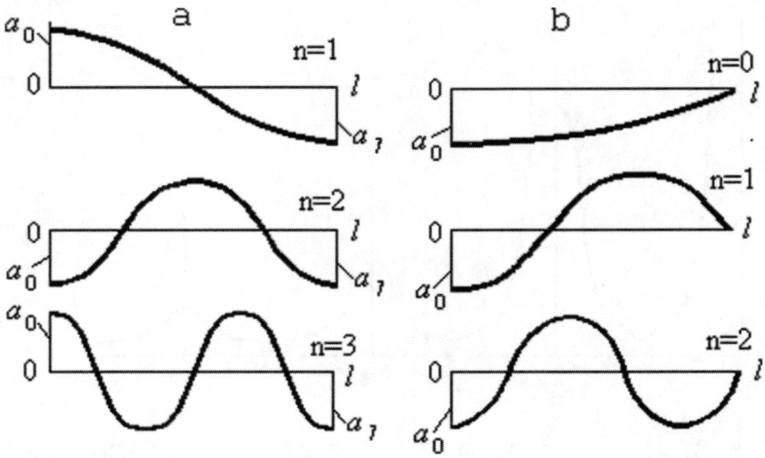

Fig. 2.13.

It is interesting that, at "antiresonance", the amplitude at the point of excitation does not depend on frequency, and the amplitude of the free end a_0^* does not depend on the absorption coefficient. As before, all calculations are limited to the first order terms of the small parameter. With an increase in frequency, the amplitude a_0^* decreases, and for

$$\xi \geq \frac{4\pi}{\psi}, \tag{2.69}$$

it transpires that $a_0^* \leq a_l^*$.

According to (2.67) and (5.31), if the excitation frequency satisfies the inequality (2.69), the antiresonant amplitude a_l^* exceeds the resonant amplitude a_l^0. This means that the vibration system moves out of resonance and there is no need to take the finite extent of the bar into account. Using the notation $\xi = \frac{\omega l}{c}$, inequality (2.69) can be rewritten in the form

$$l \geq \frac{4\pi c}{\psi\omega} = \frac{2\Lambda}{\psi},$$

coinciding with (2.56).

Another peculiarity of inequality (2.69) may now be noted. The dynamic compliance at the point of excitation $x = l$, for the resonant frequencies (2.63), may be found from (2.57),

$$L_{ll}(j\omega) = \frac{4\pi l}{jES\psi\xi^2} = \frac{4\pi c}{j\psi\omega ES\xi} = \frac{4\pi}{j\psi\omega w\xi}. \tag{2.70}$$

It follows from (2.70), that when (2.69) represents an equality, the dynamic compliance obtained $L_{ll} = (j\omega w)^{-1}$, is equal to the dynamic compliance of a semi-infinite bar (2.53) at the point of excitation.

4. Analogously to (2.57), and under the assumption of small levels of dissipation; expressions for the dynamic compliances of an ultrasonic machine's components can be written in the form,

$$L_{sx}(j\omega) = \frac{P_{sx}(j\omega)}{Q(j\omega)} = \frac{ReP_{sx}(j\omega) + jImP_{sx}(j\omega)}{ReQ(j\omega) + jImQ(j\omega)}, \tag{2.71}$$

where $P_{sx}(j\omega)$ and $Q(j\omega)$ are complex variables. Q does not depend on the coordinate x or the point of excitation s.

The real components in (2.71) characterize the elastic and inertial properties of the vibration system. The imaginary components characterize the dissipative properties and are considered to be small. Therefore, the resonant condition, providing that the amplitudes are maximal, has the form

$$ReQ(j\omega_n) = 0, \tag{2.72}$$

where ω_n are resonant frequencies.

An expression can be obtained in order to describe the amplitudes at resonance,

$$a_x^0 = \frac{F_s|P_{sx}(j\omega_n)|}{|ImQ(j\omega_n)|} = F_s\frac{\sqrt{Re^2P_{sx}(j\omega_n) + Im^2P_{sx}(j\omega_n)}}{|ImQ(j\omega_n)|}. \tag{2.73}$$

For a given ω, the equality

$$ReP_{sx}(j\omega) = 0 \tag{2.74}$$

determines the coordinates of the vibration mode's nodal points; for a given coordinate, the equality may also be used to determine the frequency at which this point becomes a node. When $x = s$, equation (2.74) gives frequencies of "antiresonances". This is because the vibration mode's node coincides with the point of excitation. Taking account of (2.74) and (2.71), the vibration amplitude at the nodal points may be written,

$$a_x^* = F_s\frac{|ImP_{sx}(j\omega)|}{|Q(j\omega)|} = F_s\frac{|ImP_{sx}(j\omega)|}{\sqrt{Re^2Q(j\omega) + Im^2Q(j\omega)}}. \tag{2.75}$$

The vibration amplitudes at the nodal points of a system at resonance are, according to (2.72) and (2.75),

$$a_x^{0*} = F_s\frac{|ImP_{sx}(j\omega_n)|}{|ImQ(j\omega_n)|}. \tag{2.76}$$

From (2.73), for the points situated at antinodes (neglecting second order terms),

$$a_x^* = F_s\frac{|ReP_{sx}(j\omega_n)|}{|ImQ(j\omega_n)|}. \tag{2.77}$$

Analogously, for the nodes at frequencies far from resonance,

$$a_x^* = F_s \frac{|ImP_{sx}(j\omega)|}{|ReQ(j\omega)|}. \tag{2.78}$$

As found in Chapter 1, the resonant amplitudes (2.77) at antinodes are the most important. They exceed the amplitudes (2.75) of the normal modes nodal points by a factor of ψ^{-1}, and the antiresonant amplitudes (2.78) by a factor of ψ^{-2}. Therefore, neglecting the small second-order terms in order to describe a process close to resonance, the following expression for the dynamic stiffness may be used. This follows from (2.71):

$$W_{sx}(j\omega) = L_{sx}^{-1}(j\omega) = U_{sx}(\omega) + jV_{sx}(\omega), \tag{2.79}$$

where,

$$\begin{aligned} U_{sx}(\omega) &= ReW_{sx}(j\omega) = \frac{ReQ(j\omega)}{ReP_{sx}(j\omega)}, \\ V_{sx}(\omega) &= ImW_{sx}(j\omega) = \\ &= \frac{ImQ(j\omega)ReP_{sx}(j\omega) - ImP_{sx}(j\omega)ReQ(j\omega)}{Re^2P_{sx}(j\omega)}. \end{aligned}$$

For the resonant frequencies and amplitudes, instead of using (2.72) and (2.77); the following expressions may be obtained by using (2.78),

$$U_{sx}(\omega_n) = 0; \quad a_x^0 = F_x/V_{sx}(\omega_n). \tag{2.80}$$

5. When designing ultrasonic machines, it can often be assumed that the system is operating under conditions of kinematic excitation, i.e., the motion of one of its components is prescribed. Such an approach is only acceptable whilst considering heterogeneous waveguides which are attached to an actuator, when the added system's influence on the actuator is small. Designing systems of this nature may be achieved by using the same values for dynamic compliance and stiffness as those used for force excitation.

For the sake of definiteness, the vibration exciter is assumed to actuate vibration $u_0(t) = \tilde{a}_0 e^{j\omega t}$ with an amplitude of $\tilde{a}_0$ to the bar vibration system at cross section $x = 0$ (Fig.2.14).

A periodic reaction of amplitude $\tilde{F}_0$, appears at the intersection between the vibration system and exciter. The following equality holds true for the amplitude of the system's cross sectional displacement,

$$\tilde{a}_x = \tilde{F}_0 L_{0x}(j\omega). \tag{2.81}$$

From (2.81) the reaction force's amplitude, at $x = 0$, may be found,

$$\tilde{F}_0 = \tilde{a}_0 / L_{00}(j\omega), \tag{2.82}$$

along with the vibration amplitude

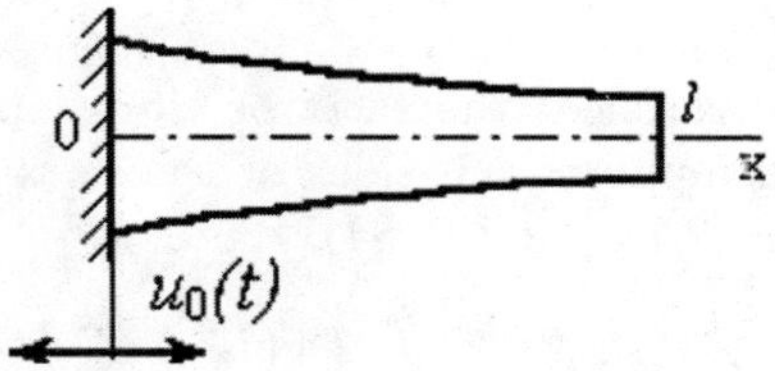

Fig. 2.14.

$$\tilde{a}_x = \tilde{a}_0 \frac{L_{0x}(j\omega)}{L_{00}(j\omega)}. \tag{2.83}$$

Equation (2.83) determines the amplitude-frequency characteristic for the vibration system under kinematic excitation.

For a homogeneous bar, after substituting values from (2.48), formula (2.83) takes the form,

$$a_x = a_0 \frac{\cosh \lambda(l-x)}{\cosh \lambda l},$$

or, assuming only small dissipation levels,

$$a_x = a_0 \frac{\cos \xi(1-\frac{x}{l}) + j\frac{\psi}{4\pi}\xi(1-\frac{x}{l}) \sin \xi(1-\frac{x}{l})}{\cos \xi + j\frac{\psi}{4\pi}\xi \sin \xi}.$$

The bar's amplitude-frequency characteristic at cross section $x = l$, under kinematic excitation, is shown in Fig.2.12, c. It is described by the equation,

$$a_l = a_0 / \sqrt{\cos^2 \xi + (\frac{\psi}{4\pi}\xi \sin \xi)^2}.$$

The frequencies

$$\xi = (2n-1)\pi/2, \quad (n = 1, 2, ...)$$

correspond to maximum vibration amplitudes, where,

$$a_l = \frac{8a_0}{\psi(2n-1)}.$$

Frequencies $\xi = \pi n$ correspond to minimum vibration amplitudes, for which $a_l = a_0$.

It must be noted, that force (2.82) determines the load which the vibration system exerts on the exciter. Taking (2.79) and (2.82) into account, it is concluded that this load is minimal under the condition,

$$U_{00}(j\omega_n) = ReQ(j\omega_n) = 0, \tag{2.84}$$

This coincides with (2.72).

If this condition is satisfied, the ratio of vibration amplitudes between the output $x = l$, and input $x = 0$ sides of the system may be written in accordance with (2.83), (2.79), and (2.84),

$$K = \frac{a_l}{a_0} = \left|\frac{V_{00}(\omega_n)}{V_{0l}(\omega_n)}\right| = \left|\frac{ReP_{0l}(j\omega_n)}{ReP_{00}(j\omega_n)}\right| = \left|\frac{U_{00}(\omega_n)}{U_{0l}(\omega_n)}\right|. \tag{2.85}$$

If the vibration system is adjusted in this manner, maximal amplitude will be seen at the output end of the waveguide, and minimal loading will be experienced by the vibration exciter. This can be achieved using concentrators - bars with a variable cross section. Parameter K characterizes the degree to which the vibration intensity is transformed as it is transmitted from the exciter to the tool. It is called the concentrator's amplification coefficient (Merkulov (1957), Teumin (1959)).

It must be noted that, the amplification coefficient is not dependant on energy losses within the concentrator's material. This can be proven using formula (2.77), and gives a more general expression for the amplification coefficient,

$$K = \frac{a_l}{a_0} = \left|\frac{ReP_{sl}(j\omega)}{ReP_{s0}(j\omega)}\right| = \left|\frac{U_{s0}(\omega)}{U_{sl}(\omega)}\right|, \quad (s = 0, l), \tag{2.86}$$

which, for $s = 0$ leads to (2.85).

The energy dissipation does, however, determine the load (2.82) which acts on the vibration exciter from the concentrator,

$$\tilde{F}_0 = jV_{00}(\omega)a_0 = jImQ(j\omega)a_0/ReP_{00}(j\omega), \tag{2.87}$$

and has a purely dissipative character.

If there is no dissipation in the concentrator, i.e. $\tilde{F}_0 = 0$, and the system is in a steady state with no working load, the vibration exciter experiences no loading from the vibration system.

6. Some ultrasonic machine concentrators may now be considered. These may be of a form as shown in Fig.2.15.

In the widely used exponential concentrator (Fig.2.15, a), the cross sectional area varies according to the law,

$$S_x = S_0 \exp(-2\gamma x/l). \tag{2.88}$$

The catenoidal concentrator (Fig.2.15, b) is described by the formula,

$$S_x = S_1 \cosh^2[\gamma(x - l)/l]. \tag{2.89}$$

The cosine concentrator (Fig.2.15, c), can be constructed using the cross sectional area given by,

$$S_x = S_0 \cos^2 \alpha x/l \quad (|\alpha| < \pi/2). \tag{2.90}$$

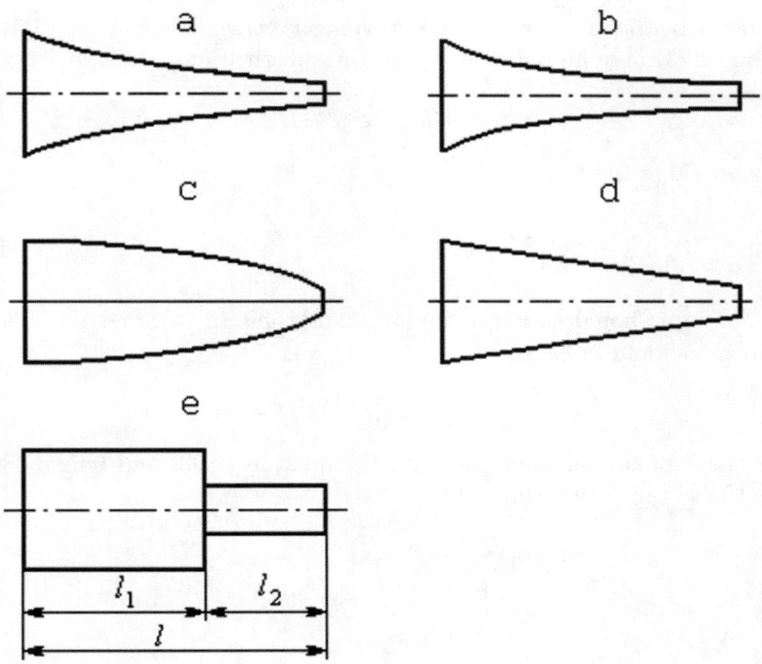

Fig. 2.15.

The cross sectional area of the conic concentrator (Fig.2.15, d) varies according to the law

$$S_x = S_0(1 - \gamma x/l)^2 \quad (\gamma < 1). \tag{2.91}$$

For $\gamma = 0$ formulae (2.88) – (2.91) describe a homogeneous waveguide $S_x = S_0 = const.$

Fig.2.15, e shows a stepped waveguide. It consists of two bars with different cross sectional areas.

It must be noted that the theory governing ultrasonic concentrators (2.88), (2.89), and (2.91) was developed in Merkulov (1957) on the basis of free vibration analysis in the absence of dissipation. This theory is widely used for designing various ultrasonic systems (Gershgal & Freedman (1976), Markov (1962, 1966, 1980), Rosenberg et al. (1964), Teumin (1959)). In this section, the results of Merkulov (1957) will be obtained as particular cases of a more general approach. The dynamic compliances of heterogeneous waveguides will be constructed, taking their dissipative properties into account. These dynamic compliances will then be used for designing concentrators which work under nonlinear elastic-dissipative loads which are generated by the intended process.

To describe the vibration in a heterogeneous waveguide (2.88) – (2.91) and equation (2.43) may be rewritten using the substitutions,

$$S_x = r_x^2, \quad \tilde{a}_x = \tilde{y}_x / r_x. \tag{2.92}$$

(2.43) now takes the form,

$$\tilde{y}_x'' + \tilde{y}_x \left(\frac{\rho \omega^2}{\tilde{E}} - \frac{r_x''}{r_x} \right) = 0. \tag{2.93}$$

Its solution should satisfy boundary conditions (2.44), and the following relationship should be used,

$$\tilde{a}_x' = (\tilde{y}_x / r_x)' = (\tilde{y}_x' r_x - \tilde{y}_x r_x')/r_x^2. \tag{2.94}$$

For the exponential waveguide (2.88), equation (2.93) and boundary conditions (2.44) take the form

$$\tilde{y}_x'' + \tilde{y}_x \left[\frac{\rho \omega^2}{\tilde{E}} - \left(\frac{\gamma}{l} \right)^2 \right] = 0, \tag{2.95}$$

$$\begin{aligned} &\left(y_x' + \frac{\gamma}{l} y_x \right) / r_x |_{x=l-s} = 0, \\ &\tilde{E} S_x \left(y_x' + \frac{\gamma}{l} y_x \right) / r_x |_{x=s} = (-1)^{1-s/l} F_s. \end{aligned} \tag{2.96}$$

Looking for a solution to equation (2.95), which satisfies boundary conditions (2.96), and takes expressions (2.92) and (2.14) into account, gives the exponential waveguide's dynamic compliance,

$$\begin{aligned} L_{sx}(j\omega) &= (-1)^{s/l} e^{\gamma \frac{x+s}{l}} \times \\ &\times \frac{\gamma (e^{\lambda x} - e^{-\lambda x} e^{2\lambda(l-s)}) + \lambda l (e^{\lambda x} + e^{-\lambda x} e^{2\lambda(l-s)})}{\omega^2 \rho S_0 l (e^{\lambda s} - e^{-\lambda s} e^{2\lambda(l-s)})}, \end{aligned} \tag{2.97}$$

where $\lambda = j\sqrt{\frac{\rho\omega^2}{\tilde{E}} - \left(\frac{\gamma}{l}\right)^2}$.

When $s = l$ and $s = 0$, from (2.97) the following dynamic compliances are obtained,

$$L_{lx}(j\omega) = \frac{\gamma \sinh \lambda x - \lambda l \cosh \lambda x}{\omega^2 \rho S_0 l \sinh \lambda} e^{\gamma (1 + \frac{x}{l})}, \tag{2.98}$$

$$L_{0x}(j\omega) = \frac{\gamma \sinh \lambda (x-l) - \lambda l \cosh \lambda (x-l)}{\omega^2 \rho S_0 l \sinh \lambda l} e^{\gamma \frac{x}{l}}. \tag{2.99}$$

Assuming that the absorption coefficient ψ is small in (2.40) and (2.97), expressions (2.98) and (2.99) may be converted into the form of (2.71) by expanding the small parameter terms and taking only the first order values:

$$L_{sx}(j\omega) = P_{sx}(j\omega)/Q(j\omega), \quad (s = l, 0) \tag{2.100}$$

where

$$Q(j\omega) = \frac{ES_0}{l}(\xi^2+\gamma^2)e^{-\gamma} \times$$
$$\times \left[\sin\xi - j\frac{\psi(\xi^2+\gamma^2)}{4\pi\xi}\cos\xi\right], \qquad (2.101)$$

$$P_{lx}(j\omega) = e^{\gamma x/l}\left\{\begin{array}{c}\gamma\sin\xi\frac{x}{l} - \xi\cos\xi\frac{x}{l}+ \\ +j\frac{\psi}{4\pi}\frac{\xi^2+\gamma^2}{\xi}[(1-\gamma\frac{x}{l})\times \\ \times\cos\xi\frac{x}{l} - \xi\frac{x}{l}\sin\xi\frac{x}{l}]\end{array}\right\}, \qquad (2.102)$$

$$P_{0x}(j\omega) = P_{l(x-l)}(j\omega). \qquad (2.103)$$

$\xi = \frac{\omega l}{c_\phi}$, and,

$$c_\phi = \frac{c}{\sqrt{1-\left(\frac{\gamma c}{l\omega}\right)^2}} = c\frac{\sqrt{\xi^2+\gamma^2}}{\xi} \qquad (2.104)$$

is the phase acoustic speed within the exponential waveguide.

It is noted that equality (2.103) simply follows from the comparison of (2.98) with (2.99). It is clear that (2.99) can be obtained from (2.98) by the substitution of $x-l$ for x, i.e., $L_{0x}(j\omega) = L_{l(x-l)}(j\omega)$.

The amplitude-frequency characteristics of the exponential waveguides cross section, $x = l$, are shown in Fig.2.16, a. They are generated using expressions (2.100) – (2.102) in which the waveguide is excited by a harmonic force $f_l = F_l e^{j\omega t}$. Using (2.72) and (2.101), the resonant frequencies of the bar can be found,

$$\xi_n = \pi n \quad (n = 1, 2, ...). \qquad (2.105)$$

These frequencies give maximum vibration amplitudes, and can be found from (2.73) using (2.101), (2.102), (2.105) and (2.88),

$$a_l^0 = \frac{4lF_l}{\psi\pi ES_l n^2[1+(\gamma/\pi n)^2]^2}. \qquad (2.106)$$

Antiresonant frequencies can be found for $x = s = l$ for condition (2.74), which, after the substitution of (2.102), leads to the equation,

$$\tan\xi_n^* = \xi_n^*/\gamma. \qquad (2.107)$$

An example of the graphical solution for equation (2.107) is shown in Fig.2.16, b. Antiresonant amplitudes are calculated using the following expression, which is obtained from (2.77), (2.101), (2.102) and (2.107),

$$a_l^* = \frac{\psi}{4\pi}\frac{F_l l}{ES_1}\left(1 - \frac{\gamma(1-\gamma)}{\xi_n^{*2}}\right). \qquad (2.108)$$

It must be noted that frequency ω is related to the dimensionless variable ξ,

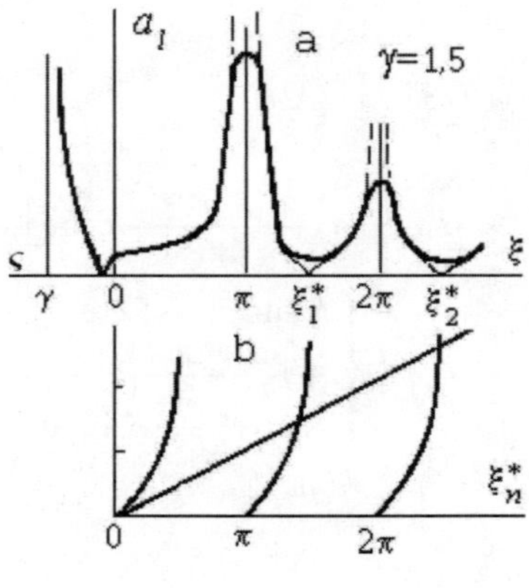

Fig. 2.16.

$$\omega = \frac{c}{l}\sqrt{\xi^2 + \gamma^2}. \tag{2.109}$$

The relations obtained above hold true for $\xi > 0$ and for frequencies,

$$\omega \geq \omega_k = |\gamma| c/l, \tag{2.110}$$

for which the phase speed (2.104) takes real values. In the frequency range below the threshold value, ω_k, the exponential bar's steady state motion does not exhibit wave behaviour. Even so, its motion can still be described by equations (2.100) – (2.104) and (2.109), in which case the following value should be used,

$$\xi = j\varsigma, \tag{2.111}$$

where, $\varsigma \leq |\gamma|$ is a real number.

Assuming $\psi = 0$, from (2.100) – (2.102), when $\omega < \omega_k$,

$$L_{lx}(j\omega) = -\frac{le^{\gamma(1+\frac{x}{l})}}{ES_0(\gamma^2 - \varsigma^2)} \, \frac{\varsigma \cosh \varsigma \frac{x}{l} - \gamma \sinh \varsigma \frac{x}{l}}{\sinh \varsigma}. \tag{2.112}$$

The amplitude-frequency characteristics of the exponential bar are shown in Fig.2.16, a, in which $0 < \omega < \omega_k$. This corresponds to $|\gamma| \geq \varsigma \geq 0$. Using (2.100) – (2.102) for $\xi \to 0$, and (2.112) for $\varsigma \to 0$, the vibration's amplitude a_l at the critical frequency, $\omega = \omega_k$, may be obtained,

$$a_l = \frac{F_l l(1+\gamma)}{ES_l \gamma^2}. \tag{2.113}$$

Interestingly, amplitude (2.113) is not dependant on the dissipation of energy within the bar's material.

The exponential bar may be attached to the vibration exciter and used as a concentrator. If it is excited at a frequency ω, its parameters, according to (2.84), should satisfy the equality (2.105). Using notations (2.104), the concentrator's length, adjusted to its working frequency is given by,

$$l = \frac{\xi_n c_\phi}{\omega} = \frac{c}{\omega}\sqrt{(\pi n)^2 + \gamma^2}. \tag{2.114}$$

Substituting the appropriate quantities from (2.101) and (2.103), the coefficient of amplification (see (2.85)), (5.78) for the exponential concentrator may be found,

$$K = e^\gamma = \sqrt{S_0/S_l}. \tag{2.115}$$

Equalities (5.77) and (5.78) coincide with those obtained in Merkulov (1957), Teumin (1959), but are found using a slightly different approach. They allow the geometrical sizes of the concentrator to be chosen in order to provide the necessary vibration amplification.

The study of vibration within other heterogeneous waveguide types is carried out in a similar manner. The main results of this analysis are now reviewed.

7. For the catenoidal bar (2.89), equation (2.93) can be converted in to the form (2.95). Its solution, which satisfies the boundary conditions obtained from (2.44) by using (2.94), allows the dynamic compliance to be found,

$$L_{lx}(j\omega) = \frac{\lambda l \cosh \lambda x - \gamma \tanh \gamma \sinh \lambda x}{\lambda \tilde{E} S_l \cosh \gamma (1 - \frac{x}{l})(\lambda l \sinh \lambda l - \gamma \tanh \gamma \cosh \lambda l)}, \tag{2.116}$$

$$L_{0x}(j\omega) = \frac{l \cosh \lambda (l - x)}{\tilde{E} S_l \cosh \gamma (1 - \frac{x}{l})(\lambda l \sinh \lambda l - \gamma \tanh \gamma \cosh \lambda l) \cosh \gamma}, \tag{2.117}$$

where λ is obtained from (2.97).

Converting expressions (5.79) and (5.80) in to the form of (2.71), it is found that,

$$Q(j\omega) = -\frac{\xi E S_l}{l} \times \times \left\{ \begin{array}{l} \xi \sin \xi + \gamma \tanh \gamma \cos \xi - \\ -j \frac{\psi}{4\pi \xi^2} \times \left[\begin{array}{l} \xi(\xi^2 + \gamma^2)((1 - \gamma \tanh \gamma) \sin \xi + \xi \cos \xi) - \\ -(\xi^2 - \gamma^2)(\xi \sin \xi + \gamma \tanh \gamma \cos \xi) \end{array} \right] \end{array} \right\}, \tag{2.118}$$

$$P_{lx}(j\omega) = \frac{\left\{ \begin{array}{l} \xi \cos \xi \frac{x}{l} - \gamma \tanh \gamma \sin \xi \frac{x}{l} - j \frac{\psi(\xi^2 + \gamma^2)}{4\pi \xi} \times \\ \times \left[(1 - \gamma \tanh \gamma) \cos \xi \frac{x}{l} - \xi \frac{x}{l} \sin \xi \frac{x}{l} \right] \end{array} \right\}}{\cosh \gamma (1 - \frac{x}{l})}, \tag{2.119}$$

$$P_{0x}(j\omega) = \frac{\xi \cos \xi \frac{l-x}{l} - j \frac{\psi(\xi^2 + \gamma^2)}{4\pi \xi} \left(\begin{array}{l} \cos \xi \frac{l-x}{l} - \\ -\xi \frac{l-x}{l} \sin \xi \frac{l-x}{l} \end{array} \right)}{\cosh \gamma \cosh \gamma (1 - \frac{x}{l})}, \tag{2.120}$$

where ξ is obtained from (2.104).

The amplitude-frequency characteristic is shown in Fig.2.17, a. The resonant and antiresonant frequencies of the catenoidal bar, subjected to a harmonic excitation at cross section $x = l$, can be found by solving equations,

$$\cot \xi_n = -\frac{\xi_n}{\gamma \tanh \gamma}, \tag{2.121}$$

$$\tan \xi_n^* = \frac{\xi_n^*}{\gamma \tanh \gamma}, \tag{2.122}$$

which are obtained from conditions (2.72) and (2.73), after the substitution of the corresponding values from (5.81) and (5.82).

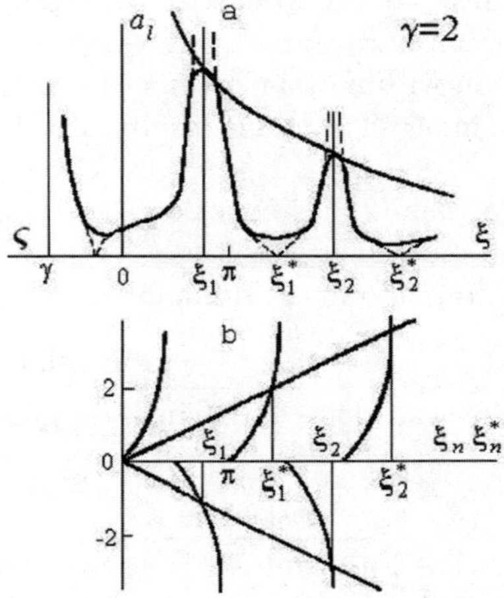

Fig. 2.17.

The graphical solution of equations (5.84) and (5.85) is shown in Fig.2.17, b. The corresponding vibration amplitudes are given by (2.76) and (2.77). Taking (5.81), (5.82), (5.84), and (5.85) into account, these equalities can be converted into the form,

$$a_l^0 = \frac{4\pi F_l l(\xi_n^2 + \gamma^2 \tanh^2 \gamma)}{\psi E S_l(\xi_n^2 + \gamma^2)(\xi_n^2 + \gamma^2 \tanh^2 \gamma - \gamma \tanh \gamma)},$$

$$a_l^* = \frac{\psi F_l l(\xi_n^2 + \gamma^2)(\xi_n^{*2} + \gamma^2 \tanh^2 \gamma - \gamma \tanh \gamma)}{4\pi E S_l \xi_n^{*2}(\xi_n^{*2} + \gamma^2 \tanh^2)}. \tag{2.123}$$

The maximum vibration amplitudes (i.e. at resonance) within the bar follow the enveloping curve, which is generated using equation (5.86). This is shown in Fig.2.17, a.

When the catenoidal bar is used as a concentrator at a frequency ω, its length is chosen, using a similar methodology to (5.77), according to the formula,

$$l = \frac{c}{\omega}\sqrt{\xi_n^2 + \gamma^2}, \tag{2.124}$$

The coefficient of amplification, (2.85) or (2.86), is given by

$$K = \frac{\cosh\gamma}{|\cos\xi_n|} = \sqrt{\frac{S_0}{S_l}}\frac{1}{|\cos\xi_n|}, \tag{2.125}$$

where ξ_n is the solution of (5.84). Relations (5.84), (5.87) and (5.88) coincide with those obtained in Merkulov (1957).

8. For the cosine waveguide (2.90), equation (2.93) can also be converted into the form of (2.95), by using the substitution,

$$\gamma = j\alpha. \tag{2.126}$$

In this instance, the dynamic compliance can be found from (5.79) or (5.80), where the indexes l and 0 should be swapped over, and $l - x$ should be substituted for x. This now gives,

$$L_{lx}(j\omega) = \frac{l\cosh\lambda x}{\tilde{E}S_0(\lambda l\sinh\lambda l + \alpha\tan\alpha\cosh\lambda l)\cos\alpha\cos\alpha\frac{x}{l}}, \tag{2.127}$$

$$L_{0x}(j\omega) = \frac{\lambda l\cosh\lambda(l-x) + \alpha\tan\alpha\sinh\lambda(l-x)}{\lambda\tilde{E}S_0(\lambda l\sinh\lambda l + \alpha\tan\alpha\cosh\lambda l)\cos\alpha\frac{x}{l}}. \tag{2.128}$$

In a similar manner, (5.81)–(5.83) may be used to obtain expressions for the dynamic compliance in the form of (2.71). Taking account of notation (5.89) for the cosine waveguide, it may be shown that relations (5.84) – (5.87) hold. These expressions are not reproduced here.

The coefficient of amplification and the length of the cosine concentrator (2.90) are calculated according to the formula,

$$K = \frac{1}{\cos\alpha|\cos\xi_n|} = \sqrt{\frac{S_0}{S_l}}\frac{1}{|\cos\xi_n|}, \quad l = \frac{c}{\omega}\sqrt{\xi_n^2 - \alpha^2}, \tag{2.129}$$

where ξ_n satisfies the equation,

$$\cot\xi_n = \frac{\xi_n}{\alpha\tan\alpha}. \tag{2.130}$$

9. For the conical bar (2.91), equation (2.93) takes the form,

$$\tilde{y}_x'' + \frac{\rho\omega^2}{\tilde{E}}\tilde{y}_x = 0.$$

Its solution, which satisfies the boundary conditions obtained from (2.44), using (2.94), allows the dynamic compliance to be found,

$$L_{lx}(j\omega) = \\ = \frac{\lambda^2 l(\gamma \sinh \lambda x - \lambda l \cosh \lambda x)}{\omega^2 \rho S_0 (1-\gamma \frac{x}{l})(\gamma^2 \lambda l \cosh \lambda l + ((\lambda l)^2(1-\gamma) - \gamma^2)\sinh \lambda l)},$$

$$L_{0x}(j\omega) = \\ = \frac{-\lambda^2 l[\gamma \sinh \lambda(l-x) + \lambda l(1-\gamma)\cosh \lambda(l-x)]}{\omega^2 \rho S_0 (1-\gamma \frac{x}{l})(\gamma^2 \lambda l \cosh \lambda l + ((\lambda l)^2(1-\gamma) - \gamma^2)\sinh \lambda l)}, \quad (2.131)$$

where $\lambda = j\omega/\sqrt{\rho/\tilde{E}} = \frac{j\omega}{c\sqrt{1+\frac{j\psi}{2\pi}}}$.

Converting expression (5.94) into the form of (2.71),

$$Q(j\omega) = \frac{ES_0}{l} \times \\ \times \left\{ \begin{array}{l} [\xi^2(1-\gamma) + \gamma^2]\sin\xi - \gamma^2 \xi \cos\xi - \\ j\frac{\psi}{4\pi}[\xi(\xi^2(1-\gamma) + 2\gamma^2)\cos\xi + \gamma^2(\xi^2 - 2)\sin\xi] \end{array} \right\}, \quad (2.132)$$

$$P_{lx}(j\omega) = \left\{ \begin{array}{l} \gamma \sin\xi\frac{x}{l} - \xi\cos\xi\frac{x}{l} + j\frac{\psi\xi}{4\pi}\times \\ \times\, [(1-\gamma\frac{x}{l})\cos\xi\frac{x}{l} - \xi\frac{x}{l}\sin\xi\frac{x}{l}] \end{array} \right\} / (1-\gamma\frac{x}{l}), \quad (2.133)$$

$$P_{0x}(j\omega) = -\frac{\left\{ \begin{array}{l} \gamma \sin\xi\frac{l-x}{l} + \xi(1-\gamma)\cos\xi\frac{l-x}{l} - \\ -j\frac{\psi\xi}{4\pi}[(1-\gamma\frac{x}{l})\cos\xi\frac{l-x}{l} - (1-\gamma)\xi\frac{l-x}{l}\sin\xi\frac{l-x}{l}] \end{array} \right\}}{(1-\gamma\frac{x}{l})}, \quad (2.134)$$

where $\xi = \omega l/c$.

The resonant ξ_n and antiresonant ξ_n^* frequencies can be found by solving equations,

$$\tan\xi_n = \frac{\gamma^2 \xi_n}{\xi_n^2(1-\gamma) + \gamma^2}, \quad (2.135)$$

$$\tan\xi_n^* = \frac{\xi_n^*}{\gamma}. \quad (2.136)$$

The corresponding amplitudes are given by,

$$a_l^0 = \frac{4\pi F_l l(\xi_n^2 + \gamma^2)}{\psi E S_0 \xi_n^2 [\xi_n^2(1-\gamma)^2 + 3\gamma^2(1-\gamma) + \gamma^4]}, \quad (2.137)$$

$$a_l^* = \frac{\psi F_l l[\xi_n^{*2} - \gamma(1-\gamma)]}{4\pi E S_0 (1-\gamma)^2(\xi_n^{*2} + \gamma^2)}. \quad (2.138)$$

When a conical bar is used as a concentrator, at a frequency of ω_n, its length may be calculated according to the formula,

$$l = c\xi_n/\omega_n, \quad (2.139)$$

Its coefficient of amplification (2.85) or (2.86) is determined by virtue of (5.96) – (5.98) as follows,

$$K = \sqrt{\frac{S_0}{S_1}} \left| \frac{\xi_n(1-\gamma)}{\gamma \sin \xi_n + \xi_n(1-\gamma)\cos \xi_n} \right|, \tag{2.140}$$

or, after a transformation with the help of (5.98),

$$K = \sqrt{\frac{S_0}{S_l}} \left| \frac{\xi_n \cos \xi_n - \gamma \sin \xi_n}{\xi_n} \right|. \tag{2.141}$$

Relations (5.98), (5.102) and (2.140) coincide with the relations obtained in Merkulov (1957) although they are derived in a different way. In Teumin (1959), the expression for the coefficient of amplification is given in the form (2.141). Taking (5.98) into account, it is easy to prove that equalities (2.140) and (2.141) coincide.

10. Waveguides composed of parts with different properties are often used in ultrasonic machine vibration systems. These parts may be made of different materials. The cross sectional areas of these different sections may vary according to different laws. Some compound parts can be represented in the form of lumped elements. Different compound concentrators (Merkulov & Kharitonov (1959)), for example vibration bar systems with a tool attached, belong to such systems. To describe the compound vibration system, it is practical to express its dynamic compliance via the dynamic compliances of its parts.

Consider, as an example, a waveguide, which consists of two steps of length l_1 and l_2 (Fig.2.18, a), where $l_1+l_2 = l$. Dividing the waveguide at the step, the interaction of its parts may be represented by a reaction force with amplitude $\tilde{F}_{12}$.

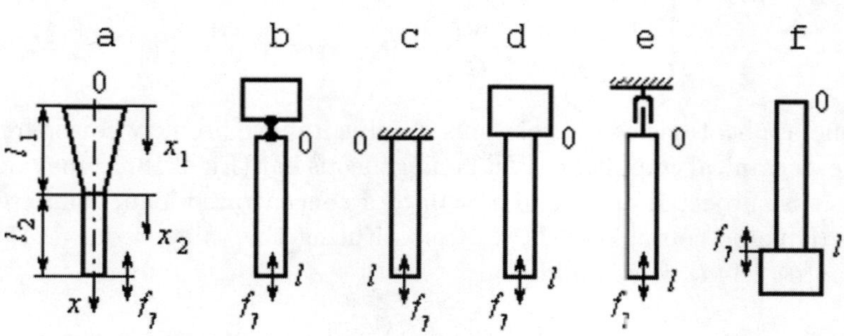

Fig. 2.18.

New coordinates $x_1 = x(x_1 \in [0, l_1])$, $x_2 = x - l_1(x_2 \in [0, l_2])$ may now be introduced, and displacement amplitude expressions may be written for the system. The system is under the action of force $\tilde{F}_l$, which is applied at end $x = l$,

$$\begin{aligned} \tilde{a}_{x_1}^{(1)} &= \tilde{F}_{12} L_{l_1 x_1}^{(1)}(j\omega), \\ \tilde{a}_{x_2}^{(2)} &= \tilde{F}_l L_{l_2 x_2}^{(2)}(j\omega) - \tilde{F}_{12} L_{0 x_2}^{(2)}(j\omega), \end{aligned} \tag{2.142}$$

where the upper index (in brackets) denotes the number of the subsystem to which the expression corresponds.

Equality $a_{l_1}^{(1)} = a_0^{(2)}$ reflects the continuity of the system at cross section $x = l_1$, and allows the reaction amplitude to be found,

$$\tilde{F}_{12} = \tilde{F}_l L_{l_2 0}^{(2)}(j\omega) / [L_{l_1 l_1}^{(1)}(j\omega) + L_{00}^{(2)}(j\omega)]. \tag{2.143}$$

Substituting (2.143) into (2.142) and using (2.14), the dynamic compliance of the compound system may be found,

$$L_{lx}(j\omega) = \begin{cases} \frac{L_{l_1 x_1}^{(1)}(j\omega) L_{l_2 0}^{(2)}(j\omega)}{L_{l_1 l_1}^{(1)}(j\omega) + L_{00}^{(2)}(j\omega)}, & 0 \le x_1 = x \le l_1, \\ L_{l_2 x_2}^{(2)}(j\omega) - \frac{L_{0 x_2}^{(2)}(j\omega) L_{l_2 0}^{(2)}(j\omega)}{L_{l_1 l_1}^{(1)}(j\omega) + L_{00}^{(2)}(j\omega)}, \\ 0 \le x_2 = x - l_1 \le l_2. \end{cases} \tag{2.144}$$

For $x_2 = l_2$, expression (2.144) is identical to expression (2.29), the only differences are that of notation.

The dynamic compliance of a compound system, excited in cross section $x = 0$, can be found in a similar manner,

$$L_{0x}(j\omega) = \begin{cases} L_{0 x_1}^{(1)}(j\omega) - \frac{L_{l_1 x_1}^{(1)}(j\omega) L_{0 l_1}^{(1)}(j\omega)}{L_{l_1 l_1}^{(1)}(j\omega) + L_{00}^{(2)}(j\omega)}, \\ 0 \le x_1 = x \le l_1, \\ \frac{L_{0 x_2}^{(2)}(j\omega) L_{0 l_1}^{(1)}(j\omega)}{L_{l_1 l_1}^{(1)}(j\omega) + L_{00}^{(2)}(j\omega)}, & 0 \le x_2 = x - l_1 \le l_2. \end{cases} \tag{2.145}$$

Some applications of the relations obtained above are now considered.

The dynamical compliance of a homogeneous bar (Fig.2.18, b) may now be found. It is subjected at its end $x = 0$, to a concentrated load, characterized by the dynamic compliance $L(j\omega)$. Substituting the values from (2.47) and (2.48) into (2.144), it is found that,

$$L_{lx}(j\omega) = -\frac{\lambda(\rho S \omega^2 \cosh \lambda x - \lambda L^{-1}(j\omega) \sinh \lambda x)}{\rho S \omega^2 (\rho S \omega^2 \sinh \lambda l - \lambda L^{-1}(j\omega) \cosh \lambda l)}. \tag{2.146}$$

Substituting the limiting factor $|L^{-1}(j\omega)| \to \infty$ into (2.146), the dynamic compliance of a bar with a fixed end (Fig.2.18, c) is found,

$$L_{lx}(j\omega) = -\frac{\lambda \sinh \lambda x}{\rho S \omega^2 \cosh \lambda l}. \tag{2.147}$$

Using (2.46), expression (2.147) is converted into the form (2.71),

$$L_{lx}(j\omega) = \frac{l}{ES\xi}\frac{\sin\xi\frac{x}{l} - j\frac{\psi}{4\pi}(\sin\xi\frac{x}{l} + \xi\frac{x}{l}\cos\xi\frac{x}{l})}{\cos\xi + j\frac{\psi}{4\pi}\xi\sin\xi}, \tag{2.148}$$

where $\xi = \frac{\omega l}{c}$.

Using (2.72) and (2.76), from (2.148), the resonant frequencies of the bar are found to be,

$$\xi_n = (2n-1)\pi/2, \tag{2.149}$$

where $n = 1, 2, ...$ is the mode number. The amplitude of any cross section may also be written, for example, $x = l$,

$$a_l^0 = \frac{16F_l l}{\pi\psi ES(2n-1)^2}. \tag{2.150}$$

If the length of the fixed bar is so small that $\xi \ll 1$, i.e., condition (2.50) is satisfied, then for $x = l$ (2.148) can be approximately written in the form,

$$L_{ll}(j\omega) = L(j\omega) = \frac{l}{\tilde{E}S} = \frac{l}{ES(1+j\psi/2\pi)} = \frac{1}{k_0 + j\omega b_0}, \tag{2.151}$$

where $k_0 = ES/l$ is the static stiffness of the bar, and $b_0 = \psi k_0/2\pi\omega$ is the equivalent coefficient of resistance.

Equality (2.151) represents the dynamic compliance of the localized elastic-viscous element (Frolov (1995)).

Expressions (2.144) and (2.145) allow all types of boundary conditions at the ends of the system to be dealt with. If the load is of an inertial type (Fig.2.18, d), then expression (2.146), by virtue of (2.51), can be written in the form,

$$L_{lx}(j\omega) = -\frac{\lambda}{\rho S\omega^2}\frac{\cosh\lambda x + \mu\lambda l\sinh\lambda x}{\sinh\lambda l + \mu\lambda l\cosh\lambda l}, \tag{2.152}$$

where $\mu = \frac{M}{\rho S l}$, M is the mass of the added body.

The influence of a linear viscous resistance, with a compliance of $L(j\omega) = (j\omega b_0)^{-1}$ on the characteristics of bar (Fig.2.18, e) may now be estimated. The energy losses in the material of the bar ($\psi = 0$) are neglected. In this case, from (2.146) and (2.46),

$$L_{lx}(j\omega) = -\frac{l}{ES\xi}\frac{\cos\xi\frac{x}{l} + j\frac{b_0}{w}\sin\xi\frac{x}{l}}{sin\xi - j\frac{b_0}{w}\cos\xi}, \tag{2.153}$$

where $w = \rho c S$, which is the wave resistance of the bar.

The case when the load resistance b_0 is equal to the wave resistance w is important from the point of view of practical applications. Taking account of the accepted notation, from (2.153) it is found that,

$$L_{lx}(j\omega) = \frac{1}{j\omega w}\frac{\cos\frac{\omega}{c}x + j\sin\frac{\omega}{c}x}{\cos\frac{\omega}{c}l + j\sin\frac{\omega}{c}l} = \frac{1}{j\omega w}e^{j\frac{\omega}{c}(x-l)}. \tag{2.154}$$

By comparison with (2.52), it is concluded, that expression (2.154) describes a travelling harmonic wave, which propagates along the bar from the excitation source. This means that, under such a load, there is no wave reflection from boundary $x = 0$, and that a finite bar works in a travelling wave regime (Teumin (1959)).

11. Finally, a system may be considered which consists of two homogeneous bars. The cross sectional areas are S_1 and S_2, and their lengths are l_1 and l_2 respectively (Fig.2.15, e). Substituting the appropriate values from (2.47) and (2.48) into (2.144), and assuming that the bars are made from the same material, it is found that,

$$L_{lx}(j\omega) = \begin{cases} -\frac{\lambda}{\rho\omega^2} \frac{\cosh \lambda x}{S_1 \sinh \lambda l_1 \cosh \lambda l_2 + S_2 \sinh \lambda l_2 \cosh \lambda l_1}, \\ 0 \leq x \leq l_1, \\ -\frac{\lambda}{S_2 \rho\omega^2} \frac{S_1 \sinh \lambda(x-l_1) \sinh \lambda l_1 + S_2 \cosh \lambda(x-l_1) \cosh \lambda l_1}{S_1 \sinh \lambda l_1 \cosh \lambda l_2 + S_2 \sinh \lambda l_2 \cosh \lambda l_1}, \\ l_1 \leq x \leq l, \end{cases} \tag{2.155}$$

where $l = l_1 + l_2$.

It is now possible to consider some particular cases, which follow from formula (2.155).

In the case when $S_1 \to \infty$, the dynamic compliance is obtained for a bar with a fixed end (2.147).

It is now considered that one of the bars has a very short length, such that $|\lambda l_i| \ll 1$ $(i = 1, 2)$. If $i = 1$, and assuming that $\rho S_1 l_1 = M, l_2 = l$, formula (2.152) is obtained. This determines the dynamic compliance of a bar with a concentrated mass at its end $x = 0$, (Fig.2.18, d). Similarly, for $i = 2$, the dynamic compliance of a bar may be found in which the mass is at the point of excitation (Fig.2.18, f),

$$L_{lx}(j\omega) = -\frac{\lambda}{\rho\omega^2 S} \frac{\cosh \lambda x}{\sinh \lambda l + \mu \lambda l \cosh \lambda l}, \tag{2.156}$$

where $\mu = M/\rho S l$.

From (2.155), calculations may be made for a stepped concentrator (Merkulov & Kharitonov (1959), Teumin (1959)), as is widely used in ultrasonic equipment. The losses in the concentrator's material are neglected, i.e., it is assumed that $\psi = 0$. Taking account of (2.46), (2.48) may be rewritten as,

$$L_{lx}(j\omega) = \begin{cases} -\frac{l}{E\xi} \frac{\cos \xi x/l}{S_1 \sin \xi_1 \cos \xi_2 + S_2 \sin \xi_2 \cos \xi_1}, \\ 0 \leq x \leq l_1, \\ \frac{l}{ES_2\xi} \frac{S_1 \sin(\xi x/l - \xi_1) \sin \xi_1 - S_2 \cos(\xi x/l - \xi_1) \cos \xi_1}{S_1 \sin \xi_1 \cos \xi_2 + S_2 \sin \xi_2 \cos \xi_1}, \\ l_1 \leq x \leq l, \end{cases} \tag{2.157}$$

where $\xi = \omega l/c, \xi_i = \omega l_i/c = \xi l_i/l, (i = 1, 2)$.

Using (2.71), (2.72) and (2.157) an equation determining the natural frequencies of a stepped concentrator may be obtained,

$$(S_1/S_2)\tan\xi_1 = -\tan\xi_2, \tag{2.158}$$

this coincides with the generally accepted equation (Merkulov & Kharitonov (1959), Teumin (1959)).

The amplification coefficient (2.86) is given by

$$K = |\cos\xi_1\cos\xi_2 - (S_1/S_2)\sin\xi_1\sin\xi_2|. \tag{2.159}$$

Taking (2.158) into account, expression (2.158) can be converted to the known form (Merkulov & Kharitonov (1959), Teumin (1959)),

$$K = \left|\frac{\cos\xi_1}{\cos\xi_2}\right| = \frac{S_1}{S_2}\left|\frac{\sin\xi_1}{\sin\xi_2}\right| = \frac{S_1}{S_2}\sqrt{\frac{1+\tan^2\xi_2}{(S_1/S_2)^2+\tan^2\xi_2}}. \tag{2.160}$$

As follows from (2.160), the maximum amplification coefficient,

$$K = S_1/S_2 \tag{2.161}$$

is achieved for a stepped concentrator when $\xi_2 = (2k+1)\pi/2, \xi_1 = (2n+1)\pi/2, (k,n = 0,1,2,...)$, i.e., the concentrator has lengths equal to,

$$l_1 = (2n+1)\pi c/2\omega,\ l_2 = (2k+1)\pi c/2\omega. \tag{2.162}$$

These lengths are an odd number of quarter-wavelength multiples, where the wavelength, $\Lambda = 2\pi c/\omega$. As shown in Bogomolov & Simson (1981), such concentrators demonstrate the maximum amplitude amplification among all known types of concentrators having the same difference in cross sectional area.

Fig.2.19 shows the dependence of the amplification coefficient K and length l, on the ratio of cross section areas S_0/S_1 for different types of concentrators. It relates to the concentrator types shown in Fig.2.15. Curve 1 corresponds to the exponential, 2 the catenoidal, 3 the cosine, 4 the Conic and 5 the stepped (for $l_1 = l_2$).

12. When designing ultrasonic vibration systems, cases often arise when forces acting on the system are applied in intermediate cross sections $x = s$ (Fig.2.20, a). In order to describe such a system, it may be separated into two parts at the position where the force acts. The assumption is made that the force acts at a single position and against only one of the parts. The interaction between the two parts can now be replaced with a reaction force having amplitude $\tilde{F}_{12}$.

Introducing the coordinates $x_1 = x(x_1 \in [0,s]), x_2 = x - s(x_2 \in [0, l-s])$, gives a relation similar to (2.142),

$$\tilde{a}^{(1)}_{x_1} = (\tilde{F}_s + \tilde{F}_{12})L^{(1)}_{sx_1}(j\omega),$$
$$\tilde{a}^{(2)}_{x_2} = -\tilde{F}_{12}L^{(2)}_{0x_2}(j\omega).$$

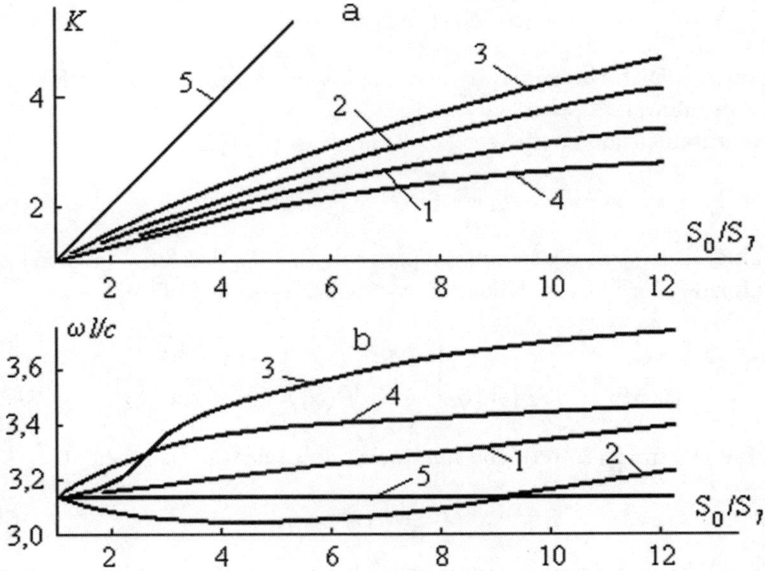

Fig. 2.19.

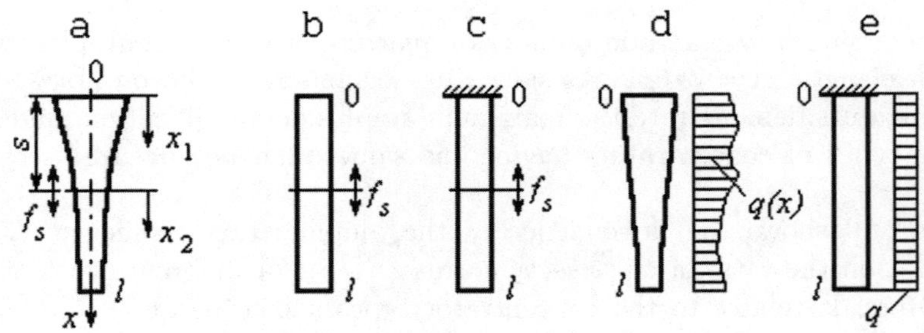

Fig. 2.20.

With the equality $\tilde{a}_s^{(1)} = \tilde{a}_0^{(2)}$, the unknown reaction $\tilde{F}_{12}$ can be excluded. Introducing the notation $L_{sx}(j\omega) = \tilde{a}_x/\tilde{F}_s$ gives,

$$L_{sx}(j\omega) = \begin{cases} \frac{L_{sx_1}^{(1)}(j\omega)L_{00}^{(2)}(j\omega)}{L_{ss}^{(1)}(j\omega)+L_{00}^{(2)}(j\omega)}, & 0 \le x_1 = x \le s, \\ \frac{L_{0x_2}^{(2)}(j\omega)L_{ss}^{(1)}(j\omega)}{L_{ss}^{(1)}(j\omega)+L_{00}^{(2)}(j\omega)}, & 0 \le x_2 = x - s \le l - s. \end{cases} \quad (2.163)$$

Some examples may now be considered.

The dynamic compliance of a homogeneous bar with free ends (Fig.2.20, b) may be found. It is excited at cross section $x = s$. Substituting the appropriate quantities from (2.47) and (2.48) into (2.163) gives,

$$L_{sx}(j\omega) = \begin{cases} -\frac{\lambda}{\rho S\omega^2}\frac{\cosh\lambda x\cosh\lambda(l-s)}{\sinh\lambda l}, & 0 \le x \le s, \\ -\frac{\lambda}{\rho S\omega^2}\frac{\cosh\lambda(l-x)\cosh\lambda s}{\sinh\lambda l}, & s \le x \le l. \end{cases} \quad (2.164)$$

Comparing (2.164) for $x = l$, and (2.47) for $x = s$ the following equality is obtained,

$$L_{sl}(j\omega) = L_{ls}(j\omega), \quad (2.165)$$

which is a manifestation of the reciprocity principle.

Taking account of (2.57), (2.63) and (2.165) the following expression can be written, giving the resonant vibration amplitude of cross section $x = l$. The bar is excited by a harmonic force $f_s = F_s e^{j\omega t}$ at cross section $x = s$.

$$\tilde{a}_l = \frac{4F_s l}{\psi\pi ESn^2}\left[\cos\pi n\frac{s}{l} - j\frac{\psi}{4\pi}\left(\cos\pi n\frac{s}{l} - \pi n\frac{s}{l}\sin\pi n\frac{s}{l}\right)\right]. \quad (2.166)$$

If force f_s acts at one of the antinodes of the bar's natural vibration mode (2.64), i.e. it is applied at the point $s = lk/n$, where $k \le n$ is an integer, then expression (2.163) leads to formula (2.67). When the force acts at one of the nodal points $s = l(2k+1)/2n$ (2.64), then (2.166) yields,

$$a_l = \frac{F_s l(2k+1)}{2\pi ESn^2}. \quad (2.167)$$

The amplitude (2.167) is an order of magnitude smaller than the amplitude given by (2.67). Therefore, if the exciting force acts at the node of the bar's natural vibration mode, the corresponding resonance is not revealed and observed experimentally.

In a similar manner to that of (2.163), the dynamic compliance of a bar with a fixed end (Fig.2.20, c) may now be found. It is subjected to excitation at cross section $x = s$. Taking (2.147) and (2.148) into account,

$$L_{sx}(j\omega) = \begin{cases} -\frac{\lambda}{\rho S\omega^2}\frac{\sinh\lambda x\cosh\lambda(l-s)}{\cosh\lambda l}, & 0 \le x \le s, \\ -\frac{\lambda}{\rho S\omega^2}\frac{\cosh\lambda(l-x)\sinh\lambda s}{\cosh\lambda l}, & s \le x \le l. \end{cases} \quad (2.168)$$

It must be noted that expressions (2.164) and (2.168) can be found in Butkovsky (1977, 1979) in which they were obtained using a different method.

Using the results obtained above, the bar system's vibration may now be considered. This results from a distributed force that acts along the system according to the law $q(x)$, (Fig.2.20, d). Force $q(s)ds$ acts on element ds at cross section s and causes the vibration of cross section x with an amplitude $d\tilde{a}_x = L_{sx}(j\omega)q(s)ds$. The amplitude of cross section x under the action of the distributed load is,

$$\tilde{a}_x = \int_0^l L_{sx}(j\omega)q(s)ds. \quad (2.169)$$

As an example, consideration shall be given to a homogeneous bar's vibration when subjected to an excitation force $q(x) = q = const$, which is uniformly distributed along its length, (Fig.2.20, e). Substituting (2.168) into (2.169) and integrating gives,

$$\tilde{a}_x = \frac{q}{\rho S \omega^2} \frac{\cosh \lambda(l - x) - \cosh \lambda l}{\cosh \lambda l}. \tag{2.170}$$

For a small value of ψ, expression (2.170) can be converted, by virtue of (2.49), to the form,

$$\tilde{a}_x = \frac{ql^2}{ES\xi^2} \frac{\cos \xi(1 - \frac{x}{l}) - \cos \xi + j\frac{\psi \xi}{4\pi}[(1 - \frac{x}{l}) \sin \xi(1 - \frac{x}{l}) - \sin \xi]}{\cos \xi + j\frac{\psi}{4\pi} \xi \sin \xi}, \tag{2.171}$$

where $\xi = \omega l / c$.

Equality (2.171) allows the amplitudes and phases of all cross sections within the bar to be determined. For example, the vibration amplitude at cross section $x = l$, at its resonant frequency (2.149), is given by,

$$a_l^0 = \frac{32ql^2}{\pi^2 \psi ES(2n - 1)^3}. \tag{2.172}$$

As follows from the comparison between (2.172) and (2.150), the excitation of a system using a concentrated force is more efficient than by using a distributed force. This is provided that the total force $F_l = ql$ is the same. This is especially noticeable during the excitation of higher vibration modes.

The approach developed in this section allows the vibration of complicated systems to be described by using known dynamic characteristics of their parts. Of course, a similar approach can be applied to describe the torsional vibration of bars or transverse vibration of strings using the same equations. The general relations obtained in this section can also be used to describe bending vibration within bar systems in cases when the interaction between the subsystems leads only to the appearance of transverse forces and does not cause bending moments at the joints of the parts. In general, the description of bending vibration gives rise to a number of specific peculiarities that will be discussed in the next section.

2.3 Bending vibration within an ultrasonic bar system

1. Consider the bending vibration of a bar system that is excited by a harmonic transverse force $f_s(t) = \tilde{F}_s e^{j\omega t}$, and a bending moment $r_s(t) = \tilde{R}_s e^{j\omega t}$. These act at cross section $x = s$ (Fig.2.21, a).

The motion of the system's arbitrary cross section x is characterized by the displacement u_x of its centre of inertia, and its rotation $\theta_x = \partial u_x / \partial x$. Consider the steady state motion of the form (2.38). The displacement amplitude and rotation angle are related as follows,

$$\tilde{\theta}_x = d\tilde{a}_x / dx. \tag{2.173}$$

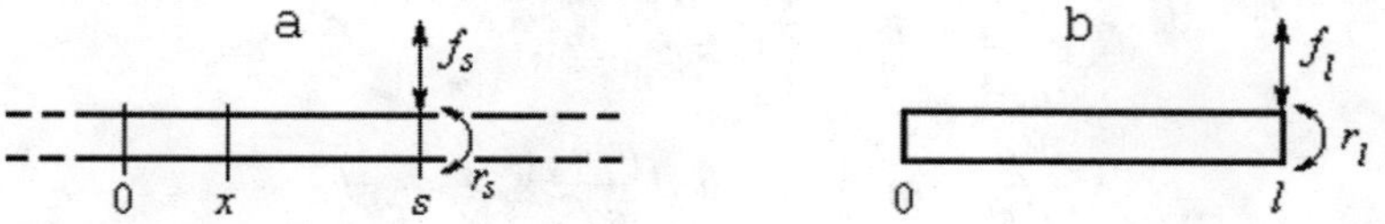

Fig. 2.21.

A notion similar to that of dynamic compliance $L_{sx}(j\omega)$ (see (2.14)) may be introduced for an action of bending moment. The dynamic compliance for a bending moment system can be written,

$$\Gamma_{sx}(j\omega) = \tilde{a}_x / \tilde{R}_s, \tag{2.174}$$

This relates the complex amplitude of the bending moment at cross section s to the displacement of cross section x. Firstly, the dynamic compliances $L_{sx}(j\omega)$ and $\Gamma_{sx}(j\omega)$ of a homogeneous bar with free ends may be found. The bar is excited in cross section $s = l$ (Fig.2.21, b), or $s = 0$. To facilitate this, the equation for bending vibration in a homogeneous bar (Biderman (1978), Timoshenko (1955)) may be written,

$$\rho \frac{\partial^2 u(x,t)}{\partial t^2} + \tilde{E}_0 I \frac{\partial^4 u(x,t)}{\partial x^2} = 0, \tag{2.175}$$

where ρ is the mass of the bar per unit length; $\tilde{E}_0 = E_0(1 + j\psi/2\pi)$; E_0 is the modulus of elasticity; ψ is the absorption coefficient; I is the moment of inertia at the bar's cross section.

If the bar is excited at cross section $x = l$, the solution of equation (2.175) should satisfy the following boundary conditions:

$$\frac{\partial^2 u(x,t)}{\partial x^2}|_{x=0} = 0, \quad \frac{\partial^3 u(x,t)}{\partial x^3}|_{x=0} = 0, \tag{2.176}$$

$$\frac{\partial^2 u(x,t)}{\partial x^2}|_{x=l} = 0, \quad \tilde{E}_0 I \frac{\partial^3 u(x,t)}{\partial x^3}|_{x=l} = -f_l(t) \tag{2.177}$$

for a force excitation, and

$$\tilde{E}_0 I \frac{\partial^2 u(x,t)}{\partial x^2}|_{x=l} = r_l(t), \quad \frac{\partial^3 u(x,t)}{\partial x^3}|_{x=l} = 0 \tag{2.178}$$

for a moment excitation.

Substituting (2.38) into (2.175) – (2.178) the following equation is obtained to determine the complex amplitude $\tilde{a}_x$,

$$\tilde{a}_x - \frac{\tilde{E}_0 I}{\rho\omega^2} \cdot \frac{d^4 \tilde{a}_x}{dx^4} = 0 \tag{2.179}$$

with boundary conditions,

$$\frac{d^2\tilde{a}_x}{dx^2}|_{x=0} = \frac{d^3\tilde{a}_x}{dx^3}|_{x=0} = 0,$$
$$\frac{d^2\tilde{a}_x}{dx^2}|_{x=l} = 0, \quad \tilde{E}_0 I\frac{d^3\tilde{a}_x}{dx^3}|_{x=l} = -\tilde{F}_l,$$
$$\tilde{E}_0 I\frac{d^2\tilde{a}_x}{dx^2}|_{x=l} = \tilde{R}_l, \quad \frac{d^3\tilde{a}_x}{dx^3}|_{x=l} = 0. \tag{2.180}$$

In the following derivation, use is made of the Krylov functions that are dependant on frequency. These are tabulated in Ananiev (1946) and use the same notation:

$$S(z) = \frac{1}{2}(\cosh z + \cos z), \quad T(z) = \frac{1}{2}(\sinh z + \sin z),$$
$$U(z) = \frac{1}{2}(\cosh z - \cos z), \quad V(z) = \frac{1}{2}(\sinh z - \sin z),$$
$$A(z) = \cosh z \sin z + \sinh z \cos z,$$
$$B(z) = \cosh z \sin z - \sinh z \cos z,$$
$$C(z) = 2\cosh z \cos z, \quad S_1(z) = 2\sinh z \sin z,$$
$$D(z) = \cosh z \cos z - 1, \quad E(z) = \cosh z \cos z + 1. \tag{2.181}$$

By looking for solutions of equation (2.179) which satisfy boundary conditions (2.180), values are found for dynamic compliances (2.14) and (2.174),

$$L_{lx}(j\omega) = \frac{2l^3}{E_0 I\tilde{\xi}^3 D(\tilde{\xi})}\left[U(\tilde{\xi})T(\tilde{\xi}\frac{x}{l}) - V(\tilde{\xi})S(\tilde{\xi}\frac{x}{l})\right], \tag{2.182}$$

$$\Gamma_{lx}(j\omega) = \frac{2l^2}{E_0 I\tilde{\xi}^2 D(\tilde{\xi})}\left[T(\tilde{\xi})T(\tilde{\xi}\frac{x}{l}) - U(\tilde{\xi})S(\tilde{\xi}\frac{x}{l})\right], \tag{2.183}$$

where $\tilde{\xi} = l\left(\frac{\rho\omega^2}{\tilde{E}_0 I}\right)^{1/4} = \xi(1 + j\frac{\psi}{2\pi})^{-1/4}$, $\xi = l\left(\rho\omega^2/E_0 I\right)^{1/4}$.

Taking account of the geometrical symmetry of a homogeneous bar (Fig.2.21, b), expressions can be written for its dynamic compliance when it is excited at cross section $x = 0$,

$$L_{0x}(j\omega) = L_{l(l-x)}(j\omega), \quad \Gamma_{0x}(j\omega) = \Gamma_{l(l-x)}(j\omega), \tag{2.184}$$

i.e. these expressions can be obtained from (2.182) and (2.183) by changing x to $l - x$.

According to (2.173), (2.14) and (2.174), the amplitude of cross section x rotation angle, when excited in cross section s by a force or moment, is given by,

$$\tilde{\theta}_x = \tilde{F}_s L'_{sx}(j\omega), \quad \tilde{\theta}_x = \tilde{R}_s \Gamma'_{sx}(j\omega), \tag{2.185}$$

where the dash denotes differentiation with respect to x.

The values of $L'_{sx}(j\omega)$ and $\Gamma'_{sx}(j\omega)$, in addition to (2.182) and (2.183), can be considered to be dynamic compliances relating the force and moment acting in the s cross section with the angular displacement of cross section x. Substituting functions (2.181) into expressions (2.182) – (2.185) allows the validity of the following equalities to be checked,

$$L'_{ll}(j\omega) = \Gamma_{ll}(j\omega),\ L'_{00}(j\omega) = \Gamma_{00}(j\omega),\ L'_{l0}(j\omega) = -\Gamma_{l0}(j\omega). \tag{2.186}$$

These equalities mean that the angular displacement of a peripheral cross section, as caused by a unitary force, is equal to the linear displacement of the same cross section under the action of a unitary moment.

As follows from (2.182) – (2.185), any dynamic compliance can be rewritten in the form (2.71). It can then be used to analyse the system's vibration in a similar way as shown in subsection 2.2.4. As an example, dynamic compliances (2.182) and (2.183) are converted in to the form of (2.71) for $x = l$,

$$L_{ll}(j\omega) = \frac{l^3}{E_0 I \xi^3} \frac{B(\xi) - j\frac{\psi}{8\pi}[3B(\xi) - \xi S_1(\xi)]}{D(\xi) - j\frac{\psi\xi}{8\pi}B(\xi)}, \tag{2.187}$$

$$\Gamma_{ll}(j\omega) = \frac{l^2}{2E_0 I \xi^2} \frac{S_1(\xi) - j\frac{\psi}{8\pi}[2S_1(\xi) - \xi A(\xi)]}{D(\xi) - j\frac{\psi\xi}{8\pi}B(\xi)}. \tag{2.188}$$

In accordance with (2.187) and (2.188), the resonant frequencies of the bar are determined by the solutions of equation $D(\xi) = 0$, or, taking into account notations (2.181),

$$\cosh\xi \cos\xi - 1 = 0. \tag{2.189}$$

With sufficient accuracy, nonzero solutions of this equation are given by,

$$\xi_n = \frac{\pi}{2}(2n+1) \quad (n = 1, 2, ..). \tag{2.190}$$

(This can easily be shown using the function (2.181) given in tables (Ananiev (1946))).

For the resonant vibration amplitudes,

$$a_l = \frac{8\pi F_l l^3}{\psi \xi_n^4 E_0 I} = \frac{128}{\psi\pi^3 (2n+1)^4} \frac{F_l l^3}{E_0 I} \tag{2.191}$$

when using force excitation, and

$$a_l = \frac{4\pi R_l l^2 S_1(\xi_n)}{\psi \xi_n^3 E_0 I B(\xi_n)} = \frac{64}{\psi\pi^2 (2n+1)^3} \frac{R_l l^2}{E_0 I} \tag{2.192}$$

when using moment excitation. In (2.192) it is assumed that $S_1(\xi_n)/B(\xi_n) = 2$ for frequencies (2.190).

Antiresonant frequencies can be obtained for force or moment excitation by solving equations,

$$B(\xi) = 0, \quad S_1(\xi) = 0$$

respectively.

The first equation gives values,

$$\xi_n^* = \frac{\pi}{4}(4n+1) \quad (n = 1, 2, ...), \tag{2.193}$$

while the second leads to,

$$\xi_n^* = \pi n \quad (n = 1, 2, ...). \tag{2.194}$$

At frequencies (2.193) under force excitation conditions, and (2.194) under moment excitation conditions, cross section $x = l$ centre of inertia remains practically static. As such, this cross section only performs angular vibration with an amplitude determinable from equalities (2.185). There are also specific frequencies at which the cross section performs purely translational vibration. When a force is used to excite vibration, purely translational vibration exists in cross sections $x = 0$ and $x = l$. This occurs at frequencies (2.194) and can be shown from equalities (2.186). In particular, it follows that the dynamic compliance (2.186) describes the angular vibration of cross section $x = l$ whilst subjected to a force $f(t)$, for which the frequencies (2.194) are antiresonant. It should be noted that, in principle, different types of motion may take place within the vibration system's cross sections. This should be taken into account when considering the motion of an ultrasonic system's working parts. This has been mentioned previously in section 2.1 (see, for example, Fig.2.8, a). It will be demonstrated later, that sections containing several modes of vibration can be efficiently realized by careful dynamic adjustment of the vibration system's components.

It can be shown, that at low frequencies of excitation, quantities (2.182) and (2.183) describe the bar's vibration as that of a rigid body. Assuming that $\xi \ll 1$, both the numerators and denominators of expressions (2.182) and (2.183) can be expanded into a power series of the small parameter ξ. Taking only the first term of the expansion into account gives,

$$L_{lx}(j\omega) = \frac{2(1-\frac{3x}{l})}{\rho l\omega^2} = -\frac{1}{M\omega^2} - \frac{lx_1}{2J\omega^2}, \tag{2.195}$$

$$\Gamma_{lx}(j\omega) = \frac{6(1-\frac{2x}{l})}{\rho l^2\omega^2} = -\frac{x_1}{J\omega^2}, \tag{2.196}$$

where, $M = \rho l$ is the mass of the bar; $J = \rho l^3/12$ is its moment of inertia with respect to the centre of mass; $x_1 = x - \frac{l}{2}$ is the coordinate under consideration and is measured from the components mass centre.

Equality (2.196) describes the displacements of the bar's cross sections, during angular vibration, as caused by a unitary moment. The dynamic compliance, in the instance of angular vibration, may be written,

$$\Gamma'_{lx}(j\omega) = \Gamma'(j\omega) = -\frac{1}{J\omega^2}. \tag{2.197}$$

The first term in the right side of(2.195) describes the displacement due to translational vibration within the bar when subjected to a unitary force, the second determines the displacement due to angular vibration with respect to the centre of mass and is subject to the moment created by a unitary force with a moment arm $l/2$.

Analogously to (2.182) and (2.183), the dynamic compliances of system components can be found. These correspond to the boundary conditions imposed at the edges of the components. A selection are given in Table 2.1.

2. Using dynamic compliance expressions for various components, as given in Table 2.1, dynamic compliances $L_{sx}(j\omega)$ and $\Gamma sx(j\omega)$ for a system being excited at an intermediate cross section $x = s$ may be calculated. The system under consideration is disjointed at the point at which the excitation force acts. It is assumed that this force acts on one of the components and the interaction between the components can be replaced by an equivalent force and moment having amplitudes $\tilde{F}_{12}$ and $\tilde{R}_{12}$ respectively. Introducing the coordinate systems $x_1 = x(x_1 \in [0,s])$ and $x_2 = x - s(x_2 \in [0, l-s])$ gives,

$$\begin{aligned}\tilde{a}^{(1)}_{x_1} &= \tilde{A}_{x_1} + \tilde{F}_{12} L^{(1)}_{sx_1}(j\omega) + \tilde{R}_{12}\Gamma^{(1)}_{sx_1}(j\omega),\\ \tilde{a}^{(2)}_{x_2} &= -\tilde{F}_{12} L^{(2)}_{0x_2}(j\omega) - \tilde{R}_{12}\Gamma^{(2)}_{0x_2}(j\omega),\end{aligned} \tag{2.198}$$

where $\tilde{A}_{x_1} = \tilde{F}_s L^{(1)}_{sx_1}(j\omega)$ for force excitation and $\tilde{A}_{x_1} = \tilde{R}_s \Gamma^{(1)}_{sx_1}(j\omega)$ for moment excitation.

Using the obvious equalities $\tilde{a}^{(1)}_s = \tilde{a}^{(2)}_0$ and $\tilde{a}'^{(1)}_s = \tilde{a}'^{(2)}_0$, the following system of equations for the unknown reactions $\tilde{F}_{12}$ and $\tilde{R}_{12}$ can be obtained:

$$\begin{aligned}\alpha_{11}\tilde{F}_{12} + \alpha_{12}\tilde{R}_{12} &= -\tilde{A}_s,\\ \alpha_{21}\tilde{F}_{12} + \alpha_{22}\tilde{R}_{12} &= -A'_s,\end{aligned} \tag{2.199}$$

where, $\alpha_{11} = L^{(1)}_{ss} + L^{(2)}_{00}$, $\alpha_{12} = \Gamma^{(1)}_{ss} + \Gamma^{(2)}_{00}$, $\alpha_{21} = \alpha_{12}$, $\alpha_{22} = \Gamma'^{(1)}_{ss} + \Gamma'^{(2)}_{00}$.

Solving (2.199) gives,

$$\tilde{F}_{12} = \frac{\alpha_{12}A'_s - \alpha_{22}A_s}{\alpha_{11}\alpha_{22} - \alpha^2_{12}}, \quad \tilde{R}_{12} = \frac{\alpha_{12}A_s - \alpha_{11}A'_s}{\alpha_{11}\alpha_{22} - \alpha^2_{21}}. \tag{2.200}$$

Substituting (2.200) into (2.198), and using (2.14) and (2.174), expressions can finally be found for the dynamic compliances.

As an example, consider the vibration within a homogeneous bar with hinged supports which is subjected to excitation in cross section $x = s$ by a force (Fig.2.22, a), and a moment (Fig.2.22, b).

Using dynamic compliance expressions from Table 2.1 (lines 3 and 4), the coefficients of equations (2.199) are found,

$$\alpha_{11} = -\frac{l^3}{\tilde{E}_0 I\tilde{\xi}^3}\left[\frac{S_1(\tilde{\xi}_l)}{B(\tilde{\xi}_l)} + \frac{S_1(\tilde{\xi}_2)}{B(\tilde{\xi}_2)}\right],$$

$$\alpha_{12} = -\frac{l^2}{\tilde{E}_0 I\tilde{\xi}^2}\left[\frac{A(\tilde{\xi}_l)}{B(\tilde{\xi}_l)} - \frac{A(\tilde{\xi}_2)}{B(\tilde{\xi}_2)}\right],$$

$$\alpha_{22} = -\frac{l}{\tilde{E}_0 I\tilde{\xi}}\left[\frac{C(\tilde{\xi}_l)}{B(\tilde{\xi}_l)} + \frac{C(\tilde{\xi}_2)}{B(\tilde{\xi}_2)}\right],$$

Table 2.1.

	Element schematic.	Dynamic compliance
1.	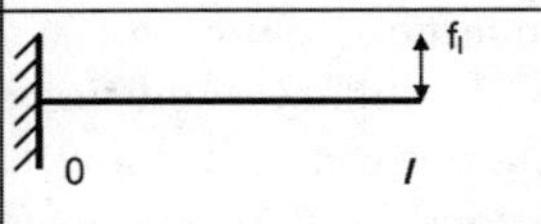	$L_{lx}(j\omega)=\frac{2l^3}{\tilde{E}_0 I}\cdot\frac{T(\tilde{\xi})\cdot U\left(\tilde{\xi}\frac{x}{l}\right)-S(\tilde{\xi})\cdot V\left(\tilde{\xi}\frac{x}{l}\right)}{\tilde{\xi}^3 E(\tilde{\xi})}$
2.	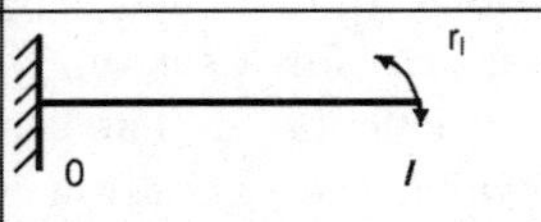	$\Gamma_{lx}(j\omega)=\frac{2l^2}{\tilde{E}_0 I}\cdot\frac{S(\tilde{\xi})\cdot U\left(\tilde{\xi}\frac{x}{l}\right)-V(\tilde{\xi})\cdot V\left(\tilde{\xi}\frac{x}{l}\right)}{\tilde{\xi}^2 E(\tilde{\xi})}$
3.	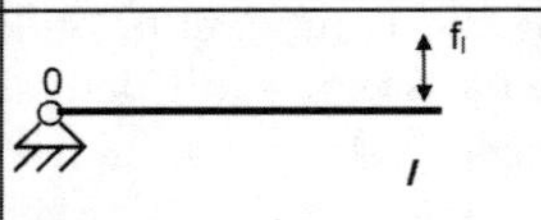	$L_{lx}(j\omega)=-\frac{l^3}{\tilde{E}_0 I}\cdot\frac{\operatorname{sh}\tilde{\xi}\cdot\sin\tilde{\xi}\frac{x}{l}+\sin\tilde{\xi}\cdot\operatorname{sh}\tilde{\xi}\frac{x}{l}}{\tilde{\xi}^3 B(\tilde{\xi})}$
4.	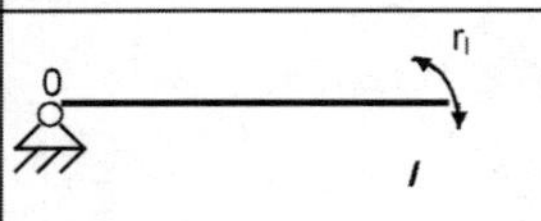	$\Gamma_{lx}(j\omega)=-\frac{l^2}{\tilde{E}_0 I}\cdot\frac{\operatorname{ch}\tilde{\xi}\sin\tilde{\xi}\frac{x}{l}+\cos\tilde{\xi}\cdot\operatorname{sh}\tilde{\xi}\frac{x}{l}}{\tilde{\xi}^2 B(\tilde{\xi})}$
5.	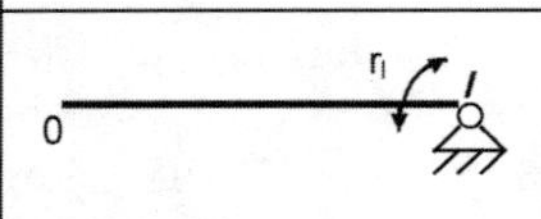	$\Gamma_{lx}(j\omega)=\frac{2l^2}{\tilde{E}_0 I}\cdot\frac{T(\tilde{\xi})\cdot S\left(\tilde{\xi}\frac{x}{l}\right)-S(\tilde{\xi})\cdot T\left(\tilde{\xi}\frac{x}{l}\right)}{\tilde{\xi}^2 B(\tilde{\xi})}$
6.	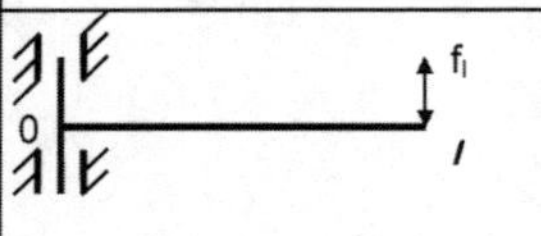	$L_{lx}(j\omega)=-\frac{l^3}{\tilde{E}_0 I}\cdot\frac{\operatorname{ch}\tilde{\xi}\cdot\cos\tilde{\xi}\frac{x}{l}+\cos\tilde{\xi}\cdot\operatorname{ch}\tilde{\xi}\frac{x}{l}}{\tilde{\xi}^3 A(\tilde{\xi})}$
7.	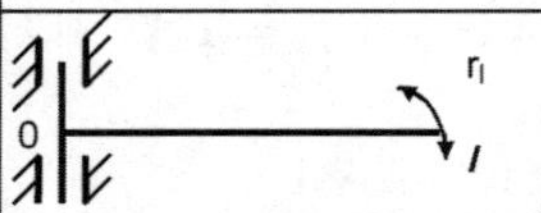	$\Gamma_{lx}(j\omega)=\frac{l^2}{\tilde{E}_0 I}\cdot\frac{\sin\tilde{\xi}\cdot\operatorname{ch}\tilde{\xi}\frac{x}{l}-\operatorname{sh}\tilde{\xi}\cdot\cos\tilde{\xi}\frac{x}{l}}{\tilde{\xi}^2 A(\tilde{\xi})}$
8.	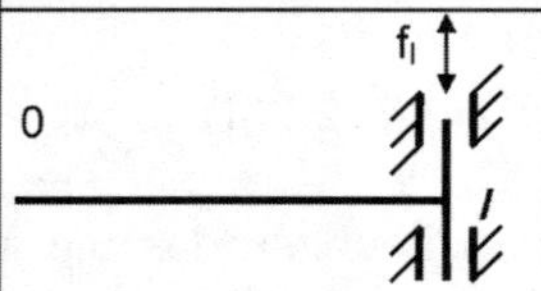	$L_{lx}(j\omega)=\frac{2l^3}{\tilde{E}_0 I}\cdot\frac{V(\tilde{\xi})\cdot T\left(\tilde{\xi}\frac{x}{l}\right)-S(\tilde{\xi})\cdot S\left(\tilde{\xi}\frac{x}{l}\right)}{\tilde{\xi}^3 A(\tilde{\xi})}$
9.	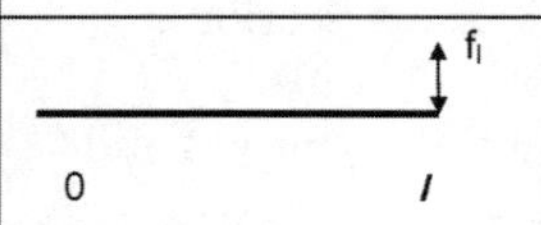	$L_{lx}(j\omega)=\frac{2l^3}{\tilde{E}_0 I}\cdot\frac{U(\tilde{\xi})\cdot T\left(\tilde{\xi}\frac{x}{l}\right)-V(\tilde{\xi})\cdot S\left(\tilde{\xi}\frac{x}{l}\right)}{\tilde{\xi}^3 D(\tilde{\xi})}$
10.	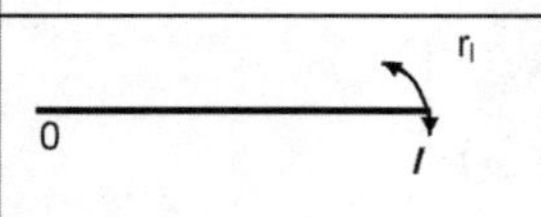	$\tilde{A}_{lx}(j\omega)=\frac{2l^2}{\tilde{E}_0 I}\cdot\frac{T(\tilde{\xi})\cdot T\left(\tilde{\xi}\frac{x}{l}\right)-U(\tilde{\xi})\cdot S\left(\tilde{\xi}\frac{x}{l}\right)}{\tilde{\xi}^2 D(\tilde{\xi})}$
11.	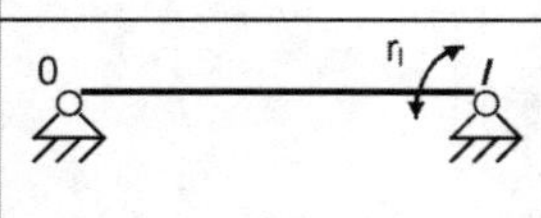	$\Gamma_{lx}(j\omega)=\frac{l^2}{\tilde{E}_0 I}\cdot\frac{\sin\tilde{\xi}\cdot\operatorname{sh}\tilde{\xi}\frac{x}{l}-\operatorname{sh}\tilde{\xi}\cdot\sin\tilde{\xi}\frac{x}{l}}{\tilde{\xi}^2 S_1(\tilde{\xi})}$

Fig. 2.22.

where, $\tilde{\xi} = l\left(\frac{\rho\omega^2}{\tilde{E}I}\right)^{1/4}$, $\tilde{\xi}_1 = \tilde{\xi}\frac{s}{l}$, $\tilde{\xi}_2 = \tilde{\xi}\frac{l-s}{l}$.

Using formulae (2.200), the reactions between the components may be found,

$$\begin{aligned}
\tilde{F}_{12} &= -\frac{F_s}{S_1(\tilde{\xi})}(\sin\tilde{\xi}_1\cos\tilde{\xi}_2\sinh\tilde{\xi} + \sinh\tilde{\xi}_1\cosh\tilde{\xi}_2\sin\tilde{\xi}),\\
\tilde{R}_{12} &= -\frac{F_s l}{\tilde{\xi}S_1(\tilde{\xi})}(\sin\tilde{\xi}_1\sin\tilde{\xi}_2\sinh\tilde{\xi} +\\
&\quad + \sinh\tilde{\xi}_1\sinh\tilde{\xi}_2\sin\tilde{\xi})
\end{aligned} \tag{2.201}$$

for force excitation, and

$$\begin{aligned}
\tilde{F}_{12} &= -\frac{\xi R_s}{lS_1(\tilde{\xi})}(\cos\tilde{\xi}_1\cos\tilde{\xi}_2\sinh\tilde{\xi} + \cosh\tilde{\xi}_1\cosh\tilde{\xi}_2\sin\tilde{\xi}),\\
\tilde{R}_{12} &= -\frac{R_s}{S_1(\tilde{\xi})}(\cos\tilde{\xi}_1\sin\tilde{\xi}_2\sinh\tilde{\xi} +\\
&\quad + \cosh\tilde{\xi}_1\sinh\tilde{\xi}_2\sin\tilde{\xi})
\end{aligned} \tag{2.202}$$

for moment excitation.

By substituting (2.201), (2.202) and the appropriate expressions from Table 2.1 into equalities (2.198), the required dynamic compliances are obtained,

$$\begin{aligned}
L_{sx}(j\omega) &= \frac{l^3}{\tilde{E}_0 I\tilde{\xi}^3}\times\\
&\times\begin{cases}
\dfrac{\sinh\tilde{\xi}\sin\tilde{\xi}\frac{l-s}{l}\sin\tilde{\xi}\frac{x}{l} - \sin\tilde{\xi}\sinh\tilde{\xi}\frac{l-s}{l}\sinh\tilde{\xi}\frac{x}{l}}{S_1(\tilde{\xi})}, & 0 \le x \le s,\\
\dfrac{\sinh\tilde{\xi}\sin\tilde{\xi}\frac{s}{l}\sin\tilde{\xi}\frac{l-x}{l} - \sin\tilde{\xi}\sinh\tilde{\xi}\frac{s}{l}\sinh\tilde{\xi}\frac{l-x}{l}}{S_1(\tilde{\xi})}, & s \le x \le l;
\end{cases}
\end{aligned} \tag{2.203}$$

$$\begin{aligned}
\Gamma_{sx}(j\omega) &= \frac{l^2}{\tilde{E}_0 I\tilde{\xi}^2}\times\\
&\times\begin{cases}
\dfrac{-\sinh\tilde{\xi}\cos\tilde{\xi}\frac{l-s}{l}\sin\tilde{\xi}\frac{x}{l} + \sin\tilde{\xi}\cosh\tilde{\xi}\frac{l-s}{l}\sinh\tilde{\xi}\frac{x}{l}}{S_1(\tilde{\xi})}, & 0 \le x \le s,\\
\dfrac{\sinh\tilde{\xi}\cos\tilde{\xi}\frac{s}{l}\sin\tilde{\xi}\frac{l-x}{l} - \sin\tilde{\xi}\cosh\tilde{\xi}\frac{s}{l}\sinh\tilde{\xi}\frac{l-x}{l}}{S_1(\tilde{\xi})}, & s \le x \le l.
\end{cases}
\end{aligned} \tag{2.204}$$

An expression similar to (2.203) is given in Butkovsky (1979). From (2.204), when $s = l$ the dynamic compliance of the component shown in line 11 of Table 2.1 is obtained.

Dynamic compliances of systems with different boundary conditions can be found in a similar manner. As an example, for a homogeneous cantilever bar, the coefficients of equation (2.199) take the form,

$$\alpha_{11} = \frac{l^3}{\tilde{E}_0 I \tilde{\xi}^3}\left[\frac{B(\tilde{\xi}_l)}{E(\tilde{\xi}_l)} + \frac{B(\tilde{\xi}_2)}{D(\tilde{\xi}_2)}\right],$$

$$\alpha_{12} = \frac{l^2}{2\tilde{E}_0 I \tilde{\xi}^2}\left[\frac{S_1(\tilde{\xi}_l)}{E(\tilde{\xi}_l)} - \frac{S_1(\tilde{\xi}_2)}{D(\tilde{\xi}_2)}\right],$$

$$\alpha_{22} = \frac{l}{\tilde{E}_0 I \tilde{\xi}}\left[\frac{A(\tilde{\xi}_l)}{E(\tilde{\xi}_l)} + \frac{A(\tilde{\xi}_2)}{D(\tilde{\xi}_2)}\right]. \tag{2.205}$$

By substituting coefficients (2.205) and the appropriate quantities from lines 1, 2, 9 and 10 of Table 2.1 into expressions (2.200) and (2.198), the dynamic compliances of a cantilever bar, subjected to the excitation of a force and a moment in cross section $x = s$, are obtained,

$$L_{sx}(j\omega) = \frac{l^3}{2\tilde{E}_0 I \tilde{\xi}^3 E(\tilde{\xi})} \times \times \begin{cases} \chi_1(\tilde{\xi}_1, \tilde{\xi}_2) U(\tilde{\xi}\frac{x}{l}) - \chi_2(\tilde{\xi}_1, \tilde{\xi}_2) V(\tilde{\xi}\frac{x}{l}), & 0 \le x \le s, \\ \chi_3(\tilde{\xi}_1, \tilde{\xi}_2) T(\tilde{\xi}\frac{l-x}{l}) - \chi_4(\tilde{\xi}_1, \tilde{\xi}_2) S(\tilde{\xi}\frac{l-x}{l}), & s \le x \le l, \end{cases} \tag{2.206}$$

where,

$$\chi_1(\tilde{\xi}_1, \tilde{\xi}_2) = 2T(\tilde{\xi}_1)E(\tilde{\xi}_2) - 2U(\tilde{\xi}_1)B(\tilde{\xi}_2) - V(\tilde{\xi}_1)S_1(\tilde{\xi}_2),$$
$$\chi_2(\tilde{\xi}_1, \tilde{\xi}_2) = 2S(\tilde{\xi}_1)E(\tilde{\xi}_2) - 2T(\tilde{\xi}_1)B(\tilde{\xi}_2) - U(\tilde{\xi}_1)S_1(\tilde{\xi}_2),$$
$$\chi_3(\tilde{\xi}_1, \tilde{\xi}_2) = 2D(\tilde{\xi}_1)U(\tilde{\xi}_2) - 2B(\tilde{\xi}_1)V(\tilde{\xi}_2) - S_1(\tilde{\xi}_1)S(\tilde{\xi}_2),$$
$$\chi_4(\tilde{\xi}_1, \tilde{\xi}_2) = 2D(\tilde{\xi}_1)V(\tilde{\xi}_2) - 2B(\tilde{\xi}_1)S(\tilde{\xi}_2) - S_1(\tilde{\xi}_1)T(\tilde{\xi}_2);$$

$$\Gamma_{sx}(j\omega) = \frac{l^2}{2\tilde{E}_0 I \tilde{\xi}^2 E(\tilde{\xi})} \times \times \begin{cases} \chi_5(\tilde{\xi}_1, \tilde{\xi}_2) U(\tilde{\xi}\frac{x}{l}) - \chi_6(\tilde{\xi}_1, \tilde{\xi}_2) V(\tilde{\xi}\frac{x}{l}), & 0 \le x \le s, \\ -\chi_7(\tilde{\xi}_1, \tilde{\xi}_2) T(\tilde{\xi}\frac{l-x}{l}) + \chi_8(\tilde{\xi}_1, \tilde{\xi}_2) S(\tilde{\xi}\frac{l-x}{l}), & s \le x \le l, \end{cases} \tag{2.207}$$

where,

$$\chi_5(\tilde{\xi}_1, \tilde{\xi}_2) = 2V(\tilde{\xi}_1)A(\tilde{\xi}_2) + 2S(\tilde{\xi}_1)E(\tilde{\xi}_2) + U(\tilde{\xi}_1)S_1(\tilde{\xi}_2),$$
$$\chi_6(\tilde{\xi}_1, \tilde{\xi}_2) = 2U(\tilde{\xi}_1)A(\tilde{\xi}_2) + 2V(\tilde{\xi}_1)E(\tilde{\xi}_2) + T(\tilde{\xi}_1)S_1(\tilde{\xi}_2),$$
$$\chi_7(\tilde{\xi}_1, \tilde{\xi}_2) = 2D(\tilde{\xi}_1)T(\tilde{\xi}_2) + 2A(\tilde{\xi}_1)S(\tilde{\xi}_2) + S_1(\tilde{\xi}_1)V(\tilde{\xi}_2),$$
$$\chi_8(\tilde{\xi}_1, \tilde{\xi}_2) = 2D(\tilde{\xi}_1)U(\tilde{\xi}_2) + 2A(\tilde{\xi}_1)T(\tilde{\xi}_2) + S_1(\tilde{\xi}_1)S(\tilde{\xi}_2);$$

Notations $\tilde{\xi}, \tilde{\xi}_1$ and $\tilde{\xi}_2$ used in (2.206) and (2.207) are the same as that used in (2.201). It is easy to prove that when $s = l$, expressions (2.206) and

(2.207) can be written in the form shown in lines 1 and 2 of Table 2.1. Using dynamic compliances in the form of (2.203), (2.204), (2.206) and (2.207), bending vibration of bars, caused by forces that are distributed along their lengths, may be considered. This can be done in a similar manner to the considerations shown in the section 2.1; vibration caused by a distributed force is described by formula (2.169). As an example, for a hinged homogeneous bar, excited by a distributed load and with a density of q_0; substituting (2.203) into (2.169) and with subsequent integration, it is found that,

$$\tilde{a}_x = \frac{q_0 l^4}{\tilde{E}_0 I \tilde{\xi}^4} \frac{\cosh\frac{\tilde{\xi}}{2}\cos(\frac{\tilde{\xi}}{2} - \tilde{\xi}\frac{x}{l}) + \cos\frac{\tilde{\xi}}{2}\cosh(\frac{\tilde{\xi}}{2} - \tilde{\xi}\frac{x}{l}) - 2C(\frac{\tilde{\xi}}{2})}{C(\frac{\tilde{\xi}}{2})}. \tag{2.208}$$

In the instance of moment excitation, the quantity $L_{sx}(j\omega)$ in (2.169) should be replaced by $\Gamma_{sx}(j\omega)$; $q(s)$ is the moment distribution density.

3. The most important task in designing ultrasonic machine vibration systems is obtaining the tool motion parameters. In most cases, the tool can be considered as a rigid body which is attached to the bar system. As mentioned previously, during the bar's bending vibration, the body to which it is attached performs complicated motion in a plane. The problem of finding this motion may now be considered. Its solution will be used later, for analysing the nonlinear processes that take place in ultrasonic vibration systems.

Bending vibration within the bar is now considered. A body is attached to its end and is subjected to an external harmonic excitation $f_c(t) = F_c e^{j\omega t}$, applied at its mass centre C (Fig.2.23).

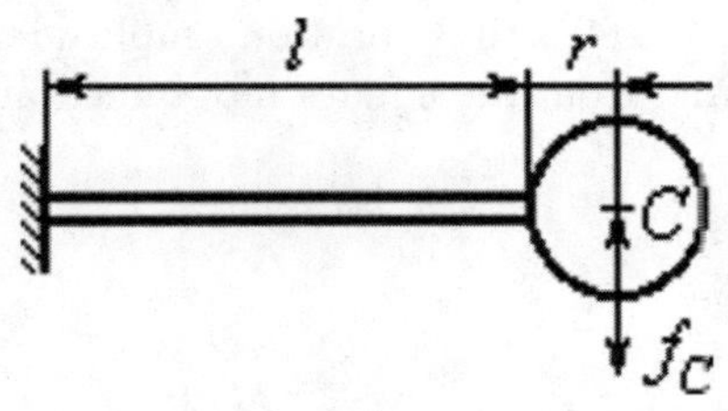

Fig. 2.23.

In order to describe the displacement of bar cross sections from their undistorted state, the function $u_x(t) = \tilde{a}_x e^{j\omega t}$ shall be used. The motion of the additional body is described by the coordinate of its mass centre $u_c(t) = \tilde{a}_c e^{j\omega t}$ and its rotation angle $\theta_c(t) = \tilde{\theta}_c e^{j\omega t}$. For small levels of vibration, these quantities are related as,

$$\tilde{a}_c = \tilde{a}_l + r\tilde{\theta}_c, \quad \tilde{\theta}_c = \tilde{\theta}_l = \frac{d\tilde{a}_x}{dx}|_{x=l}, \tag{2.209}$$

where r is the distance between the mass centre of the body and its fixing point.

Using notation $\tilde{F}_l$ and $\tilde{R}_l$ for the cutting force amplitude and bending moment in the bar's cross section $x = l$, and dynamic compliances (2.51) and (2.197) describing translational and angular vibration of the solid body, the following amplitude relations hold true,

$$-M\omega^2\tilde{a}_c = F_c - \tilde{F}_l, \quad -J\omega^2\tilde{\theta}_c = r\tilde{F}_l - \tilde{R}_l, \tag{2.210}$$

where M and J are the body's mass and moment of inertia with respect to the axis passing through its mass centre.

The motion of the bar is described using dynamic compliances $L_{lx}(j\omega)$ and $\Gamma_{lx}(j\omega)$,

$$\tilde{a}_x = L_{lx}(j\omega)\tilde{F}_l + \Gamma_{lx}(j\omega)\tilde{R}_l. \tag{2.211}$$

Substituting the quantities $\tilde{F}_l$ and $\tilde{R}_l$ from (2.210) into equation (2.211),

$$\tilde{a}_x = L_{cx}(j\omega)(\tilde{F}_l + M\omega^2\tilde{a}_c) + \Gamma_{lx}(j\omega)J\omega^2\tilde{\theta}_c, \tag{2.212}$$

where $L_{cx}(j\omega) = L_{lx}(j\omega) + r\Gamma_{lx}(j\omega)$.

When $x = l$, and by virtue of the last equality in (2.209), differentiating (2.212) leads to,

$$\tilde{\theta}_c = \frac{(\tilde{F}_l + M\omega^2\tilde{a}_c)L'_{cl}(j\omega)}{1 - J\omega^2\Gamma'_{ll}(j\omega)}, \tag{2.213}$$

where the dash denotes the operation d/dx.

The quantity $r\tilde{\theta}_c$ can be added to both parts of equality (2.212), where $x = l$, and (2.213) can be substituting into the right-hand side. Taking account of the first relation in (2.209), the vibration amplitude of the added body's mass centre can be found from the equation obtained above,

$$\tilde{a}_c = F_c/W_c(j\omega), \tag{2.214}$$

where,

$$W_c(j\omega) = \left[L_{cl}(j\omega) + L'_{cl}(j\omega)\frac{r + \omega^2 J\Gamma_{ll}(j\omega)}{1 - \omega^2 J\Gamma'_{ll}(j\omega)}\right]^{-1} - M\omega^2 \tag{2.215}$$

is the dynamic stiffness of the vibration system at point c.

The displacement amplitudes of the bar's cross sections and the angular vibration of the body can now be found from relations (2.212) and (2.213). The angular vibration amplitude is given by,

$$\tilde{\theta}_c = \tilde{a}_c/K_\theta(j\omega), \tag{2.216}$$

where,

$$K_\theta(j\omega) = \frac{L_{cl}(j\omega) + rL'_{cl}(j\omega) + J\omega^2\Gamma^2_{ll}(j\omega)\left[\frac{L_{cl}(j\omega)}{\Gamma_{ll}(j\omega)}\right]'}{L'_{cl}(j\omega)}. \tag{2.217}$$

As an example, the vibration of a cantilever bar with a body fixed on it (Fig.2.23), is considered. From the expressions shown in lines 1 and 2 of Table 2.1, and using formulae (2.181) for $x = l$,

$$\begin{aligned} L_{ll}(j\omega) &= \frac{l^3}{\tilde{E}_0 I \tilde{\xi}^3} \frac{B(\tilde{\xi})}{E(\tilde{\xi})}, \quad \Gamma_{ll}(j\omega) = \frac{l^2}{2\tilde{E} I \tilde{\xi}^2} \frac{S_1(\tilde{\xi})}{E(\tilde{\xi})}, \\ L'_{ll}(j\omega) &= \Gamma_{ll}(j\omega), \qquad \Gamma'_{ll}(j\omega) = \frac{l}{\tilde{E} I \tilde{\xi}} \frac{A(\tilde{\xi})}{E(\tilde{\xi})}, \end{aligned} \tag{2.218}$$

where, as before (see (2.182) and (2.183)),

$$\tilde{\xi} = l\left(\frac{\rho\omega^2}{\tilde{E}_0 I}\right)^{1/4} = \xi(1 + j\frac{\psi}{2\pi})^{-1/4}, \quad \xi = l(\frac{\rho\omega^2}{E_0 I})^{1/4}. \tag{2.219}$$

Substituting (2.218) into relations (2.215) and (2.217),

$$\begin{aligned} &W_c(j\omega) = \\ &= \frac{\tilde{E}_0 I}{l^3} \times \left\{ \frac{\tilde{\xi}^3 [E(\tilde{\xi}) - \varsigma \tilde{\xi}^3 A(\tilde{\xi})]}{B(\tilde{\xi}) + \eta\tilde{\xi} S_1(\tilde{\xi}) + \varsigma\tilde{\xi}^3 D(\tilde{\xi}) + \eta^2\tilde{\xi}^2 A(\tilde{\xi})} - \mu\xi^4 \right\}, \\ &K_\theta(j\omega) = \\ &= \frac{2l}{\tilde{\xi}} \times \frac{B(\tilde{\xi}) + \eta\tilde{\xi} S_1(\tilde{\xi}) + \varsigma\tilde{\xi}^3 D(\tilde{\xi}) + \eta^2\tilde{\xi}^2 A(\tilde{\xi})}{S_1(\tilde{\xi}) + 2\eta\tilde{\xi} A(\tilde{\xi})}, \end{aligned} \tag{2.220}$$

where $\mu = M/\rho l$, $\varsigma = J/\rho l^3$ and $\eta = r/l$.

Using the results obtained above, the amplitude-frequency characteristics for the translational and angular vibration within the body may be found. Assuming that the absorption coefficient ψ is small, and retaining only the linear terms in the expansions of quantities containing ψ, expressions (2.220) can be converted into a form similar to (2.79),

$$\begin{aligned} W_c(j\omega) &= U_c(\omega) + jV_c(\omega), \\ K_\theta(j\omega) &= K_l(\omega) + jK_2(\omega), \end{aligned} \tag{2.221}$$

where $U_c(\omega) = ReW_c(j\omega)$, $V_c(\omega) = ImW_c(j\omega)$, $K_1(\omega) = ReK_\theta(j\omega)$, $K_2(\omega) = ImK_\theta(j\omega)$.

The real parts of quantities (2.221) are determined by expressions (2.220), where, assuming that $\psi = 0$, from (2.219), $\tilde{\xi} = \xi$ and $\tilde{E}_0 = E_0$. The imaginary parts are given by,

$$\begin{aligned} V_c(\omega) &= \frac{\psi\xi^3 E_0 I}{8\pi l^3} \begin{bmatrix} 4\xi T^2 + BE + 2\eta\xi(4\xi ST + ES_1) + \\ +8\varsigma\xi^3 T(V - \xi U) + \eta^2\xi^2(4\xi S^2 + 3AE) + \\ +\varsigma^2\xi^6(4\xi U - AD) + \varsigma\eta\xi^4(AS_1 - 8\xi SU) \end{bmatrix} \times \\ &\times (B + \eta\xi S_1 + \varsigma\xi^3 D + \eta^2\xi^2 A)^{-2}, \\ K_2(\omega) &= \frac{\psi l}{16\pi\xi} \begin{bmatrix} BS_1 - 8\xi TV + 4\eta\xi(AB - \xi V(2\xi)) + \\ +2\varsigma\xi^3(4\xi UV - DS) + \eta^2\xi^2(AS_1 - 8\xi SU) + \\ +2\varsigma\eta\xi^4(4\xi U^2 - AD) \end{bmatrix} \times \\ &\times (S_1 + 2\eta\xi A)^{-2}. \end{aligned} \tag{2.222}$$

(In formulae (2.222), the dependence of functions (2.181) on ξ is not explicitly shown).

Fig.2.24 shows the amplitude-frequency characteristics of the body's vibration at its mass centre (Fig.2.24, a) and also those of its angular vibration (Fig.2.24, b). The equations determining the resonant curves can be found from (2.214) and (2.216). Taking account of (2.221), these equations can be written in the form,

$$a_c = \frac{F_c}{|W_c(j\omega)|} = \frac{F_c}{\sqrt{U_c^2(\omega) + V_c^2(\omega)}},$$
$$\theta_c = \frac{a_c}{|K_\theta(j\omega)|} = \frac{a_c}{\sqrt{K_1^2(\omega) + K_2^2(\omega)}}. \qquad (2.223)$$

According to (2.223), the resonant frequencies ω_i can be found from the relation $U_c(\omega_i) = 0$, which after the substitution of (2.220) yields the following equation for the natural frequencies,

$$E - \varsigma\xi_i^3 A - \mu\xi_i(B + \eta\xi_i S_1 + \varsigma\xi_i^3 D + \eta^2\xi_i^2 A) = 0. \qquad (2.224)$$

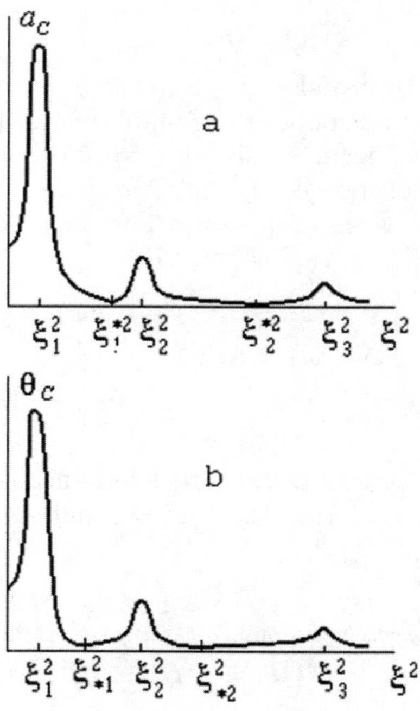

Fig. 2.24.

The amplitude of resonant vibration is then found from equality,

$$a_c = F_c/V_c(\omega_i),$$

where $V_c(\omega_i)$ is given by (2.222), and is calculated after the substitution of the solution ξ_i from equation (2.224).

To analyse the dependence of the resonant frequencies ξ_i^2 on the parameters ς, η and μ, it is convenient to rewrite equation (2 224) as follows,

$$\mu = \frac{E - \varsigma\xi_i^3 A}{\xi_i[B + \eta\xi_i(S_1 + \eta\xi_i A) + \varsigma\xi_i^3 D]}, \tag{2.225}$$

$$\varsigma = \frac{E - \mu\eta\xi_i^2(S_1 + \eta\xi_i A) - \mu\xi_i B}{\xi_i^3(\mu\xi_i D + A)}, \tag{2.226}$$

$$\eta = \frac{-S_1 \pm [S_1^2 - 4A(B + \varsigma\xi_i^3 D) + 4A(E - \varsigma\xi_i^3)\mu\xi_i]^{1/2}}{2\xi_i A}. \tag{2.227}$$

Analogously, equality $U^{-1}(\omega) = 0$ leads, by virtue of (2.220), to the following equation for the antiresonant frequencies ξ_i^* of the body's mass centre vibration,

$$\eta^2\xi_i^{*2} A + \eta\xi_i^* S_1 + \varsigma\xi_i^{*3} D + B = 0, \tag{2.228}$$

which can be rewritten as,

$$\varsigma = -\frac{B + \eta\xi_i^*(S_1 + \eta\xi_i^* A)}{\xi_i^{*3} D}, \tag{2.229}$$

$$\eta = \frac{-S_1 \pm [S_1^2 - 4A(B + \varsigma\xi_i^{*3} D)]^{1/2}}{2\xi_i^* A}. \tag{2.230}$$

Finally, from (2.223) an expression can be found for the antiresonant frequencies ξ_{*i} of the body's angular vibration, $K_1^{-1}(\omega) = 0$, or, by virtue of (2.220),

$$\eta = -S_1/2\xi_{*i} A. \tag{2.231}$$

The arguments ξ_i, ξ_i^* and ξ_{*i} of functions (2.181) are omitted in formulae (2.224) – (2.231).

The dependence on (2.226) (curves 1) and (2.229) (curves 2) for the first three natural modes are shown in Fig.2.25, a for $\mu = 0.5$ and $\eta = 0.05$. The circles on the x-axis correspond to the natural frequencies of a bar with an added point mass. These are found from equation (2.223) when $\varsigma = \eta = 0$. (In these circumstances, all the results obtained here reduce to the formulae given in Babitsky (1998), Babitsky & Tresviatsky (1976) in which the effects of the added body's geometrical and inertial parameters on the angular vibration is neglected). Fig.2.25, a demonstrates the increase in influence of the body's moment of inertia on the intrinsic properties of the vibration system when it oscillates in its higher modes.

Curves 1 and 2, when $\varsigma \to \infty$, tend to the asymptotes given by the squares of the roots of equations,

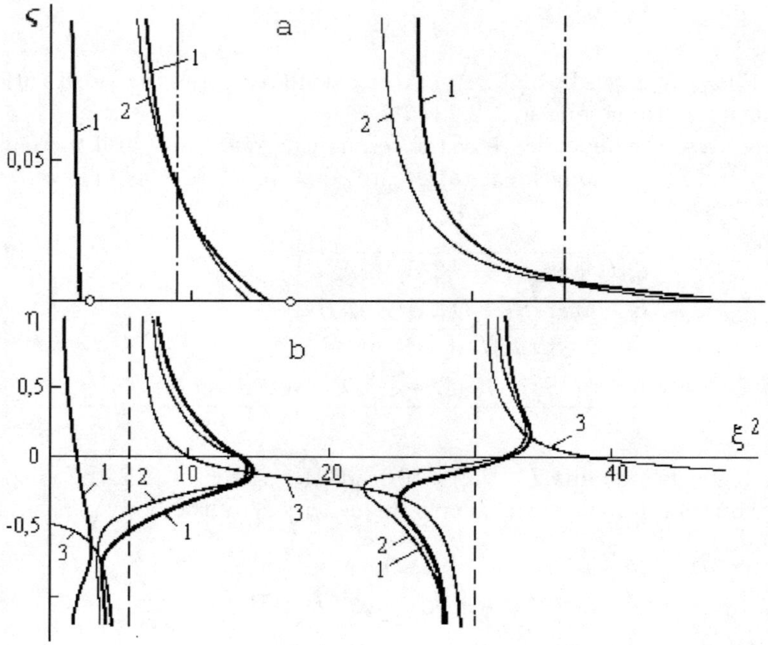

Fig. 2.25.

$$\mu\xi D + A = 0, \quad D = 0. \tag{2.232}$$

It must be noted that, when $\varsigma \to \infty$, from (2.220) $|K_\theta(j\omega)| \to \infty$, and, therefore, $\theta_c \to 0$. Thus, the dynamic stiffness,

$$W_l(j\omega) = -\frac{\tilde{E}_0 I}{l^3}\left[\frac{A(\tilde{\xi})}{D(\tilde{\xi})}\tilde{\xi}^3 + \mu\tilde{\xi}^4\right] \tag{2.233}$$

describes a system in which the added body performs translational vibration at all frequencies. The resonant and antiresonant frequencies of such a system are given by equations (2.232).

Comparing expressions (2.226) and (2.229), it can be seen that they coincide if,

$$A^2\xi_i^2\eta^2 + AS_1\xi_i\eta + AB + ED = 0. \tag{2.234}$$

According to formulae (2.181), $S_1^2 \equiv 4(AB + ED)$. Taking this into account, solutions can be found for (2.234),

$$\eta = \frac{-S_1 \pm \sqrt{S_1^2 - 4(AB + ED)}}{2\xi_i A} = -\frac{S_1}{2\xi_i A}. \tag{2.235}$$

Thus, as follows from (2.231) and (2.235), curves 1 and 2 are tangential to one another at frequencies coinciding with angular vibration antiresonant frequencies, ξ_{*i}. These are only dependant on the parameter η, and are shown in Fig.2.25, a by the dot-and-dash vertical lines.

Functions (2.227) and (2.230) can be determined by using the relationship between η, and ξ^2 under the following conditions,

$$S_1^2 - 4A(B + \varsigma\xi_i^3 D) + 4A(E - \varsigma\xi_i^3 A)/\mu\xi_i \geq 0,$$
$$S_1^2 - 4A(B + \varsigma\xi_i^{*3} D) \geq 0, \tag{2.236}$$

respectively.

Fig.2.25, b demonstrates the dependence on (2.227) (curve 1), (2.230) (curve 2) and (2.231) (curve 3) for the first three natural modes for $\mu = 0.5$ and $\varsigma = 0.01$. It can be seen that all the graphs have the same asymptotes as $\eta \to \pm\infty$. These are given by the squares of the roots of equation $A(\xi) = 0$, and are shown in Fig.2.25, b by dashed lines.

The equalities in conditions (2.236) convert the dependences (2.227) and (2.230) into a form which coincides with equation (2.231) for angular vibration antiresonant frequencies. This shows that the unique points of curves 1 and 2 for any values of parameters μ and ς are located on the lines 3. As follows from (2.226) and (2.230), curves 1 and 2 have two points of tangency located symmetrically with respect to the value (2.231). This is shown in Fig.2.25, b, provided that the condition $E - \varsigma\xi_i^3 A = 0$ holds.

The analysis carried out above shows the strong dependence of the vibration system's resonant and antiresonant frequencies on the inertial and geometrical parameters of the added body. It must be noted, that for all modes of bending vibration, there are certain parameter values which allow the resonant and antiresonant frequencies of the body's angular vibration to coincidence. These frequencies correspond to an intensive translational vibration of the tool with minimal angular vibration. Such a possibility has already been mentioned in section 2.1.3. The calculations shown above allow devices to be designed which realize a given tool motion within an ultrasonic system.

This problem has been considered in detail, since the questions that have arisen are not covered well enough by the existing literature. In the known papers, the vibration of concentrated masses, attached to systems, are usually considered (Babitsky (1998), Babitsky & Tresviatsky (1976)). The estimation of the added body's effect on the natural frequencies and vibration modes of the bar is based on the modelling of an added body using a concentrated mass with an equivalent moment of inertia.

4. To conclude this section, consider briefly the bending vibration of a bar system under kinematic excitation. Known values for dynamic compliance may be used. The vibration system is kinematically actuated in a specific cross section, $x = s$. Using notation $\tilde{F}_s$ and $\tilde{R}_s$ for the amplitudes of force and moment at the point of excitation, the following expression for the vibration amplitudes at the arbitrary cross section x can be obtained,

$$\tilde{a}_x = L_{sx}(j\omega)\tilde{F}_s + \Gamma_{sx}(j\omega)\tilde{R}_s. \tag{2.237}$$

Thus, for cross section $x = s$, using (2.173),

$$\begin{aligned} \tilde{a}_s &= L_{ss}(j\omega)\tilde{F}_s + \Gamma_{ss}(j\omega)\tilde{R}_s, \\ \tilde{\theta}_s &= L'_{ss}(j\omega)\tilde{F}_s + \Gamma'_{ss}(j\omega)\tilde{R}_s. \end{aligned} \tag{2.238}$$

As can be seen from (2.238), the kinematic excitation of bending vibration can be realized by initiating translational vibration with an amplitude $\tilde{a}_s$ in cross section s, or by using angular vibration with an amplitude $\tilde{\theta}_s$. If one of these quantities is given, the other three unknowns may be found by using system of equations (2.238) supplemented by an appropriate constraint which is usually known from the physical context of the problem.

A few examples may be considered. Consider the vibration of a homogeneous bar $(-l \leq x \leq l)$, whose average cross section $x = s = 0$ is forced to perform alternating motion $u_0(t) = a_0 e^{j\omega t}$. Due to symmetry, the systems shown in Fig.2.8, a, b lead to such an arrangement. In this case, $\tilde{\theta}_0 = 0, \Gamma_{00}(j\omega) = 0$, and from (2.238) and (2.237), when $s = 0$,

$$\tilde{F}_0 = a_0 / L_{00}(j\omega), \tag{2.239}$$

$$\tilde{a}_x = a_0 L_{0x}(j\omega) / L_{00}(j\omega). \tag{2.240}$$

Substituting the appropriate values of dynamic compliance from table 2.1 (line 8) into (2.240) and taking (2.181) into account,

$$\tilde{a}_x = 2a_0 \frac{S(\tilde{\xi})S(\tilde{\xi}\frac{x-l}{l}) - V(\tilde{\xi})T(\tilde{\xi}\frac{x-l}{l})}{E(\tilde{\xi})}. \tag{2.241}$$

Further analysis can be carried out in a similar manner to that in section 2.2.5. The maximal amplitudes are achieved at frequencies given by the equation $E(\xi) = 0$. These coincide with the natural frequencies of the cantilever bar (see lines 1 and 2 of Table 2.1 and formulae (2.206) and (2.207)). For the first two natural modes, the handbook's tables Ananiev (1946) give,

$$\xi_1 = 1.87; \quad \xi_2 = 4.69. \tag{2.242}$$

In practice, in order to find the resonant frequencies, the following relation can be used with sufficient accuracy,

$$\xi_n = (2n-1)\pi/2, \quad n = 2, 3, ... \tag{2.243}$$

Assuming, as before, that the absorption coefficient ψ is small, from (2.241) the amplitude of resonant vibration at cross section $x = l$ is found,

$$a_l = \frac{16\pi a_0}{\psi \xi_n} \left| \frac{S(\xi_n)}{B(\xi_n)} \right|. \tag{2.244}$$

For the first two natural modes of vibration, using tables Butkovsky (1979), $a_l = 10.02a_0/\psi$ and $a_l = 5.45a_0/\psi$ are obtained. For higher modes ($n > 1$), taking (2.181) and (2.243)into account, expression (2.244) can be approximated to,

$$a_l = 8\pi a_0/\psi\xi_n = 16a_0/\psi(2n-1), \quad n = 2, 3, ... \tag{2.245}$$

It is noted that relations (2.239) and (2.240) coincide with expressions (2.82) and (2.83) describing a bar's longitudinal vibration under kinematic excitation. Therefore, in this situation, the results of section 2.2.5, in particular those devoted to designing ultrasonic concentrators, hold. When using a bar for concentrating bending vibration, its parameters should satisfy equality (2.84), which, after the substitution of the appropriate quantities, can be converted into the form $A(\xi) = 0$. This yields, with sufficient accuracy for practical needs,

$$\xi_n = \frac{(4n-1)\pi}{4}, \quad n = 1, 2, ... \tag{2.246}$$

Taking account of the notation used in (2.182) and (2.183), the length of a concentrator tuned to a working frequency of ω can be found,

$$l = \xi_n \left(\frac{E_0 I}{\rho\omega^2}\right)^{1/4} = \frac{(4n-1)\pi}{4} \left(\frac{E_0 I}{\rho\omega^2}\right)^{1/4}. \tag{2.247}$$

The amplification coefficient of a bending vibration concentrator can be found by using (2.86) after the substitution of the appropriate quantities from line 6 of Table 2.1,

$$K = \frac{a_l}{a_0} = \frac{\cosh\xi_n \cos\xi_n}{S(\xi_n)} \approx \sqrt{2} \approx 1.41. \tag{2.248}$$

To conclude this section, consider bending vibration of the same system when its $x = s = 0$ cross section is subjected to an angular vibration $\Theta_0(t) = \theta_0 e^{j\omega t}$. In this case, $L_{00}(j\omega) = L'_{00}(j\omega) = 0$, and from (2.237) and (2.238), when $s = 0$,

$$\begin{aligned} \tilde{R}_0 &= \theta_0/\Gamma'_{00}(j\omega), \\ \tilde{a}_x &= \theta_0\Gamma_{0x}(j\omega)/\Gamma'_{00}(j\omega). \end{aligned} \tag{2.249}$$

Substituting the appropriate quantities from table 2.1 (line 5) into (2.249),

$$a_x = 2\theta_0 \frac{l}{\tilde{\xi}} \frac{T(\tilde{\xi})S(\tilde{\xi}\frac{l-x}{l}) - S(\tilde{\xi})T(\tilde{\xi}\frac{l-x}{l})}{E(\tilde{\xi})}. \tag{2.250}$$

As follows from (2.250), the resonant frequencies of the system are given by the values (2.242) and (2.243) obtained above. The resonant amplitudes a_l at $x = l$ are found analogously to those in (2.244) and give the following expression,

$$a_l = \frac{16\pi l\theta_0}{\psi\xi_n}\left|\frac{T(\xi_n)}{B(\xi_n)}\right|. \tag{2.251}$$

For the first two natural modes (2.242), using tables Ananiev (1946), $a_l = 13.51 l\theta_0/\psi$ and $a_l = 5.49 l\theta_0/\psi$. For higher modes ($n > 2$), expression (2.251) can be approximated to the form (2.245), where it should be assumed that $a_0 = l\theta_0$.

These results will be used in the study of resonant vibration in ultrasonic bar systems which are subjected to a nonlinear load (section 3.2).

2.4 Dynamic behaviour of electroacoustic transducers

1. Within the generalized arrangement of ultrasonic machines (Fig.2.9), the dynamic behaviours of typical ultrasonic system components have been considered. The quantities characterizing the dynamic behaviour of vibration actuators will now be determined. Considerations will focus on magnetostrictive and piezoelectric transducers, which are the most frequently used in ultrasonic applications.

Consider a magnetostrictive transducer (Rimsky-Korsakov (1973)), with a core taking the form of a closed loop, and with a winding of N turns (Fig.2.26). The transducer is being fed from an external source, which generates a voltage,

$$\nu = \nu_m + \tilde{\nu}_0 e^{j\omega t}. \tag{2.252}$$

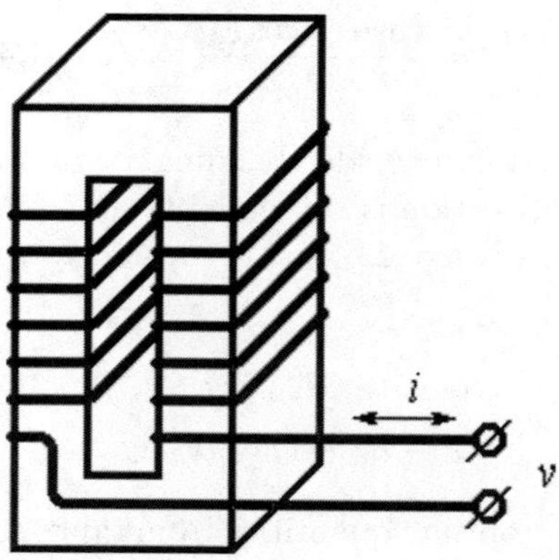

Fig. 2.26.

Consequently, a current,

$$i = i_m + \tilde{i}_0 e^{j\omega t} \tag{2.253}$$

is induced in the winding.

In (2.252) and (2.253), the quantities represented by ν_m and i_m represent constant components, while $\tilde{\nu}_0$ and $\tilde{i}_0$ denote the complex amplitudes of the alternating components of voltage and current.

The current passing through the winding induces a magnetic field in the core with an intensity (Kharkevich (1973)),

$$H_i = \frac{N}{l} i, \tag{2.254}$$

where l is the length of the core (Fig.2.26).

The direct magnetostrictive effect results in the appearance of mechanical stresses in the material of the core as caused by the magnetic field,

$$\sigma_h = -\gamma B = -\gamma \tilde{\mu} H. \tag{2.255}$$

B is the magnetic flux density, γ is the magnetostriction constant, $\tilde{\mu}$ is the complex magnetic permeability of the medium.

Equality (2.255) means that a magnetic field causes a compressional stress in the mounted workpiece.

The reverse magnetostrictive effect takes place when the stress-strain state of the core is physically changed. This results in a change of the magnetic field surrounding the core.

The change in the magnetic field caused by a change in strain ε is given by the equality,

$$H_\varepsilon = \gamma \varepsilon. \tag{2.256}$$

If the material of the core is simultaneously subjected to both mechanical and magnetostrictive deformations, the stress induced in the material is given by,

$$\sigma = \tilde{E}_0 \varepsilon - \gamma \tilde{\mu} H, \tag{2.257}$$

and the total field in the core, as caused by both the current and the deformation, is, according to (2.254) and (2.256),

$$H = \frac{N}{l} i + \gamma \varepsilon. \tag{2.258}$$

Substituting (2.258) into (2.257), the relation between the mechanical stresses and strains of the magnetostrictive transducer's core and its magnetic and electric parameters is obtained:

$$\sigma = \tilde{E}_1 \varepsilon - \frac{\tilde{A}}{S} i, \tag{2.259}$$

where $\tilde{A} = \gamma \tilde{\mu} S N / l$ is the coefficient of electromechanical coupling;

$$\tilde{E}_1 = \tilde{E}_0 - \gamma^2 \tilde{\mu} = E_1 (1 + j \frac{\psi_1}{2\pi}). \tag{2.260}$$

$E_1 = E_0 - \gamma^2\mu$ is the elastic modulus of the magnetostrictive material for a constant magnetic bias field $H_m = \frac{N}{l}i_m$; ψ_1 is the adjusted absorption coefficient, which takes into account the mechanical and magnetic hysteresis.

Taking into account the specifics of the process under consideration, analogous relations for the constant components and complex amplitudes of the periodic components of the quantities used in (2.259) can be obtained. It must be noted that quantities γ and μ should be considered as specific linearized values in the vicinity of a constant magnetic field - as caused by the bias current i_m.

Relation (2.259) describes the stress present in element dx of the core. The change in its magnetic state induces an opposing electromotive force,

$$de = -\frac{N}{l}S\frac{\partial B}{\partial t}dx = -\frac{N}{l}S\tilde{\mu}\frac{\partial H}{\partial t}dx \tag{2.261}$$

in the corresponding element of the winding, or, taking (2.258) into account,

$$de = -\frac{\partial}{\partial t}\left(\frac{\tilde{L}}{l}i + \tilde{A}\varepsilon\right)dx, \tag{2.262}$$

where $\tilde{L} = \tilde{\mu}S\frac{N^2}{l}$ is the inductivity of the transducer.

The total opposing electromotive force can be found by integrating (2.262) along the full length of the transducer,

$$e = -\frac{d}{dt}[\tilde{L}i + \tilde{A}(u_1 - u_0)], \tag{2.263}$$

where u_1 and u_0 are the displacements of the end cross sections of the core.

Using (2.263) and the relation $i = dq/dt$ (where q is the electric charge), an equation for the transducer's electric circuit can be written,

$$\tilde{L}\frac{d^2q}{dt^2} + r_0\frac{dq}{dt} + \frac{1}{C}q + \tilde{A}\frac{d}{dt}(u_1 - u_0) = \nu(t), \tag{2.264}$$

where r_0 and C are the resistance and capacitance of the circuit.

Equations (2.259) and (2.264) allow the vibration of a magnetostrictive transducer to be described in full.

2. Some typical cases shall be considered. Suppose that the length of the core is small in comparison to the length of the elastic wave within the material, i.e., it satisfies condition (2.50). Let the core be fixed at cross section $x = 0$ and loaded by a force f_{12} at cross section $x = l$ (Fig.2.27). In this case $\sigma = f_{12}/S, \varepsilon = u_1/l$ and $u_0 = 0$.

Taking into account the specifics of the process under consideration, from (2.259) and (2.264) the following equations are obtained:

$$\begin{aligned} \tilde{F}_{12} &= \tilde{k}_1\tilde{a}_1 - j\omega\tilde{A}\tilde{q}_0, \\ \tilde{\nu}_0 &= j\omega\tilde{A}\tilde{a}_1 + j\omega Z(j\omega)\tilde{q}_0, \end{aligned} \tag{2.265}$$

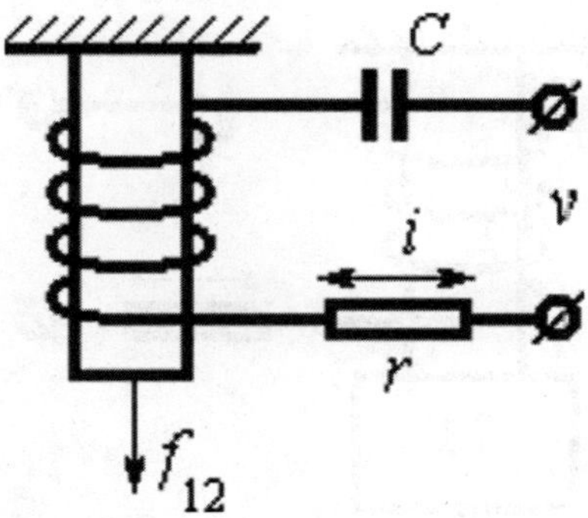

Fig. 2.27.

where $\tilde{k}_1 = \tilde{E}_1 S/l$ is the dynamic stiffness of the transducer's core when $\tilde{i}_0 = 0$; $Z(j\omega) = r + j[\omega L - (\omega C)^{-1}]$ is the complex impedance of the electric circuit when $\tilde{a}_1 = 0$; $r = r_0 + r_1$; $r_1 = \frac{SN^2\omega^2}{l} I_m \tilde{\mu}$ is the equivalent series resistance which takes into account the magnetic hysteresis.

Equations (2.265) are of the same type as the four-pole circuit equations (2.19). The coefficients are given by,

$$a_{11} = \tilde{k}_1, \ -a_{12} = a_{21} = j\omega \tilde{A}, \ a_{22} = j\omega Z(j\omega). \tag{2.266}$$

Substituting these quantities into relations (2.21), the unloaded transducer's vibration amplitude ($\tilde{F}_{12} = 0$) can be found. The transducer is supplied by a current source,

$$\tilde{a}_1^* = \tilde{A}\tilde{i}_0/\tilde{k}_1. \tag{2.267}$$

Suppose that the transducer excites vibration in a mechanical system with a dynamic stiffness $W(j\omega)$. To find the vibration amplitude of such a transducer, expression (2.26) can be used, where $\tilde{F}_l = 0, L_{00}(j\omega) = W^{-1}(j\omega)$ and $L_n(j\omega) = \tilde{k}_1^{-1}$. Using (2.23) and (2.266), it is found that,

$$\tilde{a}_l = \frac{\tilde{i}_0 \tilde{A}}{\tilde{k}_1 + W(j\omega)}. \tag{2.268}$$

As follows from (2.268), a system excited by a magnetostrictive transducer with a current source power supply, demonstrates the same behaviour as a mechanical component with a dynamic stiffness of $W_0(j\omega) = \tilde{k}_1 + W(j\omega)$ when subjected to a periodic force of amplitude $\tilde{F} = \tilde{A}\tilde{i}_0$. Further analysis of complete vibrating systems follow the method described in subsection 2.2.4.

For the sake of definiteness, the transducer is supposed to excite the simplest vibration system (Fig.2.28), with a dynamic stiffness,

$$W(j\omega) = (k_2 - M\omega^2) + j\omega b_2, \tag{2.269}$$

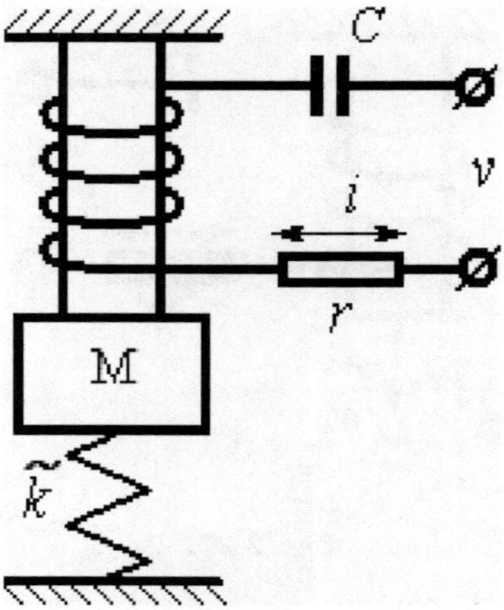

Fig. 2.28.

where M, k_2 and b_2 are the mass, stiffness and coefficient of resistance of the system.

Representing coefficient $\tilde{A}$ in the form

$$\tilde{A} = A(1 - j\frac{\chi}{4\pi}), \tag{2.270}$$

it is supposed, as before, that the quantities ψ_1, b_2, r and χ, describing the dissipative properties of the mechanical, electrical and magnetic circuits of the system, are small. From (2.268), using (2.260), (2.269) and the notations introduced in (2.265),

$$a_1 = \frac{\tilde{i}_0 A}{k} \frac{1 - j\frac{\chi}{4\pi}}{(1 - \xi^2) + j\xi\eta}, \tag{2.271}$$

where $\xi = \omega/\varsigma, \varsigma = \sqrt{k/M}, \eta = b/\sqrt{kM}, k = k_1 + k_2, b = b_1 + b_2$ and $b_1 = \frac{k_1\psi_1}{2\pi\omega}$.

This allows, analogously to (2.72) and (2.77), the resonant frequency and amplitude to be found,

$$\xi = 1, \quad a_1 = i_0 A/k\eta. \tag{2.272}$$

Substituting (2.271) into the second equality (2.265), and retaining quantities of the first order, the transducer's supply voltage is found,

$$\tilde{\nu}_0 = \frac{\tilde{i}_0 L\varsigma}{j\xi[(1 - \xi^2) + j\xi\eta]}\{(1 - \xi^2)(\xi_1^2 - \xi^2) - \xi_M^2\xi^2 + \\ + j\xi[\eta_1(1 - \xi^2) + \eta(\xi_1^2 - \xi^2) + \xi_M^2\xi\frac{\chi}{4\pi}]\}, \tag{2.273}$$

where $\xi_1 = 1/\varsigma\sqrt{LC}, \xi_M = A/\sqrt{kL}$ and $\eta_1 = r/L\varsigma$.

Fig.2.29 shows the amplitude-frequency characteristics of the mass's (M) vibration and the transducer's supply voltage when excited by a current source. The graphs are plotted using relations (2.271) and (2.273) for $\eta = \eta_1 = 0.2, \chi = 0$ and $\xi_M = 0.71$.

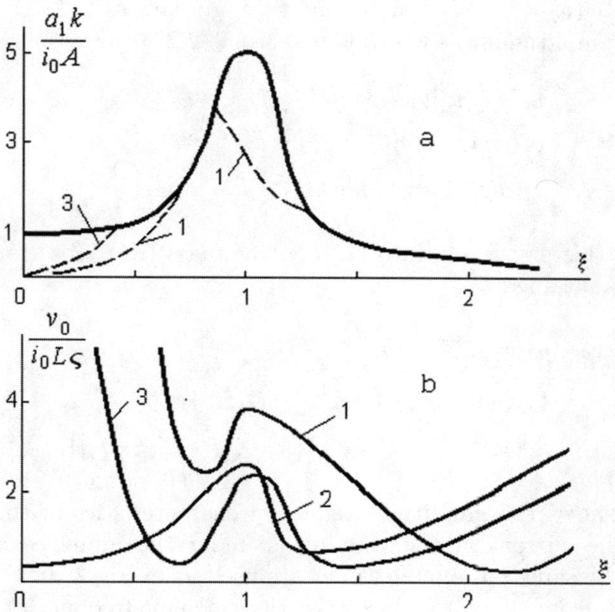

Fig. 2.29.

In general, when $\chi_1 \neq 0$ (curve 1 in Fig.2.29, b), the supply voltage $\nu_0 \to \infty$ when $\xi \to 0$ and $\xi \to \infty$ due to the presence of capacitance and inductance in the circuit. The curve has a local maximum at the system's mechanical resonant frequency ($\xi = 1$), and two minimums at frequencies determined by equation,

$$(1 - \xi^2)(\xi_1^2 - \xi^2) - \xi_M^2 \xi^2 = 0. \tag{2.274}$$

Solutions of this equation have the form,

$$\xi_{1,2}^{*2} = \frac{1 + \xi_1^2 + \xi_M^2}{2} \pm \sqrt{\left(\frac{1 + \xi_1^2 + \xi_M^2}{2}\right)^2 - \xi_1^2}. \tag{2.275}$$

Curve 1 is plotted for $\xi_1 = 2$.

Curve 2 in Fig.2.29, b corresponds to $\xi_1 = 0$.

Curve 3 is when $\xi_1 = 1$. In this case, the natural frequencies of the partial mechanical and electrical systems coincide, i.e. $\sqrt{k/M} = 1/\sqrt{LC}$. It can be shown that such a tuning of the partial systems reduces the supply voltage required in the vicinity of resonance $\xi = 1$ and frequencies (2.274) to a minimum. It must be noted that in the complete absence of dissipation ($\eta_1 = \eta_2 = \chi = 0$), $a_1(1) = \infty, \nu_0(1) = \infty, \nu_0(\xi^*_{1,2}) = 0$.

Parameters describing the vibration of a system which is excited by a voltage supply ($\nu_0 = const$) can be found using relations (2.271) and (2.273). The current amplitude may be evaluated from (2.273),

$$\begin{aligned} \tilde{i}_0 = & j\xi\tilde{\nu}_0[(1-\xi^2) + j\xi\eta]/ \\ & (\varsigma L\{(1-\xi^2)(\xi_1^2 - \xi^2) - \xi_M^2\xi^2 + \\ & +j\xi[\eta_1(1-\xi^2) + \eta(\xi_1^2 - \xi^2) + \xi_M^2\xi\frac{\chi}{4\pi}]\}) \end{aligned} \qquad (2.276)$$

Substituting this expression into (2.271), the mass (M) vibration amplitude (Fig.2.28), is obtained,

$$\begin{aligned} \tilde{a}_1 = & j\xi\xi_M^2\nu_0(1 - j\frac{\chi}{4\pi})/ \\ & \{A\varsigma[(1-\xi^2)(\xi_1^2 - \xi^2) - \xi_M^2\xi^2 + \\ & +j\xi(\eta_1(1-\xi^2) + \eta(\xi_1^2 - \xi^2) + \xi_M^2\xi\frac{\chi}{4\pi})]\}. \end{aligned} \qquad (2.277)$$

Fig.2.30 shows the amplitude-frequency characteristics of the mass (M) vibration. The current in the winding is under the influence of a voltage supply and the same parameter values are used as in Fig.2.29.

As follows from (2.276) and (2.277), the resonant frequencies, which provide maximum displacement amplitudes and current intensity, are given by expression (2.275). It is easy to show that a minimum amplitude (2.277) occurs in the vicinity of the frequency determined by equality,

$$\xi = \sqrt{(1 + \xi_1^2 + \xi_M^2)/2}.$$

This frequency is located between the resonant frequencies (2.275).

These simple examples show the dependence of the vibration system's behaviour on both the parameter relation between the mechanical and electrical partial systems, and the type of power supply used. This situation is similar to that arising in mechanics, when a system is excited by both a force, and kinematically. This was considered in previous sections.

The relations obtained above can also be applied to the situation when the power source can not be considered to be an ideal current or voltage supply. For example, if the voltage or current is limited, as shown by the dashed lines in Fig.2.29, b and Fig.2.30, b, then the corresponding parts of the amplitude-frequency characteristics should be replaced by the branches shown by the dashed lines in Fig.2.29, a and Fig.2.30, a.

3. Consider a magnetostrictive transducer, whose core length is comparable to, or exceeds the vibration wavelength in the transducer's material. For the

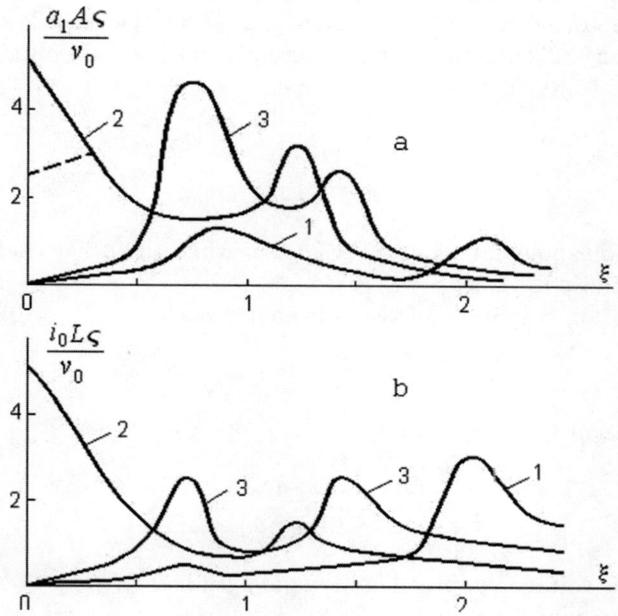

Fig. 2.30.

sake of definiteness, it is still assumed that the $x = 0$ end of the core is fixed, while the $x = l$ end is subjected to an external force, f_{12}. In this situation, the core strain is not homogeneous and is related to the displacement u_x of cross section x by the formula $\varepsilon_x = \partial u_x/\partial x$. Writing the equation of motion for the core's dx element as $\rho S dx \frac{\partial^2 u_x}{\partial t^2} = S \frac{\partial \sigma_x}{\partial x} dx$ and using (2.259), the following equation for longitudinal vibration within the core is obtained,

$$\rho \frac{\partial^2 u_x}{\partial t^2} - \tilde{E}_1 \frac{\partial^2 u_x}{\partial x^2} = 0. \tag{2.278}$$

Solutions are sought for (2.278), satisfying the following boundary conditions,

$$u_x|_{x=0} = 0, \quad \tilde{E}_1 S \frac{\partial u_x}{\partial x}|_{x=l} = f_{12} + \tilde{A} i = f_{12} + f_l. \tag{2.279}$$

Equation (2.278) with conditions (2.279) coincides with equation (2.41) describing bar vibration with one fixed end and subjected to periodic forces applied at the free end. Its solution can be given in terms of the dynamic compliance operators obtained in section 2.2. The vibration amplitudes of the core cross sections are,

$$\tilde{a}_x = L_{lx}(j\omega)(\tilde{F}_{12} + \tilde{F}_l) = L_{lx}(j\omega)(\tilde{F}_{12} + \tilde{A}\tilde{i}_0). \tag{2.280}$$

From (2.280), when $x = l$, equation (2.264) for the electric circuit, and also considering (2.279), the four-pole circuit equations are obtained. These are analogous to (2.265).

$$\tilde{F}_{12} = W_{ll}^{(1)}(j\omega)\tilde{a}_1 - j\omega\tilde{A}\tilde{q}_0,$$
$$\tilde{\nu}_0 = j\omega\tilde{A}\tilde{a}_l + j\omega Z(j\omega)\tilde{q}_0. \tag{2.281}$$

For this four-pole circuit $a_{11} = W_{ll}(j\omega)$, while the other coefficients are the same as those in (2.266).

The vibration amplitude of the unloaded transducer ($F_{12} = 0$) is,

$$\tilde{a}_l^* = \tilde{A}i_0/W_{ll}(j\omega) \tag{2.282}$$

when using a current supply, and,

$$\tilde{a}_l^* = \frac{j\omega\tilde{A}\tilde{\nu}_0}{j\omega Z(j\omega)W_{ll}(j\omega) - \omega^2 A^2} \tag{2.283}$$

when using a voltage supply. The dynamic stiffness $W_{ll}(j\omega) = L_{ll}^{-1}(j\omega)$ is determined by expression (2.148).

Using equations (2.281), the vibration of the system (Fig.2.31) can be described, in which the bar waveguide 2 is attached to the magnetostrictive transducer 1. Quantities referring to the transducer and waveguide will be denoted by the indexes 1 and 2, as was done before when considering the composite bar systems in section 2.2.10. The vibration amplitude at the connecting end of the transducer is given by,

$$\tilde{a}_l^{(1)} = \tilde{a}_0^{(2)} = \frac{\tilde{A}\tilde{i}_0}{W_{ll}^{(1)}(j\omega) + W_{00}^{(2)}(j\omega)}. \tag{2.284}$$

Taking (2.103) into account, the vibration amplitude at the free end of the waveguide can be found,

$$\tilde{a}_l^{(2)} = \frac{L_{0l}^{(2)}(j\omega)}{L_{00}^{(2)}(j\omega)}\frac{\tilde{A}\tilde{i}_0}{W_{ll}^{(1)}(j\omega) + W_{00}^{(2)}(j\omega)}. \tag{2.285}$$

Direct calculations using (2.28) yield the same result.

Substituting this expression into the second equation in (2.281), the voltage of the magnetostrictor's supply is found,

$$\tilde{\nu}_0 = \tilde{i}_0\left[Z(j\omega) + j\omega A^2\frac{L_{0l}^{(2)}(j\omega)}{L_{00}^{(2)}(j\omega)[W_{ll}^{(1)}(j\omega) + W_{00}^{(2)}(j\omega)]}\right]. \tag{2.286}$$

Parameters for a transducer being fed from a voltage supply can be found using expressions (2.285) and (2.286) as was done in the subsection 2.

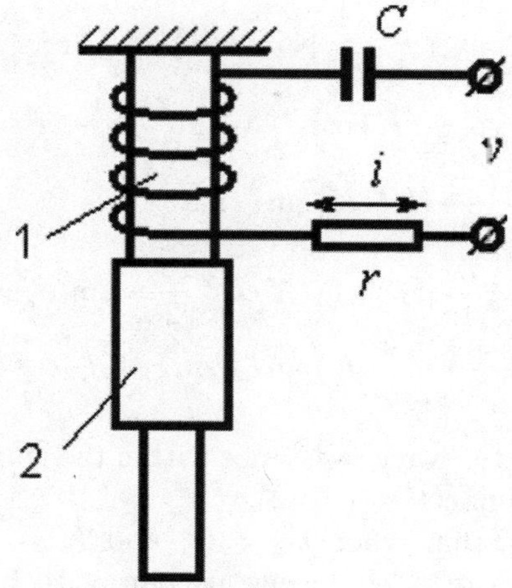

Fig. 2.31.

As an example, consider a vibration system consisting of a magnetostrictive transducer with a stepped waveguide attached to it. The waveguide is assumed to consist of cylindrical bars of equal length. The energy dissipation in the material of the waveguide is negligible in comparison with the dissipation rates found in the transducer. For further convenience the appropriate quantities from (2.284) and (2.285), using (2.148) and (2.157) can be written,

$$W_{ll}^{(1)}(j\omega) = \frac{E_1 S_1 \xi_1}{l_1} \frac{\cos\xi_1 + j\frac{\psi_1}{4\pi}\xi_1 \sin\xi_1}{\sin\xi_1 - j\frac{\psi_1}{4\pi}(\sin\xi_1 + \xi_1\cos\xi_1)},$$
$$W_{00}^{(2)}(j\omega) = \frac{E_2 S_2 \xi_2}{2l_2} \frac{(1+K)\sin\xi_2}{\sin^2\frac{\xi_2}{2} - K\cos^2\frac{\xi_2}{2}},$$
$$L_{00}^{(2)}(j\omega) = 1/W_{00}^{(2)}(j\omega),$$
$$L_{0l}^{(2)}(j\omega) = -\frac{2l_2}{E_2 S_2 \xi_2} \frac{K}{(1+K)\sin\xi_2}, \tag{2.287}$$

where $\xi_i = \omega l_i / c_i, i = 1, 2; K = S_2/S_3$; and S_2, S_3 are the cross sectional areas of the waveguide's cylindrical bars.

Substituting (2.287) and (2.270) into (2.284) and (2.285),

$$\tilde{a}_l^{(2)}/\tilde{a}_0^{(2)} = K/(K\cos^2\frac{\xi_2}{2} - \sin^2\frac{\xi_2}{2}),$$
$$a_l^{(2)} = \frac{K\delta}{\xi_1}\frac{P(j\omega)}{Q(j\omega)}, \tag{2.288}$$

where $\delta = Ai_0 l_1 / E_1 S_1$ is the static deformation of the transducer's core as caused by the force $F = Ai_0$;

$$
\begin{aligned}
P(j\omega) &= \sin\xi_1 - j\frac{1}{4\pi}[\psi_1\xi_1\cos\xi_1 + (\psi_1+\chi)\sin\xi_1],\\
Q(j\omega) &= \cos\xi_1(K\cos^2\frac{\xi_2}{2} - \sin^2\frac{\xi_2}{2}) -\\
&-\frac{w_2}{2w_1}(1+K)\sin\xi_1\sin\xi_2 +\\
&+j\frac{\psi_1}{4\pi}[\xi_1\sin\xi_1(K\cos^2\frac{\xi_2}{2} - \sin^2\frac{\xi_2}{2}) +\\
&+\frac{w_2}{2w_1}(1+K)\sin\xi_2(\sin\xi_1 + \xi_1\cos\xi_1)],
\end{aligned}
$$

$w_i = S_i\sqrt{E_i\rho_i}$ denotes the wave resistance within the transducer's core ($i = 1$) and the waveguide's connection ($i = 2$).

It is easy to check, that when $w_1 = w_2$ and $K = 1$, expression (2.288) describes the vibration of a homogeneous bar, with an effective length of $l = l_1 + l_2c_1/c_2$, which is excited in the $x = l_1$ cross section by a force of amplitude $F = Ai_0$ (see subsection 2.2.12). The resonant frequencies of such system are given by the formula (2.149),

$$
\xi = \xi_1 + \xi_2 = (2n-1)\pi/2, \quad n = 1, 2, \tag{2.289}
$$

and the resonant amplitudes of the free end and the end connecting it to the transducer are given by

$$
\begin{aligned}
a_0^{(2)} &= a_l^{(2)}|\cos\xi_2|,\\
a_l^{(2)} &= \left|\frac{4\pi\delta\sin\xi_1}{\psi_1\xi_1[\sin\xi_1\sin\xi_2 - (-1)^n\xi_1]}\right|.
\end{aligned} \tag{2.290}
$$

If the transducer satisfies the resonant condition,

$$
\xi_1 = (2n_1-1)\pi/2, \quad n_1 = 1, 2, ... \tag{2.291}
$$

then the resonance of the whole system is achieved at,

$$
\xi_2 = \xi - \xi_1 = (n - n_1)\pi \tag{2.292}
$$

and equalities (2.290) yield,

$$
a_0^{(2)} = a_l^{(2)} = \frac{16\delta}{\pi\psi_1(2n_1-1)^2}. \tag{2.293}
$$

Thus, if the attached bar is tuned in accordance with (2.292), it does not distort the resonant vibration of the transducer. This also holds true in the general case (2.288). Indeed, resonant frequencies of this system can be found from equation,

$$
\frac{w_1}{w_2(1+K)}\cot\xi_1 = \frac{\tan(\xi_2/2)}{K - \tan(\xi_2/2)}, \tag{2.294}
$$

and obtained in accordance with the condition (2.72).

In this case, the waveguide works as a stepped concentrator, providing an increase in the amplitude of vibration which has been transferred from the transducer (see (2.161)). In this analysis, the concentrator has been considered to be without dissipative loses. It must be noted, that any vibration system without dissipation that is attached to the transducer, does not distort its working regime at frequencies equal to the natural frequencies of the attached system. At these frequencies, $W_{00}^{(2)}(j\omega) = 0$, and (2.284) coincides with (2.282). The influence of dissipation within the concentrator will be estimated below in Chapter 4.

It should be emphasized that the values given by (2.291) and (2.292) don't determine all of the system's resonance frequencies, only those at which the waveguide acts as a vibration concentrator. The complete spectrum of resonant frequencies is given by equation (2.294). An example of its graphical solution is shown in Fig.2.32 for $l_2c_1/l_1c_2 = 2$, i.e. $\xi_2 = 2\xi_1$. The function on the right-hand side of (2.294) is plotted for different values of K, and the function on the left-hand side, for different values of the parameter $\alpha = w_1/w_2(1+K)$.

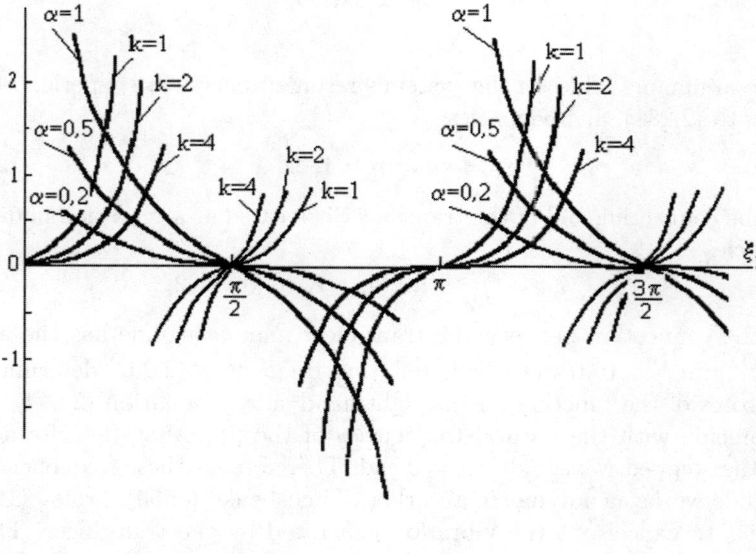

Fig. 2.32.

Fig.2.33 shows the amplitude-frequency characteristics of the waveguide's free end for $K = 4, \alpha = 0.2$ and $\psi_1 = 0.1$. The maximum amplitudes, reached at the resonant frequencies, are obtained according to (2.288) and (2.294) from the expression,

$$a_l^{(2)} = \left| \frac{4\pi K \delta \sin^2 \xi_1}{\psi_1 \xi_1 (\xi_1 + 0.5 \sin 2\xi_1)(K \cos^2 \xi_2/2 - \sin^2 \xi_2/2)} \right|. \tag{2.295}$$

At resonant frequencies (2.291) and (2.292), this formula yields values (2.293) for the vibration amplitudes at the end cross sections of the concentrator.

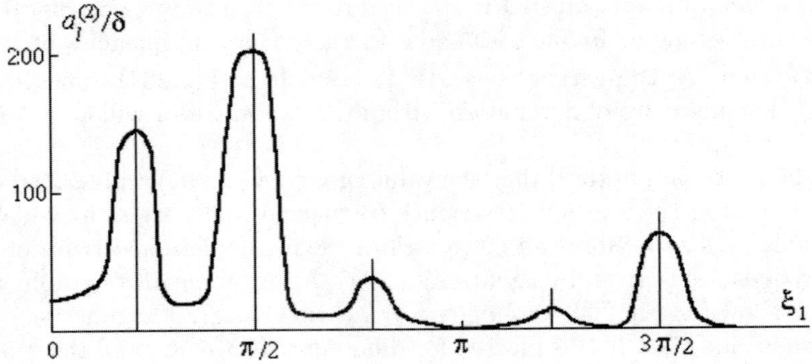

Fig. 2.33.

The minimum values of the system's resonant curves are experienced, according to (2.288), at frequencies

$$\xi_1 = \pi n, \quad n = 1, 2, ..., \tag{2.296}$$

when the connecting end of the transducer is located at a vibration node, and frequencies

$$\xi_2 = (2n-1)\frac{\pi}{2} \pm \arctan\sqrt{K}, \quad n = 1, 2, ..., \tag{2.297}$$

when the connection between the transducer and waveguide has the amplitude $a_l^{(1)} = a_0^{(2)} = 0$. It must be noted, that frequencies (2.297) determine the asymptotes of the function on the right-hand side of equation (2.294). They also coincide with the natural frequencies of the partial system, formed by fixing the stepped waveguide's $x = 0$ end. Therefore, at these frequencies, the waveguide works as a dynamic absorber (Timoshenko (1955), Frolov (1995)), and tries to cancel out the vibration generated by the transducer. The vibration amplitudes at the waveguide's free end, at frequencies (2.296) and (2.297), are small compared to the resonant ones: at frequencies (2.296) the amplitudes,

$$a_l^{(2)} = \left| \frac{\psi_1 K \delta}{4\pi(K\cos^2\xi_2/2 - \sin^2\xi_2/2)} \right|$$

are of the same small order as ψ_1, while at frequencies (2.297) the amplitudes,

$$a_l^{(2)} = \frac{\sqrt{K} A i_0 l_2}{E_2 S_2 \xi_2} = \frac{\sqrt{K} w_1}{w_2} \frac{\delta}{\xi_1}$$

are of an order similar to the static strain δ.

Finally, when $\omega = 0$, i.e. $\xi_1 = \xi_2 = 0$, from (2.287), $a_1^{(2)} = a_0^{(2)} = \delta$ is found.

Fig.2.33 shows the advantage of using a heterogeneous waveguide as a concentrator at frequencies (2.291). The results of this example are in accordance with the general theory of concentrators as discussed in subsection 2.2.5. The interaction of the concentrator with the vibration actuator, and a nonlinear elastic-dissipative loading in the presence of internal concentrator losses, will be discussed in detail in Chapter 4.

4. The dynamic characteristics are now found for piezoelectric transducers. A typical piezoelectric transducer is shown in Fig.2.34. Element 1 is manufactured from a piezoelectric material. The metalized plates 2 of the piezoelectric element are subjected to an alternating voltage ν from a power supply.

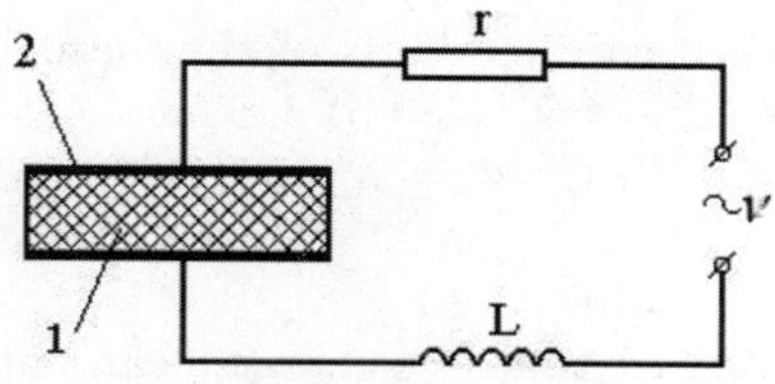

Fig. 2.34.

Longitudinal vibration of the element is considered in the direction perpendicular to these plates. The action of the transducer is based on the piezoelectric effect (Kharkevich (1973), Rimsky-Korsakov (1973), Kikuchi (1969)). The direct piezoelectric effect results from the electric polarization of crystals within the material when subjected to a mechanical stress σ. The polarization gives rise to an electric field with a dielectric displacement D, which is related to the mechanical stress, σ as

$$\left.\frac{D}{\sigma}\right|_{\Sigma=0} = -d, \tag{2.298}$$

where d is a piezoelectric material constant. The ratio in (2.298) is calculated for $\Sigma = 0$, i.e. in the absence of an electric field.

The inverse piezoelectric effect results in the appearance of strain within the material when subjected to an electric field in the absence of mechanical stress ($\sigma = 0$). It is given by the relation,

$$\varepsilon/\Sigma|_{\sigma=0} = -d, \tag{2.299}$$

where, ε is the relative strain in crystal and Σ is the electric field intensity.

Using (2.298) and (2.299), equations can be written for a piezoelectric transducer when subjected to both a mechanical stress and an electric field,

$$D = \xi\Sigma - d\sigma,$$
$$\varepsilon = -d\Sigma + \sigma/\tilde{\Sigma}' \tag{2.300}$$

ξ is the permittivity coefficient of the crystal in the absence of mechanical stress, and $\tilde{\Sigma}$ is the elastic modulus in the absence of an electric field.

It is convenient to rewrite linear relations (2.300) in a different form, where σ and Σ are given as functions of D and ε,

$$\sigma = \Sigma_1\varepsilon + \Phi D,$$
$$E = \Phi\varepsilon + \xi_1^{-1}D' \tag{2.301}$$

$\Sigma_1 = \Sigma/(1-\lambda^2)$ is the elastic modulus of the piezoelectric element in the absence of dielectric displacement ($D = 0$), i.e. for short-circuited plates, $\lambda^2 = d^2\Sigma/\xi < 1$ (Kharkevich (1973)); $\Phi = \lambda^2/d(1-\lambda^2)$ is Mason's piezoelectric modulus (Mason (1939)); $\xi_1 = \xi(1-\lambda^2)$ is the materials permittivity in the absence of deformation ($\varepsilon = 0$).

Taking account of the fact that free charge q can only appear on the crystal's plates, the dielectric displacement must be constant along the ceramics length,

$$D = q/S, \tag{2.302}$$

from the first equation in (2.301) a relation is obtained which describes the stress state within the crystal,

$$\sigma = \Sigma_1\varepsilon + \frac{\Phi}{S}q. \tag{2.303}$$

To describe the electrical state of the piezoelectric element, the equality $\Sigma = \frac{\partial\varphi_x}{\partial x}$ is used (φ_x is the potential in the x cross-section), $\nu_1 = \varphi_1 - \varphi_0$, and $\varepsilon = \partial u_x/\partial x$. Integrating the second equation in (2.301) along the element's length l,

$$\nu_1 = \frac{q}{C_1} + (u_1 - u_0)\Phi, \tag{2.304}$$

where $C_1 = \xi_1 S/l$ is the capacitance of the piezoelectric element in the absence of deformation.

The electric circuit, with a resistance r and an inductivity L (Fig.7.9) is described by the equation,

$$L\frac{d^2q}{dt^2} + r\frac{dq}{dt} + \frac{q}{C_1} + (u_l - u_0)\Phi = \nu. \tag{2.305}$$

Equations (2.303) and (2.305) fully describe the vibration of a piezoelectric transducer. The structure of these equations is similar to that of equations (2.259) and (2.264), describing a magnetostrictive transducer. Therefore, the four-pole circuit equations for a piezoelectric transducer can be obtained from (2.265) and (2.281) by the substitution of $j\omega A \to \Phi$ and by changing the sign in front of the last term in the first equation.

5. As an example, consider the vibration of a piezoelectric element (Fig.2.35, a), when one end ($x = 0$) is fixed, and the other ($x = l$), is loaded by a force f_{12}.

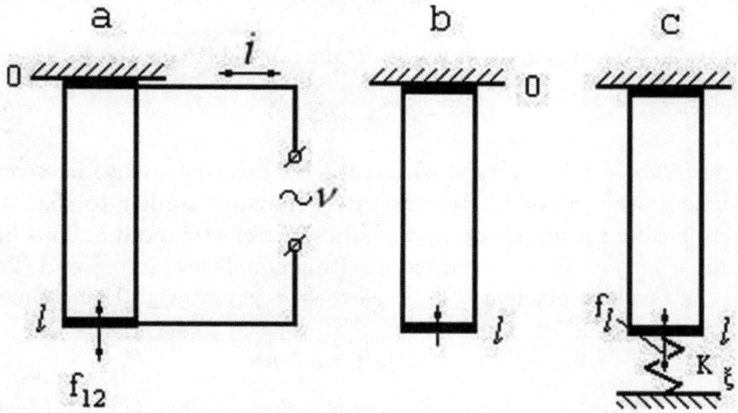

Fig. 2.35.

It is assumed that the resistance and inductance in the electric circuit is negligible, and the piezoelectric element is fed from a current or voltage source. The four-pole circuit equations take the form,

$$\tilde{F}_{12} = W_{ll}(j\omega)\tilde{a}_l + \Phi\tilde{q}_0,$$
$$\tilde{\nu}_0 = \Phi\tilde{a}_l + \frac{1}{C_1}\tilde{q}_0. \tag{2.306}$$

When $\tilde{F}_{12} = 0$ (no load at end $x = l$), the vibration amplitude at idle is found to be,

$$\tilde{a}_l^* = -\Phi\tilde{q}_0/W_{ll}(j\omega) \tag{2.307}$$

when a current supply is used, and,

$$\tilde{a}_l^* = -\Phi C_1\nu_0/[W_{ll}(j\omega) - \Phi^2 C_1] \tag{2.308}$$

when a voltage supply is used.

As follows from (2.307) and (2.308), the transducer can be described using the following models. The first (Fig.2.35, b) consists of a homogeneous bar, and is excited at end $x = l$ by a force with amplitude $F_l = -\Phi q_0$. The second (Fig.2.35, c) consists of a homogeneous bar with a dummy negative stiffness

$K_g = -C_1\Phi^2$ at the end which is subjected to a force $F_l = -\Phi C_1\nu_0$. The dynamic stiffness $W_{ll}(j\omega)$ in (2.306)–(2.308) is given by the relation (2.148), and the resonant frequencies in these two instances transpire to be different.

The physical meaning of this difference can be understood by considering low frequencies of excitation. In this case, the dynamic stiffness is given by $W_{ll}(j\omega) = \Sigma_1 S/l$. Equality (2.307) now takes the form,

$$a_l = -\Phi q_0 l/\Sigma_1 S, \tag{2.309}$$

and equality (2.308), using the notation from (2.301), takes the form,

$$a_l = -\Phi C_1 \nu_0 l/\Sigma S. \tag{2.310}$$

As such, piezoelectric crystals, when subjected to equivalent forces but fed from different power supplies, demonstrate behaviour similar to that of elements manufactured from materials with different elastic moduli. This apparent contradiction is easily resolved by rewriting equalities (2.309) and (2.310), and using the initial coefficients of the piezoelectric material. Using the notation from (2.301),

$$a_l = \delta = -dq_0/C = -d\nu_0, \tag{2.311}$$

where $C = \xi S/l$ is the capacitance of the piezoelectric element with free ends.

These equalities can be obtained directly from equations (2.300) by using (2.302), making $\sigma = 0$, and assuming that $\Sigma = \nu/l$ and $\varepsilon = u_l/l$. This is the case due to deformation homogeneity. It is worth noting that the element's deformation, as caused by the piezoelectric effect, does not depend on the elastic properties of the material; it resembles the kinematic tension-compression of the crystal. This reflects the process of rebuilding the atomic structure of the piezoelectric element's crystal lattice when subjected to an electric field.

6. To conclude this section, the vibration system (Fig.2.36) is considered. Homogeneous bars 2 symmetrically surround the piezoelectric element 1.

Due to symmetry, vibration can be considered for only half of the system. A nodal point exists in the middle cross section of the transducer. In order to describe the vibration, the relations obtained above can be used. As in section 2.4.3, similar variables for the transducer and bar are denoted by the indexes 1 and 2. From the first system equation (2.306), the vibration amplitude present at the joining cross section, when using a current source power supply, is found,

$$\tilde{a}_l^{(1)} = \tilde{a}_0^{(2)} = -\Phi\tilde{q}_0/[W_{ll}^{(1)}(j\omega) + W_{\infty}^{(2)}(j\omega)]. \tag{2.312}$$

The variable $\tilde{q}_0$ may be eliminated from the second equation in (2.306), and equalities (2.312). The following can now be written for a voltage supply,

$$\tilde{a}_l^{(1)} = \tilde{a}_0^{(2)} = -\frac{C_1\Phi\tilde{\nu}_0}{W_{ll}^{(1)}(j\omega) + W_{00}^{(2)}(j\omega) - C_1\Phi^2}. \tag{2.313}$$

The vibration amplitude at the bar's free end can be found using (2.103),

$$\tilde{a}_l^{(2)} = \tilde{a}_0^{(2)} L_{0l}^{(2)}(j\omega)/L_{00}^{(2)}(j\omega). \tag{2.314}$$

The values of dynamic stiffness and compliance in (2.312)– (2.314) are determined by the expressions obtained in section 2.2.

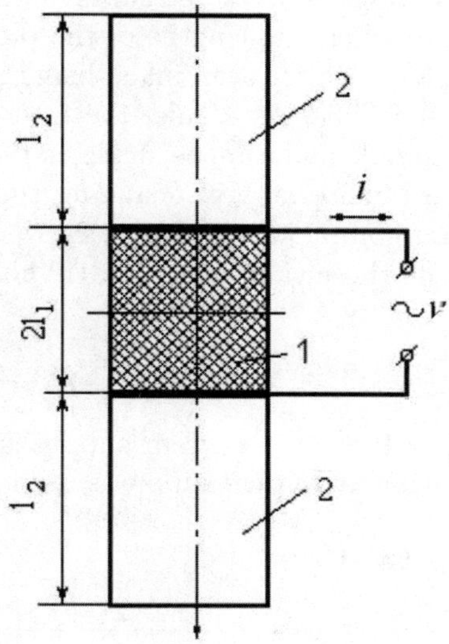

Fig. 2.36.

Consider some typical cases. Suppose that the lengths of the piezoelectric element l_1, and bar l_2, are small enough to satisfy the condition (2.50). Taking (2.151) and (2.51) into account, it is found that $\tilde{a}_l^{(1)} = \tilde{a}_l^{(2)} = \tilde{a}$, and expressions (2.312) and (2.313) take the form,

$$\begin{aligned} \tilde{a} &= -\frac{\Phi q_0}{(K_1 - M\omega^2) + j\omega b_1}, \\ \tilde{a} &= -\frac{C_1 \Phi \nu_0}{(K - M\omega^2) + j\omega b}, \end{aligned} \tag{2.315}$$

where $K_1 = \Sigma_1 S/l_1$, b_1 are the static stiffness and the resistance coefficient of a short-circuited transducer; $K = K_1 - \Phi^2 C_1 = \Sigma S/l_1$, b are the same parameters, but for an open-circuited transducer; M is the mass of the bar.

As follows from (2.315), the vibration system's resonance is achieved at frequency $\omega_q = \sqrt{K_1/M}$ when a current supply is used, and at $\omega_\nu = \sqrt{K/M}$ when a voltage supply is used.

Using the notation from (2.301) and (2.311), equalities (2.315) can be rewritten in a different form,

$$\tilde{a}_l = \frac{\delta}{(1-\omega^2/\omega_0^2) + j\eta_0\omega/\omega_0}, \tag{2.316}$$

where $\omega_0 = \omega_q$ and $\eta_0 = b_1/\sqrt{K_1 M}$ when a current supply is used, and $\omega_0 = \omega_\nu$ and $\eta_0 = b/\sqrt{KM}$ when a voltage supply is used.

Consider now the case when the lengths of the piezoelectric element and the bar are commensurable with, or even much bigger, than the length of the acoustic wave in their materials. As a rule, the internal losses in piezoelectric materials are very small, and can be neglected when compared to the losses present in the bar's material. By changing the form of the formulae for dynamic stiffness and compliance (2.148), (2.57) and (2.58) into that of (2.79), and by substituting these values into (2.312) and (2.313), the following is obtained,

$$\tilde{a}_l^{(1)} = \tilde{a}_0^{(2)} = \frac{F}{W_0(j\omega) - K_g}, \tag{2.317}$$

where $F = \Phi q_0$ and $K_g = 0$ when a current supply is used; $F = C_1\Phi\nu_0$ and $K_g = C_1\Phi^2 = \lambda^2\Sigma_1 S/l$ when a voltage supply is used;

$$W_0(j\omega) = W_{ll}^{(1)}(j\omega) + W_{00}^{(2)}(j\omega) = \\ = \frac{\Sigma_1 S_1 \xi_1}{l_1}[\cot\xi_1 - \frac{w_2}{w_1}(\tan\xi_2 - j\frac{\psi_2}{4\pi}\cdot\frac{\xi_2\cos 2\xi_2 + 0,5\sin 2\xi_2}{\cos^2\xi_2})], \tag{2.318}$$

$\xi = \omega l_i/c_i$; $c_i = \sqrt{\Sigma_i/\rho_i}$; $w_i = \sqrt{\Sigma_i\rho_i}S_i$ is the wave resistance of elements; $i = 1, 2$.

The vibration amplitude (2.314) at the bar's free end is given by the relation,

$$\tilde{a}_l^{(2)} = \tilde{a}_0^{(2)}/\cos\xi_2,$$

which coincides with (2.288) for $K = 1$.

Substituting $w_1 = w_2 = w$ into (2.318), and neglecting energy losses ($\psi_2 = 0$),

$$W_0(j\omega) = w\omega\frac{\cos\xi}{\sin\xi_1\cos\xi_2}, \tag{2.319}$$

where $\xi = \xi_1 + \xi_2$.

Expression (2.168), giving the homogeneous bar's dynamic compliance when subjected to a force concentrated in the middle cross section, has a similar form if $l = l_1 + l_2$ and $s = l_1$.

Thus, in this case, the vibration system can be described by a bar dynamic model, subjected to an equivalent excitation force. Further analysis is analogous to that in subsection 2.4.3.

3

Processes in ultrasonic systems

There are no things, only processes.

David Bohm (1917-1992)

3.1 Nonlinear load

1. The interaction between an ultrasonic machine's tool and the workpiece or medium being treated during an ultrasonic process gives rise to a load on the vibration system. As shown in Chapter 1, the interaction force can be described by the working process' dynamic characteristic $f_l(u_l, \dot{u}_l)$. This relates the force f_l acting on the workpiece to the tool's displacement u_l, and speed $\dot{u}_l$.

In the simplest of cases, this characteristic is comprised of a linear force dependence on the tool displacement,

$$f_l(u_l) = k_0 u_l \tag{3.1}$$

or speed,

$$f_l(\dot{u}_l) = b_0 \dot{u}_l. \tag{3.2}$$

Dependence (3.1) describes a linear elastic load. This appears, for example, during the continuous deformation of an elastic specimen with stiffness k_0. In this instance, the deformation takes place in the elastic zone of the material's stress-strain characteristic (see Fig.1.10, b). Dependence (3.2) describes a linear dissipative load, and has a coefficient of resistance b_0. This arises when vibration occurs in a viscous fluid (in the absence of cavitation) (Kikuchi (1969)), or when a wave is radiated into an unbounded waveguide system (see subsection 2.2.2).

The complete characteristic

$$f_l(u_l, \dot{u}_l) = k_0 u_l + b_0 \dot{u}_l \tag{3.3}$$

describes a linear elastic-dissipative load, which may be encountered, for example, during the deformation of elastic-viscous materials.

Taking into account the nature of the motions (2.4) and forces (2.7) under consideration, equalities (3.1)–(3.3) can be rewritten in terms of complex displacements and forces. In particular, (3.3) yields the relation,

$$\tilde{F}_l(\tilde{a}_l) = W_H(j\omega)\tilde{a}_l.$$

The complex vibration amplitude and force are related by a dynamic characteristic describing the elastic-dissipative load,

$$W_H(j\omega) = k_0 + j\omega b_0. \tag{3.4}$$

When designing an ultrasonic machine, the inertial load generated by a detachable tool's mass M, or, the mass added to the system during the processing of a workpiece, often needs to be considered. An inertial load characteristic takes the form,

$$f_l(\ddot{u}_l) = M\ddot{u}_l,$$

and its dynamic stiffness is given by,

$$W_H(j\omega) = -M\omega^2. \tag{3.5}$$

As shown in subsection 2.1.7, in which a typical ultrasonic machine's arrangement is considered, a machine tool's vibration amplitude, whilst subjected to a linear load, is given by formula (2.37). The dynamic characteristics of vibration systems and electroacoustic transducers, as found in sections 2.2-2.4, should be used.

2. The real working process, as realized by an ultrasonic machine, cannot usually be represented as a linear load. Therefore, consideration should be paid to nonlinear models of these processes. Some of these have been presented in Chapter 1. In the instance under consideration, the interaction force between the working tool and the workpiece, or medium, is described by a nonlinear dynamic characteristic $f_l = f_l(u_l, \dot{u}_l)$. In order to find an ultrasonic system's periodic motion, in which a nonlinear load is present, the approximate method of harmonic linearization will be employed. This is widely used for the analysis of nonlinear automatic control systems (Popov & Paltov (1960)), nonlinear vibration protection systems (Kolovsky (1999)), vibro-impact systems (Babitsky (1998)), etc.

The harmonic linearization method is based on the use of the harmonic balance principle.

Suppose, as before, that the motion of the machine's tool is periodic and has a frequency, ω. It may be represented as a Fourier series,

$$u_l(t) = m_l + \sum_{n=1}^{\infty}(A_{cn}\cos n\omega t + A_{sn}\sin n\omega t). \tag{3.6}$$

This can be written in an equivalent form,

$$u_l(t) = m_l + \frac{1}{2}\sum_{n=\pm 1}^{\pm\infty} \tilde{a}_{ln}e^{nj\omega t}, \tag{3.7}$$

where, m_l is a constant component and $\tilde{a}_{ln}$ are the complex amplitudes of the motion's harmonic components. The latter are related to the coefficients of series (3.6) by the equalities,

$$\tilde{a}_{ln} = A_{cn} - jA_{sn}\mathrm{sgn}n \tag{3.8}$$

As follows from these equalities, amplitudes $\tilde{a}_{ln}$ for $n = k$ and $n = -k$ are complex conjugates.

By substituting periodic function (3.7) into the nonlinear dynamic characteristic, a periodic function with the same frequency ω is obtained. Expanding it in the Fourier series gives,

$$f_l[u_l(t), \dot{u}_l(t)] = P_l + \frac{1}{2}\sum_{n=\pm 1}^{\pm\infty} \tilde{F}_{ln} e^{nj\omega t}, \tag{3.9}$$

where the constant component P_l, and complex amplitudes $\tilde{F}_{ln}$, of the force's f_l harmonic components are given by the formulae,

$$P_l = \frac{1}{T}\int_0^T f_l[u_l(t), \dot{u}_l(t)]dt = P_l(m_l, A_{cl}, \ldots, A_{sl}, \ldots) \tag{3.10}$$

$$\tilde{F}_{ln} = \frac{2}{T}\int_0^T f_l[u_l(t), \dot{u}_l(t)]e^{-nj\omega t}dt = \tilde{F}_{ln}(m_l, A_{cl}, \ldots, A_{sl}, \ldots). \tag{3.11}$$

$T = 2\pi/\omega$ is the time period of the process under consideration.

It is still assumed that the vibration system and the ultrasonic machine's transducer constitute a linear system and its properties are completely defined by the values of dynamic compliance found in the previous chapter. The system's vibration during its interaction with the nonlinear load can be described by the following operator equation,

$$u_l(t) = u_l^*(t) - L_M(j\omega) f_l[u_l(t), \dot{u}_l(t)]. \tag{3.12}$$

$u_l^*(t)$ is the steady-state vibration, at frequency ω, during idle motion, i.e. in the absence of load ($f_l = 0$), $L_M(j\omega)$ is the dynamic compliance of the vibration system and transducer as determined by formula (2.29).

During idling conditions, the motion of the system, $u_l^*(t)$, can be found using the methods described in the previous chapter. For the sake of generality, $u_l^*(t)$ is taken to be a periodic function which can be expanded in the Fourier series, analogously to (3.7),

$$u_l^*(t) = \frac{1}{2}\sum_{n=\pm 1}^{\pm\infty} \tilde{a}_{ln}^* e^{nj\omega t}. \tag{3.13}$$

Formulae (2.21) and (2.28) can be used to determine the complex amplitudes $\tilde{a}_{ln}^*$. This may be done by using the dynamic characteristics for each of the vibration system's components and holds true when either a voltage or current source is used to supply the transducer. Therefore, within equation (3.12), function (3.13) acts as an external periodic excitation.

Substituting series (3.7), (3.9), and (3.13) into equation (3.12) yields,

$$m_l + \frac{1}{2}\sum_{n=\pm 1}^{\pm\infty} \tilde{a}_{ln} e^{nj\omega t} = \frac{1}{2}\sum_{n=\pm 1}^{\pm\infty} \tilde{a}^*_{ln} e^{nj\omega t} - L_M(0)P_l -$$

$$\frac{1}{2}\sum_{n=\pm 1}^{\pm\infty} L_M(nj\omega)\tilde{F}_{ln} e^{nj\omega t}. \tag{3.14}$$

By equating the constant components and coefficients in front of the same power exponents on the left and right-hand sides of this equality, the following equations are found,

$$m_l = -L_M(0)P_l(m_l, A_{cl}, \ldots, A_{sl}, \ldots), \tag{3.15}$$

$$\tilde{a}_{ln} = \tilde{a}^*_{ln} - L_M(nj\omega)\tilde{F}_{ln}(m_l, A_{cl}, \ldots, A_{sl}, \ldots),$$
$$(n = 1, 2, \ldots). \tag{3.16}$$

These equations, along with equalities (3.8), complete the information required to determine the unknown parameters of motion $m_l, A_{cl}, \ldots, A_{sl}, \ldots$ and complex amplitudes $\tilde{a}_{ln}$.

It is noted that when negative values of n $(n = -1, -2, \ldots)$ are considered in the system of equations following from (3.14), they are equivalent to system (3.16). This is because the complex numbers in this equation are complex conjugates of those with positive values of n.

In general, it is impossible to find a solution for the infinite system of transcendental equations (3.15) and (3.16). This is due to the fact that P_l and $\tilde{F}_{ln}$ are dependant on all of the coefficients of the expanded function $u_l(t)$, when substituted into series (3.16). When a linear load is considered, as for example given by expression (3.16), an exception is found. In this instance, by substituting (3.7) into (3.11) and integrating, it is found that,

$$\tilde{F}_{ln} = \tilde{a}_{ln}(k_0 + nj\omega b_0)$$

and system of equations (3.16), along with (3.4), take the form:

$$\tilde{a}_{ln} = \tilde{a}^*_{ln} - \tilde{a}_{ln} L_M(nj\omega)W_H(nj\omega)$$

Thus, the complex amplitude $\tilde{a}_{ln}$ is given by the expression,

$$\tilde{a}_{ln} = a^*_{ln}\left[1 + \frac{W_H(nj\omega)}{W_M(nj\omega)}\right]$$

which coincides with (2.37).

When a nonlinear load is considered, series (3.7) and (3.9) must be truncated, retaining only a finite number of terms from each sequence. This reduces the problem of solving the transcendental equations and an approximate solution is found. As a rule, in order to construct such a solution, only the constant components, and the first harmonics of the periodic solution, are accounted for, i.e. an approximate solution of the form,

$$u_l(t) \approx m_l + \tilde{a}_l e^{j\omega t}, \tag{3.17}$$

is sought. The constant component m_l and complex amplitude $\tilde{a}_l$ of this solution can be found, according to (3.15) and (3.16), from the equations,

$$m_l = -L_M(0)P_l(m_l, a_l), \tag{3.18}$$

$$\tilde{a}_l = \tilde{a}_l^* - L_M(j\omega)\tilde{F}_l(m_l, a_l) \tag{3.19}$$

The method of finding periodic solutions in this form is called the harmonic balance method. It must be noted, that equation (3.19) has the same form as (2.27). The difference, when a nonlinear load is considered, arises from the nonlinear dependence of the complex interaction force on the tool motion amplitude a_l, and constant component m_l.

3. Using (3.11) for the first harmonic gives,

$$\tilde{F}_l = \frac{2}{T}\int_0^T f_l[u_l(t), \dot{u}_l(t)]e^{-j\omega t}dt =$$

$$\frac{2}{T}\int_0^T f_l[u_l(t), \dot{u}_l(t)](\cos\omega t - j\sin\omega t)dt. \tag{3.20}$$

The complex amplitude $\tilde{F}_l$ can be rewritten in a different form,

$$\tilde{F}_l = a_l(k + j\omega b), \tag{3.21}$$

where,

$$k = \frac{2}{Ta_l}\int_0^T f_l[u_l(t), \dot{u}_l(t)]\cos\omega t\, dt,$$

$$b = -\frac{2}{T\omega a_l}\int_0^T f_l[u_l(t), \dot{u}_l(t)]\sin\omega t\, dt \tag{3.22}$$

At this point it must be restated, that to find an approximate solution of the form (3.17), it is assumed that,

$$f_l[u_l(t), \dot{u}_l(t)] \approx P_l + \tilde{F}_l e^{j\omega t}$$

or, taking (3.21) and (3.17) into account,

$$f_l[u_l(t), \dot{u}_l(t)] \approx P_l + k u_l^0(t) + b\dot{u}_l(t), \tag{3.23}$$

where $u_l^0 = \tilde{a}_l e^{j\omega t}$.

Thus, the initial nonlinear function $f_l(u_l(t), \dot{u}_l(t))$ is substituted with a linearized expression. Such a substitution differs from the usual linearization by the dependence of coefficients P_l, k and b on the unknown solution parameters. This is called a harmonic linearization (Kolovsky (1999), Popov & Paltov (1960)).

To calculate the coefficients of the harmonic linearization, it is convenient to rewrite (3.10) and (3.22) in a different form. It is assumed that,

$$u_l(t) = m_l + a_l \cos \omega t, \dot{u}_l(t) = -a_l \omega \sin \omega t$$

and variable $\tau = \omega t$ is substituted. As a result,

$$P_l = \frac{1}{2\pi} \int_0^{2\pi} f_l(m_l + a_l \cos \tau, -a_l \omega \sin \tau) d\tau = P_l(m_l, a_l), \tag{3.24}$$

$$k = \frac{1}{\pi a_l} \int_0^{2\pi} f_l(m_l + a_l \cos \tau, -a_l \omega \sin \tau) \cos \tau d\tau = k(m_l, a_l), \tag{3.25}$$

$$b = -\frac{1}{\pi a_l \omega} \int_0^{2\pi} f_l(m_l + a_l \cos \tau, -a_l \omega \sin \tau) \sin \tau d\tau = b(m_l, a_l). \tag{3.26}$$

Variable $P_l(m_l, a_l)$ determines the constant component of nonlinear load. By comparing (3.21) and (3.4), it is convenient to write the nonlinear load's alternating component by using its linearized dynamic stiffness,

$$W_H(j\omega, m_l, a_l) = k(m_l, a_l) + j\omega b(m_l, a_l). \tag{3.27}$$

Coefficients $k(m_l, a_l)$ and $b(m_l, a_l)$ depend on the unknown motion parameters, and characterize the equivalent elastic and dissipative components of the nonlinear load.

By substituting the working process' nonlinear dynamic characteristic with the harmonic linearization (3.23), the equation describing the complex amplitude a_l is converted to the form,

$$\tilde{a}_l = \tilde{a}_l^* - L_M(j\omega) W_H(j\omega, m_l, a_l) \tilde{a}_l.$$

This yields the relation,

$$\tilde{a}_l = \tilde{a}_l^* \left[1 + \frac{W_H(j\omega, m_l, a_l)}{W_M(j\omega)} \right], \tag{3.28}$$

and has the same form as expression (2.37) which was obtained earlier for a linear load. Unlike (2.37), relation (3.28), in conjunction with equation (3.18) (which relates the constant component to the vibration amplitude), provides a closed system of equations and may be used to determine the unknown parameters m_l, a_l of solution $u_l(t)$.

In the forthcoming analysis of vibration systems under the influence of a nonlinear load, the method of harmonic linearization will be used as in the form presented above.

4. Expressions may now be found in order to obtain harmonic linearization coefficients. These reflect the dynamic characteristics of ultrasonic processes and will be used in the following analysis.

As shown in section 1.2, when deforming an elastic specimen using a vibrating tool, the loading may take the form of a periodic impulse (Fig.1.10, a). Such a process is described using a single-sided elastic limiter characteristic (Fig.3.1, a),

$$f_l(u_l) = k_0(u_l - \Delta)\eta(u_l - \Delta). \tag{3.29}$$

k_0 is the limiter's static stiffness and $\eta(u)$ is the unit step function.

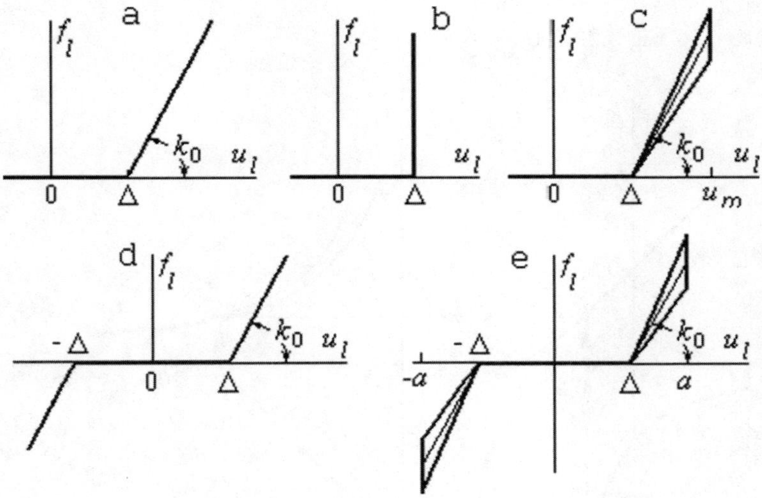

Fig. 3.1.

By substituting (3.29) into (3.24) and (3.25), it is found that,

$$P_l = \frac{1}{\pi}\int_0^{\tau_l}(m_l - \Delta + a\cos\tau)d\tau, k = \frac{2k_0}{\pi a_l}\int_0^{\tau_l}(m_l - \Delta + a\cos\tau)\cos\tau\ d\tau,$$

where, $\tau_l = \arccos(\Delta - m_l)/a_l$.

Integrating gives,

$$P_l = \frac{k_0 a_l}{\pi}\left(\sqrt{1-\left(\frac{\Delta - m_l}{a_l}\right)^2} - \frac{\Delta - m_l}{a_l}\arccos\frac{\Delta - m_l}{a_l}\right), \qquad (3.30)$$

$$k = \frac{k_0}{\pi}\left(\arccos\frac{\Delta - m_l}{a_l} - \frac{\Delta - m_l}{a_l}\sqrt{1-\left(\frac{\Delta - m_l}{a_l}\right)^2}\right). \qquad (3.31)$$

For the load's dissipative component, expression (3.26) reveals that $b = 0$. Introducing the notation,

$$(\Delta - m_l)/a_l = \alpha, -1 \leq \alpha \leq 1, \qquad (3.32)$$

from (3.30) and (3.31), an approximate formula for the linearization coefficient $k(a_l)$ is deduced.

Graphs of $k(a)$, as deduced from (3.31), and $a_l(\alpha)$, as deduced from the equation (following from (3.30)),

$$a_l = \pi\frac{P_l}{k_0}\left(\sqrt{1-\alpha^2} - \alpha\arccos\alpha\right)^{-1} \qquad (3.33)$$

are plotted in Fig.3.2, a, b.

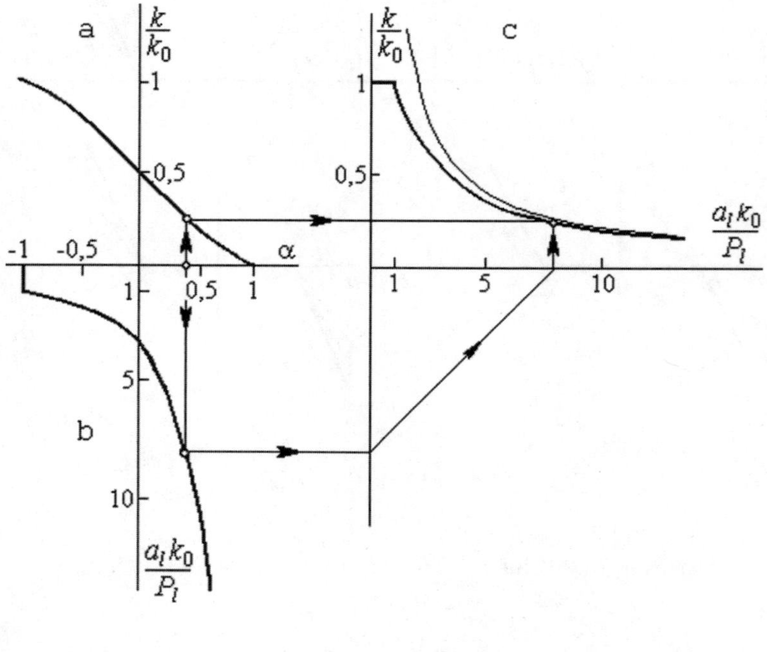

Fig. 3.2.

These curves are used to plot dependence of the coefficient k on amplitude a_l. Amplitude a_l is shown in Fig.3.2, c. The dependence can be approximated by the formula,

$$k = \frac{2P_l}{a_l + P_l/k_0}. \tag{3.34}$$

The graph of (3.34) practically coincides with the curve shown in Fig.3.2, c.

It must be noted, that for amplitudes $a_l \leq P_l k_0$, the vibration does not exceed the limits of the limiter's elastic deformation. Therefore, the load is linear and is described by $k = k_0$.

Coefficients (3.30), (3.31) and (3.34) will be used for the analysis of vibro-impact processes, which are of great importance when considering ultrasonic systems. These coefficients, however, describe only the elastic component of the load. To find the dissipative component, different models of impact interaction should be used.

The simplest model is based on stereomechanical impact theory (Babitsky (1998)). This theory completely excludes any consideration of the impact process due to its short duration. It uses general mechanical theorems to find estimates of the resulting colliding bodies' kinematic characteristics. When considering stereomechanical impact theory, the effect of the impact is estimated by the velocity restorability factor $R (0 \leq R \leq 1)$, known as restitution coefficient. This is equal to the ratio of the colliding bodies' relative speeds, after, and prior to the impact. Thus, a body of mass M, moving with speed $\dot{u}_-$, acquires the following speed after an impact with an immovable wall,

$$\dot{u}_+ = -R\dot{u}_-. \tag{3.35}$$

In (3.35), $R = 1$ corresponds to a perfectly elastic collision, while $R = 0$ corresponds to a purely plastic collision.

An impact is characterized by an impulse $J = \int_0^\tau f(t)dt$, of force $f(t)$, which appears in the contact zone during the impact time τ. The impulse's action is assumed to be instantaneous. Using the momentum theorem, and taking (3.35) into account,

$$J = M(\dot{u}_+ - \dot{u}_-) = -M\dot{u}_-(1 + R). \tag{3.36}$$

The impact is accompanied by a partial loss of energy. The dissipated energy is given by,

$$A = \frac{M}{2}(\dot{u}_-^2 - \dot{u}_+^2) = 1/2M\dot{u}_-^2(1 - R^2) \tag{3.37}$$

Stereomechanical impact theory describes a body's collision with a rigid limiter. Its characteristics are shown in Fig.3.1, b. Applying the limit $k_0 \rightarrow \infty$ to (3.34), the linearization coefficient of a rigid limiter (dashed line in Fig.3.1, c) is obtained,

$$k = 2P_l/a_l. \tag{3.38}$$

It is noted, that this expression can be obtained from formulae (3.30) and (3.31). In order to do this, it is noticed that as $k_0 \rightarrow \infty$, $\alpha = (\Delta - m_l)/a_l \rightarrow 1$.

k_0 may be excluded from (3.30) and (3.31) by taking the ratio of these expressions,

$$\frac{P_l}{k} = a_l \frac{\sqrt{1-\alpha^2} - \alpha \cdot \arccos\alpha}{\arccos\alpha - \alpha\sqrt{1-\alpha^2}} \tag{3.39}$$

Applying the limit $\alpha \to 1$ to this ratio, and resolving the indeterminacy, it is found that $P_l/k = a_l/2$. This yields relation (3.38).

In order to quantify energy losses, the linearization coefficient of the impact characteristic's dissipative component is constructed using energy considerations. If P_l is the average value of the impact interaction force over the time period, the total impact impulse acting on the body during the periodic collisions is,

$$J = -\frac{2\pi}{\omega} P_l. \tag{3.40}$$

By equating the values of impulse given by (3.36) and (3.40), the speed of the body is found as it approaches the limiter,

$$\dot{u}_- = \frac{2\pi P_l}{M\omega(1+R)}.$$

Using (3.37) the energy losses during the impact are calculated:

$$A = \frac{1-R}{1+R} \cdot \frac{2\pi^2 P_l^2}{M\omega^2}. \tag{3.41}$$

The energy dissipated over the time period T, as given by the equivalent linear resistance, is,

$$A = \frac{1}{2} b (a_l \omega)^2 T = \pi b \omega a_l^2. \tag{3.42}$$

Equating values (3.41) and (3.42), the dissipative component of the impact force's linearization coefficient is obtained,

$$b = 2\pi \frac{1-R}{1+R} \cdot \frac{P_l^2}{M\omega^3 a_l^2}. \tag{3.43}$$

Using (3.38), this expression can be rewritten in the form:

$$b = \frac{\pi k^2(a)}{2M\omega^3} \cdot \frac{1-R}{1+R}. \tag{3.44}$$

5. In cases when the impact interaction duration is not negligible, the stereomechanical model of impact is not applicable, and different models must be used. One of the alternative models is constructed in subsection 1.3.4 (Fig.1.33). Its dynamic characteristic (Fig.3.1, c) is determined by the equation,

$$f_l(u_l, \dot{u}_l) = k_0(u_l - \varDelta)\eta(u_l - \varDelta)(1 + \vartheta \operatorname{sgn}\dot{u}_l) \tag{3.45}$$

The general characteristic of this type of model can be represented in the form,

$$f_l(u_l, \dot{u}_l) = f_l^{(1)}(u_l) + f_l^{(2)}(u_l, \dot{u}_l), \tag{3.46}$$

where the first term $f_l^{(1)}(u_l)$ determines the elastic forces, and the second, which satisfies the condition $f_l^{(2)}(u_l, \dot{u}_l) = -f_l^{(2)}(u_l, -\dot{u}_l)$, describes the dissipative forces. It is easy to show that for characteristics of the type (3.46), the linearization coefficients, as given by formulae (3.24)–(3.26), can be simplified, taking the form (Kolovsky (1999))

$$P_l = \frac{1}{2\pi} \int_0^{2\pi} f_l^{(1)}(m_l + a_l \cos\tau) d\tau = P_l(m_l, a_l), \tag{3.47}$$

$$k = \frac{1}{\pi a_l} \int_0^{2\pi} f_l^{(1)}(m_l + a_l \cos\tau) \cos\tau d\tau = k(m_l, a_l), \tag{3.48}$$

$$b = -\frac{1}{\pi\omega a_l} \int_0^{2\pi} f_l^{(2)}(m_l + a_l \cos\tau, -a_l\omega \sin\tau) \sin\tau d\tau = b(m_l, a_l). \tag{3.49}$$

The last formula can be transformed,

$$b = \frac{1}{\pi\omega a_l^2} \int_0^{2\pi} f_l^{(2)}(m_l + a_l \cos\tau, -a_l\omega \sin\tau) d(a_l \cos\tau) =$$

$$= \frac{1}{\pi\omega a_l^2} \oint f_l(u_l, \dot{u}_l) du = \frac{A}{\pi\omega a_l^2}$$

in order to allow the linearization coefficient b to be expressed as a function of the energy A dissipated over the time period. It is equal to the area inside the hysteresis loop of characteristic $f_l(u_l, \dot{u}_l)$. Therefore, in order to study vibration that is close to harmonic vibration, the shape of the hysteresis loop is not essential.

Comparing characteristic (3.45) with (3.29) shows that the linearization coefficients P_l and k may be determined using expressions (3.30) and (3.31). Their approximation, (3.34), may also be used. For the dissipative component (3.48),

$$b = \frac{k_0\vartheta}{\pi\omega}\left(1 - \frac{\Delta - m_l}{a_l}\right)^2 = \frac{k_0\vartheta}{\pi\omega}(1-\alpha)^2. \tag{3.50}$$

When studying vibro-impact processes, bilateral impact interactions (Babitsky (1998)) with dynamic characteristics as shown in Fig.3.1, d, e, often have to be considered. The characteristic shown in Fig. 3.1, e is described by the equation,

$$f_l(u_l) = k_0(|u_l| - \Delta)\eta(|u_l| - \Delta)\mathrm{sign}u_l(1 + \vartheta\mathrm{sign}\dot{u}_l), \tag{3.51}$$

and that in Fig.3.1, d, by the same equation, in which $\vartheta = 0$.

In symmetrical processes, i.e. when $u_l(t+T/2) = -u_l(t)$, the constant components of the process are $m_l = 0$ and $P_l = 0$. In this case, by comparing (3.51) with (3.44), it is demonstrated that the harmonic linearization coefficients of dynamic characteristic (3.51) are obtained by doubling values (3.31) and (3.50) in which $m_l = 0$. They have the form:

$$k = \frac{2k_0}{\pi}\left(\arccos\frac{\Delta}{a_l} - \frac{\Delta}{a_l}\sqrt{1-\left(\frac{\Delta}{a_l}\right)^2}\right), \tag{3.52}$$

$$b = \frac{2k_0\vartheta}{\pi\omega}\left(1-\frac{\Delta}{a_l}\right)^2 . \tag{3.53}$$

The dependence of coefficients (3.52) and (3.53) on amplitude a_l is shown in Fig.3.3,a,b. When vibration occurs inside the gap ($a_l < \Delta$) the vibration system is unloaded. When the amplitude $a_l \to \infty$, the elastic load $k \to k_0$, while the dissipative load $b \to b_\infty = 2k_0\vartheta/\pi\omega$.

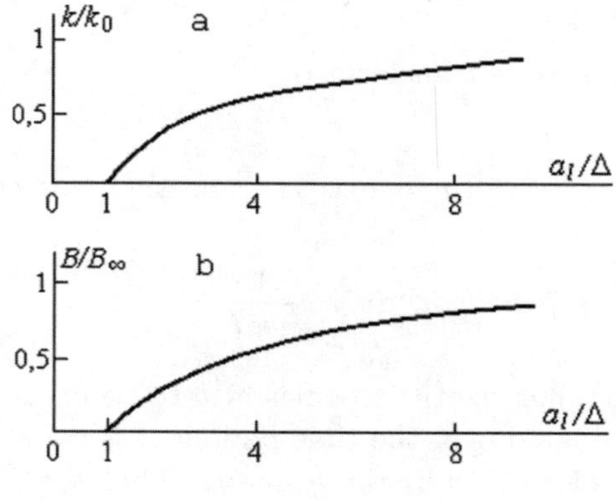

Fig. 3.3.

6. Chapter 1 included consideration of processes in which the tool was allowed to progress slowly in a specified direction under the action of a constant force. The average velocity of the tool's motion determines the process speed. These models describe the processes taking place in systems with dry friction, plastic deformation, material fracture using a vibrating instrument, etc. The dynamic characteristics of some of these processes are shown in Fig.3.4.

Considering such processes, an approximate solution is sought in the form,

$$\begin{aligned} u_l(t) &\approx vt + u_l^0(t) = vt + a_l\cos\omega t, \\ \dot{u}_l(t) &\approx v + \dot{u}_l^0(t) = v - a_l\omega\sin\omega t, \end{aligned} \tag{3.54}$$

where v is the speed at which the tool progresses, u_l^0 is a periodic component of the solution with amplitude a_l and frequency ω. The nonlinear dynamic

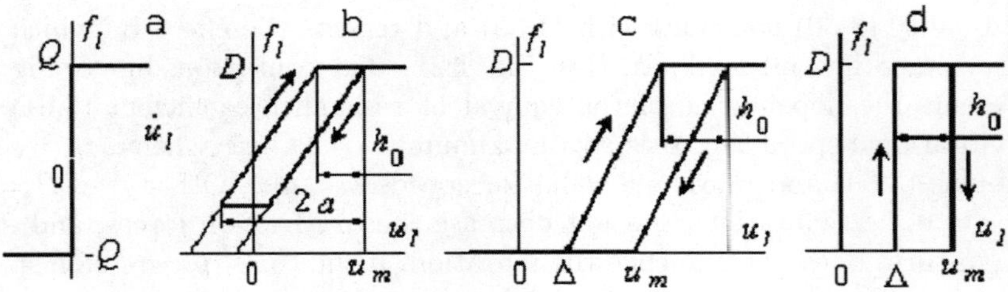

Fig. 3.4.

characteristic $f_l(u_l, \dot{u}_l)$ is substituted by the linear function,

$$f_l(u_l, \dot{u}_l) \approx P_l + ku_l^0 + b\dot{u}_l^0 \tag{3.55}$$

analogously to (3.23).

The linearization coefficients in (3.55) can be calculated using formulae similar to (3.33)–(3.26):

$$P_l = \frac{1}{2\pi} \int_0^{2\pi} f_l \left(\frac{v}{\omega}\tau + a_l \cos\tau, v - a_l\omega \sin\tau \right) d\tau = P_l(v, a_l), \tag{3.56}$$

$$k = \frac{1}{\pi a_l} \int_0^{2\pi} f_l \left(\frac{v}{\omega}\tau + a_l \cos\tau, v - a_l\omega \sin\tau \right) \cos\tau \, d\tau = k(v, a_l), \tag{3.57}$$

$$b = -\frac{1}{\pi a_l \omega} \int_0^{2\pi} f_l \left(\frac{v}{\omega}\tau + a_l \cos\tau, v - a_l\omega \sin\tau \right) \sin\tau \, d\tau = b(v, a_l). \tag{3.58}$$

As shown in Chapter 1, the tool's vibration has a considerable influence on the static forces necessary to overcome the threshold force of material deformation. Equality (3.56) reflects this influence. Relations (3.57) and (3.58) also demonstrate the inverse influence of the tool's progression speed on the coefficients of harmonic linearization k and b. It consequently affects the vibration parameters of the system.

Fig.3.4, a shows the characteristic of a "dry friction" type nonlinear element. It may be described by the equation,

$$f_l(\dot{u}_l) = Q\text{sgn}\dot{u}. \tag{3.59}$$

Substituting (3.59) and (3.54) into (3.56)–(3.58) and integrating gives,

$$P = \frac{2Q}{\pi} \arcsin \frac{v}{a_l \omega}, \tag{3.60}$$

$$k = 0, \quad b = \frac{4Q}{\pi \omega a_l} \sqrt{1 - \left(\frac{v}{\omega a_l} \right)^2} \tag{3.61}$$

Equality (3.60) coincides with (1.39) and reflects the effect by which vibration smoothes out nonlinearities (Fig. 1.21). The continuous line in Fig.3.5 represents the dependence of the equivalent resistance coefficient (3.61) on the vibrational speed ωa_l. It is at a maximum $b_m = 2Q/\pi v$ when $\omega a_l = \sqrt{2}v$. Formulae (3.60) and (3.61) are valid for speeds $v < a_l\omega$. Otherwise, $P = Q$ and $b = 0$, i.e., vibration does not decrease the dry friction forces, and friction, in turn, does not influence the vibration. If the tools' progression speed is very small, $v \ll \omega a_l$, and terms higher than the first order are neglected, from (3.61),

$$b = 4Q/\pi\omega a_l. \tag{3.62}$$

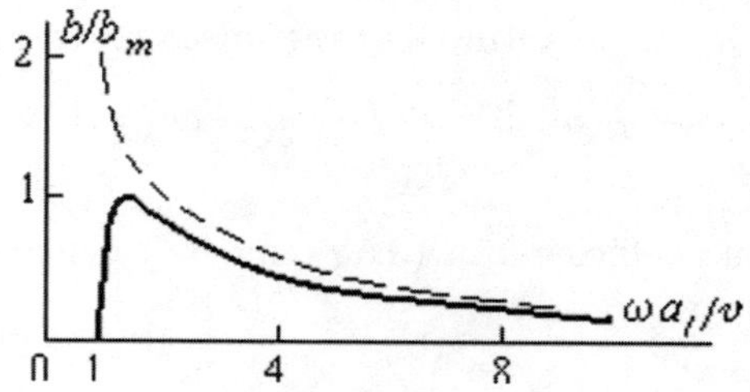

Fig. 3.5.

Dependence (3.62) is plotted in Fig.3.5 as a dashed line. As shown in Chapter 1 and following from (3.60), the advantages of using ultrasound for decreasing static force at low speeds $v \ll \omega a_l$ are clear. Therefore, an acceptable and practical form of the linearization coefficient is given by (3.62).

Fig.3.4, b, c shows the dynamic characteristic of a plastic deformation process during continuous (Fig.1.13, a) and impulsive (Fig.1.13, b) regimes. The arrows show the directions of loading and load relief. Expressions for determining the harmonic linearization coefficients may be written for these characteristics, assuming that the average deformation speed is small ($v \ll \omega a_l$). These formulae are obtained using the same assumptions made during the derivation of equalities (1.20) and (1.21) which determined the constant load components P_l.

For the characteristic shown in Fig.3.4, b,

$$k = \frac{k_0}{2\pi}\left[2\pi - \arccos\left(1 - \frac{2\pi v}{a_l\omega}\right) + \left(1 - \frac{2\pi v}{a_l\omega}\right)\sqrt{1 - \left(1 - \frac{2\pi v}{a_l\omega}\right)^2}\right], \tag{3.63}$$

$$b = \frac{2k_0}{\pi\omega}\left[1 - \left(1 - \frac{2\pi v}{a_l\omega}\right)^2\right]. \tag{3.64}$$

It must be noted that during continuous plastic deformation in the process under consideration, coefficients (3.63) and (3.64) are independent of the material's yield limit D. This is because the coefficients are determined from the difference between the material's yield limit and the static force P_l, i.e. $D - P_l$ which, according to (1.21), depends only on the average speed of vibration ωa_l, the tool's progression speed v, the vibration amplitude a_l and the workpiece's static stiffness k_0.

When $v = 0$, formulae (3.63) and (3.64) yield $k = k_0$ and $b = 0$, i.e. any deformation taking place, does so, inside the elastic zone of the materials dynamic characteristic. The material's yield limit is only attained at the point at which the tool reaches its maximum displacement.

The characteristic shown in Fig.3.4, c can be described, where,

$$k = \frac{k_0}{2\pi}[\arccos\left(1 - \frac{D}{k_0 a_l} - \frac{2\pi v}{a_l \omega}\right) + \arccos\left(1 - \frac{D}{k_0 a_l}\right) - \\ - \arccos\left(1 - \frac{2\pi v}{a_l \omega}\right) - \left(1 - \frac{D}{k_0 a_l} - \frac{2\pi v}{a_l \omega}\right) \times \\ \times \sqrt{1 - \left(1 - \frac{D}{k_0 a_l} - \frac{2\pi v}{a_l \omega}\right)} - \left(1 - \frac{D}{k_0 a_l}\right)\sqrt{1 - \left(1 - \frac{D}{k_0 a_l}\right)^2} + \\ + \left(1 - \frac{2\pi v}{a_l \omega}\right)\sqrt{1 - \left(1 - \frac{2\pi v}{a_l \omega}\right)^2}], \tag{3.65}$$

$$b = \frac{2vD}{(a_l \omega)^2}. \tag{3.66}$$

The last formula can easily be derived using energy considerations. Indeed, the work due to deformation, over the loading cycle, is equal to the area inside the hysteresis loop $A = Dh_0$, where the residual strain h_0 is related to the average deformation speed by (1.13) and is equal to $h_0 = vT = 2\pi v/\omega$. The substitution of these values into (3.49) leads to formula (3.66). It must be noted that when $v = 0$, expression (3.65) takes the form (3.31), where $(\Delta - m_l) = a_l - D/k_0$, i.e. the limiter deforms to the full extent of the workpiece material's elastic zone. If, in addition, $a_l \to D/2k_0$, then $k \to k_0$ and the process moves from an impulsive elastic deformation regime to a harmonic regime of continuous deformation.

Fig.3.4, d demonstrates the deformation characteristic of a rigid-plastic material. For small speeds of deformation ($v \ll \omega a_l$), the linearization coefficients of this nonlinear characteristic take the form,

$$P = \frac{D}{\pi}\arcsin\sqrt{\frac{\pi v}{a_l \omega}}, \tag{3.67}$$

$$k = \frac{D}{\pi a_l}\sqrt{1 - \left(1 - \frac{2\pi v}{a_l \omega}\right)^2}, \tag{3.68}$$

$$b = \frac{2Dv}{(a_l \omega)^2}. \tag{3.69}$$

Expressions (3.67), which coincide with (1.26) and (3.68) can be obtained by applying the limit $k_0 \to \infty$ to (1.20) and (3.65). Equality (3.69) coincides with (3.56). For practical convenience, these formulae can be rewritten in a different form. Firstly, from (3.67), the speed v may be expressed by alternative parameters,

$$v = \frac{a_l \omega}{\pi} \sin^2 \frac{\pi P}{D}. \tag{3.70}$$

Substituting this value into (3.68) and (3.69) gives,

$$\begin{aligned} k &= \frac{D}{\pi a_l} \sin \frac{2\pi P}{D}, \\ b &= \frac{2D}{\pi a_l \omega} \sin^2 \frac{\pi P}{D}. \end{aligned} \tag{3.71}$$

A number of ultrasonic processes take place under the influence of small static forces $P \ll D$. In these cases, expressions (3.70) and (3.71) can be simplified, and take the form,

$$v = \pi a_l \omega \left(P/D\right)^2, \tag{3.72}$$

$$k = 2P/a_l, b = \frac{2\pi}{a_l \omega} \cdot \frac{P^2}{D}. \tag{3.73}$$

It is worth noting that expression (3.73), determining the linearization coefficient k, is not dependant on the yield limit D of the rigid-plastic material. It coincides with formula (3.38), which was obtained for the characteristic of a rigid, single-sided limiter.

Thus, for a number of ultrasonic processes, the coefficients characterizing the nonlinear load have been found. A typical feature of these coefficients is their dependence on the tool's vibration characteristics. Such a dependency has been repeatedly observed, in different processes, when experimental measurements of load resistance are taken. The dependence of load resistance on vibration amplitude when radiating vibration into a fluid, after passing its cavitation threshold, was revealed in Kikuchi (1969). In Holopov (1972), a similar dependence was revealed whilst studying an ultrasonic welding process. It was also found in Ganeva et al. (1981), in which the ultrasonic machining of brittle materials was investigated.

3.2 Resonant vibration of ultrasonic bar systems with a nonlinear loading

1. A load's nonlinearity causes specific effects to occur in an ultrasonic bar system's vibration characteristic. The load significantly changes the bar system's resonant properties, causing nonlinear distortion of the amplitude-frequency characteristic. It is dependant on the bar's parameters, the way in which it

is attached, and on the load's elastic and dissipative properties. These effects are now considered in a bar system that performs longitudinal and bending vibration.

Consider a bar system, with length l, in which vibration is excited by a force $f_s(t) = F_s e^{j\omega t}$ acting at cross section $x = s$. The displacement of the arbitrary cross section x shall be described by the function $u_x(t)$. When vibrating, cross section $x = l$ interacts with the nonlinear load. The load's dynamical characteristic is dependant on the displacement function u_l of the system's end $x = l$,

$$f_l = f_l(u_l, pu_l) \qquad (p = \partial/\partial t). \tag{3.74}$$

The bar system's vibration shall be described using a dynamic compliance operator $L_{sx}(p)$, which relates the force acting at cross section s with the resulting displacement x. The displacement of the system's cross sections are determined by the expression,

$$u_x(t) = L_{sx}(p)F_s e^{j\omega t} - L_{lx}(p) f_l[u_l(t), pu_l(t)]. \tag{3.75}$$

The approximate periodic solution of equation (3.75) is sought in the form,

$$u_x(t) \approx m_x + u_x^0(t), \qquad u_x^0(t) = \tilde{a}_x e^{j\omega t} = a_x e^{j(\omega t - \varphi_x)}. \tag{3.76}$$

Taking account of the nature of the solutions sought, the harmonic linearization of function (3.74) may be undertaken,

$$f_l(u_l, pu_l) \approx P_l(m_l, a_l) + [k(m_l, a_l) + pb(m_l, a_l)]u_l^0. \tag{3.77}$$

Here, $P_l(m_l, a_l)$ is a constant component; $k(m_l, a_l)$ and $b(m_l, a_l)$ are the load's equivalent elastic and dissipative components. They are dependant on the solution's parameters and should be calculated using the method described in section 3.1.

The following is obtained by substituting (3.77) into (3.75), separating the constant and periodic components, and taking account of (3.76),

$$m_x = -L_{lx}(0)P_l(m_l, a_l), \tag{3.78}$$

$$u_x^0 = L_{sx}(p)F_s e^{j\omega t} - L_{lx}(p)[k(m_l, a_l) + pb(m_l, a_l)]u_l^0. \tag{3.79}$$

From (3.78) and (3.79), for cross section $x = l$,

$$m_l = -L_{ll}(0)P_l(m_l, a_l), \tag{3.80}$$

$$u_l^0 = L_{sl}(p)F_s e^{j\omega t} - L_{ll}(p)[k(m_l, a_l) + pb(m_l, a_l)]u_l^0. \tag{3.81}$$

From (3.76), when $x = l$, the following equality can be written,

$$e^{j\omega t} = \frac{u_l^0}{a_l} e^{j\varphi_l} = \frac{u_l^0}{a_l}(\cos\varphi_l + j\sin\varphi_l). \tag{3.82}$$

By substituting (3.82) into equation (3.81), and transforming the result into an equivalent homogeneous equation (Babitsky (1998), Popov & Paltov (1960)),

$$\{1 + [k(m_l, a_l) + pb(m_l, a_l)]L_{ll}(p) - \\ -\frac{F_s}{a_l}(\cos\varphi_l + j\sin\varphi_l)L_{sl}(p)\}u_l^0 = 0. \tag{3.83}$$

The periodic solutions (3.76) of equation (3.83) exist when its characteristic equation possesses a pair of purely imaginary roots, $p = \pm j\omega$. Substituting $p = j\omega$ into the characteristic equation,

$$\frac{F_s}{\tilde{a}_l} = \frac{F_s}{a_l}(\cos\varphi_l + j\sin\varphi_l) = W(j\omega, m_l, a_l), \tag{3.84}$$

where, $W(j\omega, m_l, a_l) = \{1 + [k(m_l, a_l) + j\omega b(m_l, a_l)]L_{ll}(j\omega)\}/L_{sl}(j\omega)$ is the complete dynamic stiffness of both the vibration system and load at cross section l.

Analogously to (2.79), the dynamic stiffness expression can be rewritten in the form,

$$W(j\omega, m_l, a_l) = U(\omega, m_l, a_l) + jV(\omega, m_l, a_l), \tag{3.85}$$

where, $U(\omega, m_l, a_l) = ReW(j\omega, m_l, a_l)$ and $V(\omega, m_l, a_l) = ImW(j\omega, m_l, a_l)$.

By substituting (3.85) into (3.84) and separating the real and imaginary parts,

$$\cos\varphi_l = a_l U(\omega, m_l, a_l)/F_s, \tag{3.86}$$

$$\sin\varphi_l = a_l V(\omega, m_l, a_l)/F_s. \tag{3.87}$$

Excluding the trigonometric functions from these equalities gives,

$$a_l = F_s/\sqrt{U^2(\omega, m_l, a_l) + V^2(\omega, m_l, a_l)} = F_s/|W(j\omega, m_l, a_l)| \tag{3.88}$$

Relations (3.80) and (3.88) constitute a system of equations for the unknown motion parameters, m_l and a_l, of cross section $x = l$. Values of m_l and a_l allow, using equations (3.86) and (3.87), the phase φ_l to be found, while equations (3.78)–(3.79) determine the periodic motion parameters of the bar system's other cross sections.

In order to study the stability of the solutions obtained with respect to small perturbations, the techniques suggested in Babitsky (1971, 1998) shall be used. These are based on the ideas of the slowly varying amplitude method. Due to the continuous dependence of the characteristic equation's (3.83) roots on the system's parameters, small perturbations in the system's motion give rise to a small deviation of the roots from the imaginary axes. Taking account of this, it is assumed that the disturbed motion in the vicinity of the steady-state periodic solution has the form,

$$\breve{u}_x(t) = \breve{m}_x(t) + \breve{a}_x(t) \exp j[\omega t - \breve{\varphi}_x(t)], \tag{3.89}$$

where $\breve{m}_x(t), \breve{a}_x(t), \breve{\varphi}_x(t)$ are the slowly varying functions of time.

The system's dissipative properties are taken into account by means of the function $V(\omega, m_l, a_l)$ in dynamic stiffness expression (3.84). Therefore, equality (3.87) constitutes the effective dissipative and driving force energy balance of the steady-state motion under consideration. Using (3.84), the motion's stability condition may be formulated with respect to a small amplitude deviation,

$$\frac{\partial}{\partial \breve{a}_l} \left[V(\omega, \breve{m}_l, \breve{a}_l) - \frac{F_s}{\breve{a}_l} \sin \breve{\varphi}_l \right]_{\breve{a}_l = a_l} > 0 \tag{3.90}$$

Inequality (3.90) means that an amplitude deviation of $\breve{a}_l(t)$ from its stationary value $a_l = const$ causes a violation of the energy balance in such a way that it will compensate for the deviation.

Three unknown functions instead of one were introduced into (3.89), and may be related by two arbitrary conditions. It is assumed that in the perturbed motion, functions $\breve{m}_l(t)$ and $\breve{a}_l(t)$ are related by equality (3.80), and $\breve{\varphi}_l(t)$ and $\breve{a}_l(t)$ by equality (3.86). Differentiating (3.90), and taking account of these relations gives,

$$\left[\frac{\partial V}{\partial \breve{a}_l} + \frac{F_s}{\breve{a}_l^2} \left(\sin \breve{\varphi}_l - \breve{a}_l \frac{\partial \breve{\varphi}_l}{\partial \breve{a}_l} \cos \breve{\varphi}_l \right) \right]_{\breve{a}_l = a_l} > 0. \tag{3.91}$$

The derivative $\partial \breve{\varphi}_l / \partial \breve{a}_l$ can be found by differentiating equality (3.86) as an implicit function of $\varphi_l(a_l)$,

$$\frac{\partial \breve{\varphi}_l}{\partial \breve{a}_l} = -\frac{1}{F_s \sin \breve{\varphi}_l} \left(U + \breve{a}_l \frac{\partial U}{\partial \breve{a}_l} \right). \tag{3.92}$$

By substituting (3.92) into inequality (3.91), and taking expressions (3.86) and (3.87) into account, the stability condition is finally obtained,

$$\left[U \left(U + \breve{a}_l \frac{\partial U}{\partial \breve{a}_l} \right) + V \left(V + \breve{a}_l \frac{\partial V}{\partial \breve{a}_l} \right) \right]_{\breve{a}_l = a_l} > 0. \tag{3.93}$$

In the absence of the system's dissipative losses, $V = 0$, condition (3.93) gives the following stability region boundaries,

$$U(\omega, m_l, a_l) = 0, \qquad U(\omega, m_l, a_l) + a_l \frac{\partial U(\omega, m_l, a_l)}{\partial a_l} = 0, \tag{3.94}$$

where m_l and a_l are related by (3.80).

2. As an example, consider longitudinal vibration within a homogeneous bar (Fig.3.6, a). Its upper end, $x = 0$, is fixed, and its lower end, $x = l$, interacts with an elastic limiter. When in static equilibrium, the limiter is positioned at a distance Δ from the end of the bar. A negative value of Δ indicates that the limiter is subjected to a pre-compression.

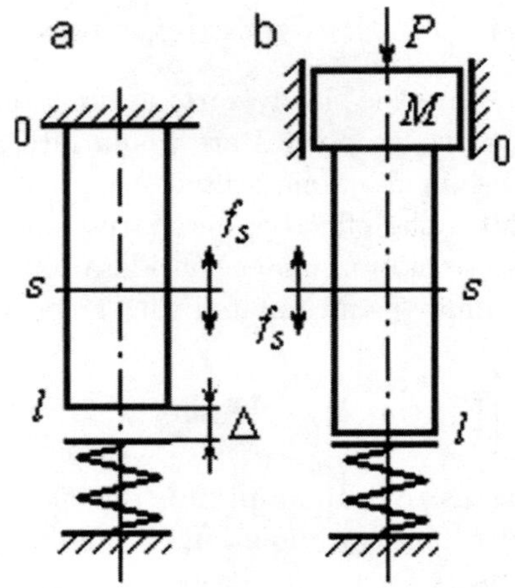

Fig. 3.6.

Harmonic linearization coefficients for the elastic limiter's characteristic are given by (3.30) and (3.31). The dynamic stiffness expression in (3.84) shall be written in the form,

$$W(j\omega, m_l, a_l) = [W_{ll}(j\omega) + k(m_l, a_l)]K(j\omega), \tag{3.95}$$

where, $K(j\omega) = W_{sl}(j\omega)/W_{ll}(j\omega)$.

By using the fixed bar's dynamic compliance (2.168), accounting for (2.49), and by limiting the hyperbolic function expansions containing the small parameter ψ to linear terms, the following can be written,

$$W_{ll}(j\omega) = \frac{ES}{l}\left[\xi \cot \xi + j\frac{\psi\xi}{4\pi}\frac{(\xi + 0.5 \sin 2\xi)}{\sin^2 \xi}\right], \tag{3.96}$$

$$K(j\omega) = \frac{\sin \xi}{\sin \chi\xi} +$$

$$+j\frac{\psi\xi}{4\pi}\frac{(\cos \xi \sin \chi \cdot \xi - \chi \sin \xi \cos \chi\xi)}{\sin^2 \chi\xi}, \tag{3.97}$$

where, $\chi = s/l$ $(0 \leq \chi \leq 1)$.

Note that if the excitation acts at the bar's end $s = l$ $(\chi = 1)$, then coefficient $K(j\omega) = 1$. Value (3.97) reflects the influence of the driving force's application point on the amplitude and phase of vibration at cross section $x = l$.

After substituting (3.95) and (3.96) into equation (3.88), (3.88) takes the form,

$$a_l = \frac{\delta}{K(\xi)}\left\{(\xi \cot \xi + q)^2 + \left[\frac{\psi\xi}{4\pi} \cdot \frac{(\xi + 0,5 \sin 2\xi)}{\sin^2 \xi}\right]^2\right\}^{-1/2}, \tag{3.98}$$

where $\delta = \frac{F_s l}{ES}$, $q = \frac{k(m_l, a_l)l}{ES}$ and $K(\xi) = |K(j\xi)|$.

In order to determine the motion parameters of the bar's lower end, equation (3.98) should be supplemented by equality (3.80), which, by virtue of (3.30) and (3.96) can be converted into the form,

$$a_l = \Delta \left[\alpha + \left(\alpha \arccos \alpha - \sqrt{1-\alpha^2}\right) q_0/\pi\right]^{-1}. \tag{3.99}$$

The expression for coefficient (3.31) can be rewritten as,

$$q = \left(\arccos \alpha - \alpha\sqrt{1-\alpha^2}\right) q_0/\pi, \tag{3.100}$$

where, $q_0 = k_0 l/ES$ and $\alpha = (\Delta - m_l)/a_l \quad (-1 \le \alpha \le 1)$.

Using these expressions, the amplitude - frequency characteristic for the bar's lower end can be constructed. According to (3.98), the equality

$$\xi_0 \cot \xi_0 + q = 0, \tag{3.101}$$

where $\xi_0 = \omega_0 l/c$, determines the configuration of the backbone curves which reflect the dependence of the system's free vibration frequency ω_0 on the amplitude.

In order to construct the backbone curves, a graphical method as shown in Fig.3.7, a, b, c shall be used. Dependencies $a_l(\alpha)$, $q(\alpha)$ and $q(\xi_0)$ are plotted in accordance with formulae (3.99)–(3.101) respectively. By choosing a value of amplitude a_l in Fig.3.7, a, and by tracing the curve in the direction shown by the arrows, with use of Fig.3.7, c, to determine the natural frequency ξ_0, the corresponding point in the backbone curve, Fig.3.7, d, is obtained.

It may be noted that there is a finite set of natural frequencies ξ_0 (each relating to a different vibration mode) corresponding to each given value of a_l. The backbone curves for the two lowest oscillation modes are shown in Fig.3.7, d. Due to the load's nonlinearity, the vibration frequency for each oscillation mode is dependant on amplitude. When a clearance is present in a system ($\Delta > 0$) the natural frequency increases with an increase in amplitude, and the system has a stiff type backbone curve. When a system's limiter is initially under compression ($\Delta < 0$), the natural frequency decreases with an increase in amplitude, and the system has a soft type backbone curve (Kolovsky (1999)).

It is possible, however, that linear vibration can occur in both cases. When $\Delta > 0$ linear vibration exists within the clearance space Δ. The amplitude is $a_l < \Delta$ and the natural frequency of the bar's vibration are as that of a bar with a free end. When $\Delta < 0$ it is possible that the vibration is limited in amplitude to the initial interference between the bar and limiter i.e. the bar's end does not separate from the limiter. In this case the amplitude is $a_l \le -\Delta/(1+q_0)$ and the bar's free vibration frequency is equal to that of a bar with an elastically fixed end. The vertical parts of the backbone curves in Fig.3.7, d correspond to linear vibration.

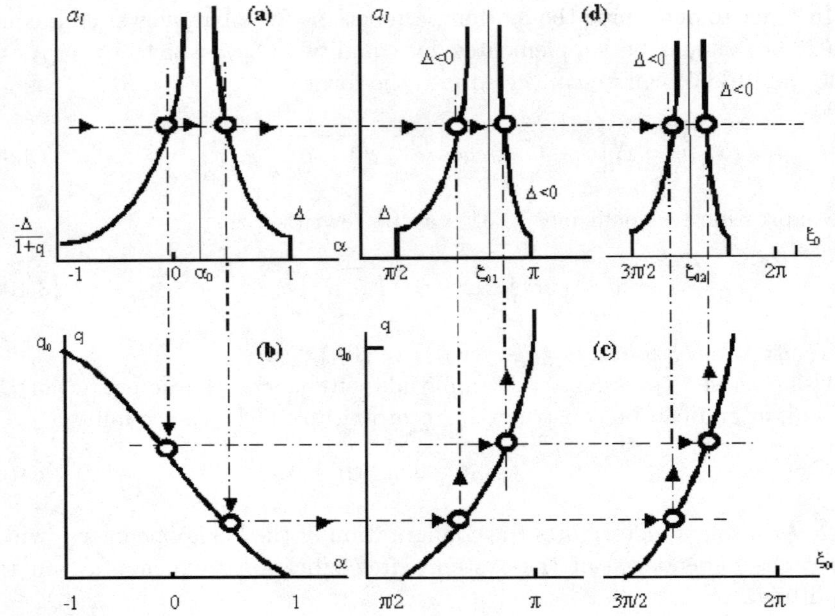

Fig. 3.7.

When $\Delta = 0$, the system has a constant spectrum of natural frequencies that differ from those of linear vibration. These natural frequencies are independent of the systems amplitude.

Due to these trends, backbone curves are determined in the following intervals,

$$\begin{aligned} &\xi_n \leq \xi_0 < \xi_{0n}, &&\text{for } \Delta > 0, \\ &\xi_0 = \xi_{0n}, &&\text{for } \Delta = 0, \\ &\xi_{0n} \leq \xi_0 < \xi_{Hn}, &&\text{for } \Delta < 0, \end{aligned} \tag{3.102}$$

where ξ_n and ξ_{Hn} are the bar's natural frequencies with either a free end, or when subjected to a linear elastic load k_0. The bar's natural frequencies can be found from equation (3.101) when $q = 0$ and $q = q_0$ respectively. ξ_{0n} are the natural frequencies of the nonlinear system when $\Delta = 0$; the index $n = 1, 2, ...$ denotes the natural mode number. Frequencies ξ_{0n} can also be obtained from equation (3.101), after the substitution of q as calculated from (3.100) when $\alpha = \alpha_0$. The latter is given, according to (3.99), by solving the transcendental equation,

$$\alpha_0 + \left(\alpha_0 \arccos \alpha_0 - \sqrt{1 - \alpha_0^2} \right) q_0 / \pi = 0 \tag{3.103}$$

3. The forced vibration amplitude of cross section $x = l$ can be found by simultaneously solving the system of equations (3.98)–(3.100). This can be done graphically. Expression (3.100), giving the linearization coefficient q, should be substituted into the amplitude-frequency characteristic equation (3.98). The obtained dependence $a_l(\alpha)$, for different frequencies $\xi = const$, should be marked on curve (3.99). This has already been used for plotting the backbone curves. The intersection points of these curves determine the values of α and vibration amplitude a_l, corresponding to the given frequency ξ.

The tracing operation is again performed in Fig.3.8, a, c and is used to reveal the possible configuration of the system's resonant curves. An initial clearance is assumed ($\Delta > 0$), and Fig.3.8, b, d shows the results. The number 1 is used to denote the backbone curves (Fig.3.8, b, d) and their mapping (3.99) onto plane (a_l, α) (Fig.3.8, a, c). Parameter ξ_i $(i = 1, 2)$ in Fig.3.8, a, c denotes the curves plotted according to equations (3.98) and (3.100) for a given value of frequency $\xi = \xi_i$. The solutions obtained are marked by points. Fig. 3.8, a, c demonstrates that the system under consideration has either one, or three solutions. In the latter case (Fig.3.8, a), one of the solutions is determined by the intersection of curve ξ_i with the vertical line $\alpha = 1$. This corresponds to linear vibration within the clearance Δ, and has an amplitude $a_l \leq \Delta$. Accompanying the realizable branches, the nonrealizable $(a_l > \Delta)$ parts of the linear system's resonant curves are shown in Fig.3.8, b, d using dashed lines. These are determined from equation (3.98) when $q = 0$. The other solutions determine branches of the resonant curve that are divided by the backbone curve and correspond to nonlinear vibration with an amplitude $a_l > \Delta$. From equations (3.98)–(3.100) in which $\alpha = 1$, and the tracing shown in Fig.3.8, it can be seen that the resonant curve's linear and nonlinear branches meet at the intersection points between the linear branches and the limiter value $a_l = \Delta$.

By setting the amplitude-frequency characteristic $\xi = \xi_0$ in equation (3.98), and taking account of equality (3.101), the following is obtained. The small denominator terms are only considered to their first order,

$$a_l = \frac{4\pi\delta}{\psi\xi_0} \cdot \left| \frac{\sin\xi_0 \sin\chi\xi_0}{\xi_0 + 0,5\sin 2\xi_0} \right|. \tag{3.104}$$

Equation (3.104) determines the limiting amplitude curve (2) (Fig.3.8, b, d), which intersects backbone curve 1 at the same point as the system's resonant curve. A limiting amplitude line is formed from the geometrical locations of the intersecting points between the resonant and backbone curves. As follows from the comparison between (3.98) and (3.104), all points on the resonant curve are situated below line (3.104). With the help from the graphical tracing of frequency ξ_2, as shown by the arrows in Fig.3.8, b, the mapping of limiting amplitude line 2 onto plane (a_l, α) is obtained (Fig.3.8, a, c). This line shows the maximum values of curves ξ_i, attained at values of α which correspond to the backbone curves.

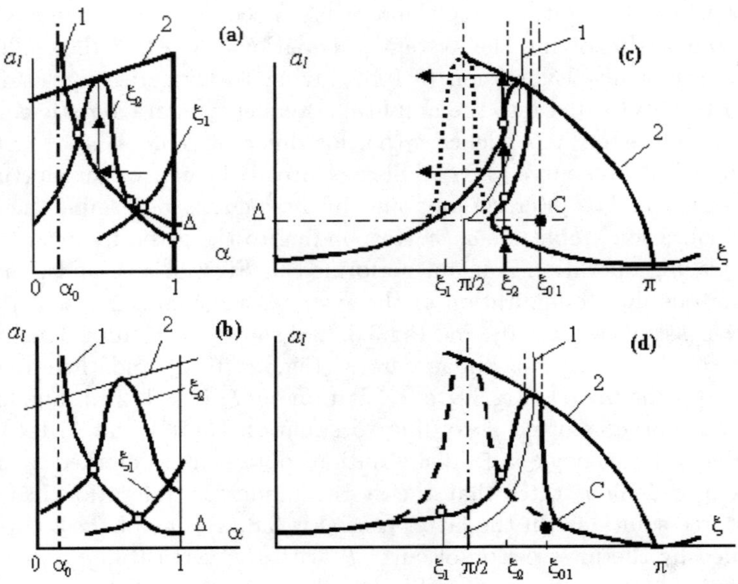

Fig. 3.8.

The possible resonant curve configurations for a system with an initial interference ($\Delta < 0$), are obtained by the analogous tracings shown in Fig.3.9.

Within the limiter's elastic deformation limits, the linear system's resonant curve branches, when working on the linear elastic load k_0 and vibrating with an amplitude $a_l \leq -\Delta/(1+q_0)$, are determined (again, the unrealizable part of the resonant curve is shown by a dashed line). These branches are described by equation (3.98) when $q = q_0$. The ambiguity between the resonant curve and the nonlinear branches enveloping the backbone curve 1, is shown in Fig.3.9, a, b. The resonant curve plotted in Fig.3.9, c, d is single-valued for the whole frequency interval under consideration. The considerations described demonstrate that ambiguity is present if the curve preceding the linear system's resonance branch passes below the intersection point C between the backbone curve's asymptote and the limiter's elastic deformation level $a_l \leq -\Delta/(1+q_0)$. The resonant curve's peak values are determined, as before, from the intersection between the backbone curve 1 and limiting amplitude line 2 (Fig.3.9, b, d) which is described by equation (3.104).

Fig.3.10 shows the amplitude-frequency characteristic of a system with no clearance ($\Delta = 0$). In this case, the vibro-impact regime exists throughout the whole frequency interval. Maximum vibration amplitude is obtained at the system's free vibration frequencies ξ_{0n} (see (3.102)) in which an elastic limiter with no clearance $\Delta = 0$ is present. The amplitude curve is determined

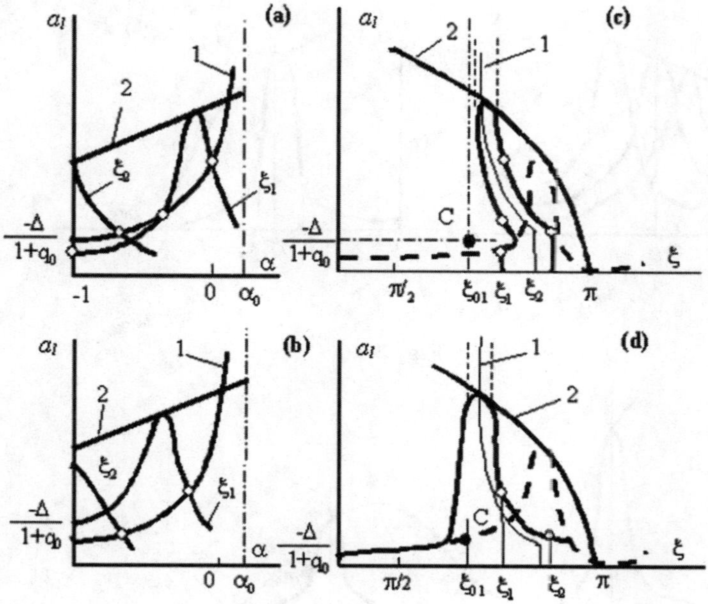

Fig. 3.9.

from the intersecting points between backbone curve 1 and limiting amplitude curve 2. From the curves in Fig.3.10, a, it can be seen that for any excitation frequency ξ, there is always a unique vibration regime.

It is noted that, in the absence of energy dissipation ($\psi = 0$), all resonant curves are continued, without restriction, along the backbone curve. This is shown by dashed lines in Fig.3.8, b, d and Fig.3.9, b. These figures also show examples of a linear system's resonant curve in the vicinity of its first vibration mode. The amplitude-frequency characteristics in the vicinity of higher vibration modes are constructed analogously.

4. The manifestation of nonlinear system properties rises with an increase in limiter stiffness. In the limiting case of having a rigid limiter ($k_0 \to \infty$), the elastic characteristic's linearization coefficient $k(a)$ relates to a constant component P_l of the limiter's reaction by the dependence (3.38). From equation (3.80), which gives the relation between the constant force component and displacement, using equality (3.32), and taking into account that $\alpha = (\Delta - m_l)/a_l \to 1$ as $k_0 \to \infty$, it is found that,

$$P_l = (a_l - \Delta) L_{ll}^{-1}(0). \tag{3.105}$$

By substituting this equality into formula (3.38), the linearization coefficient's dependence on the single unknown parameter a_l is obtained,

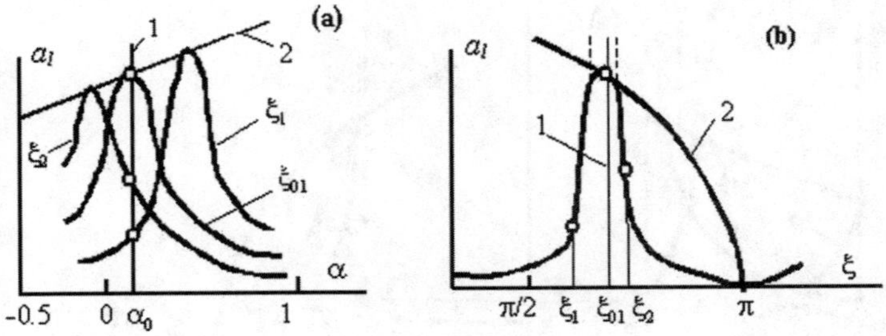

Fig. 3.10.

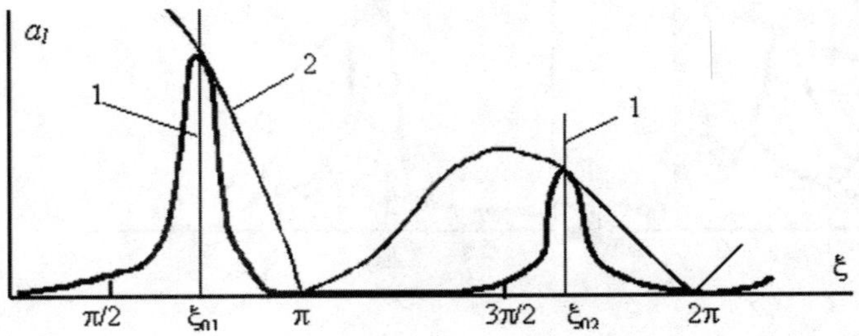

Fig. 3.11.

$$k(a_l) = \frac{2}{L_{ll}(0)}\left(1 - \frac{\Delta}{a_l}\right). \tag{3.106}$$

According to (3.96), for the system under consideration, the static stiffness can be written as $W_{ll}(0) = L_{ll}^{-1}(0) = ES/l$. Taking account of the notation used in (3.98), expression (3.106) takes the form,

$$q(a_l) = 2(1 - \Delta/a_l). \tag{3.107}$$

By substituting linearization coefficient (3.107) into equality (3.98), and solving with respect to the unknown amplitude a_l, the following expression is found for the system's amplitude-frequency characteristic:

$$a_l = \frac{2\Delta C(\xi) \pm \sqrt{[\delta(\xi)C(\xi)]^2 + \Psi^2(\xi)[\delta^2(\xi) - 4\Delta^2]}}{C^2(\xi) + \Psi^2(\xi)}, \tag{3.108}$$

where $\delta(\xi) = \delta/K(\xi)$; $C(\xi) = \xi \cot \xi + 2$; $\Psi(\xi) = \frac{\psi\xi}{4\pi} \cdot \frac{\xi + 0{,}5 \sin 2\xi}{\sin^2 \xi}$.

The resonant curves for the system's first two vibration modes with no clearance ($\Delta = 0$), an initial clearance ($\Delta > 0$), and an initial interference ($\Delta < 0$) are shown in Fig.3.11–3.13. An index 1 denotes a backbone curve, and is described by the equation,

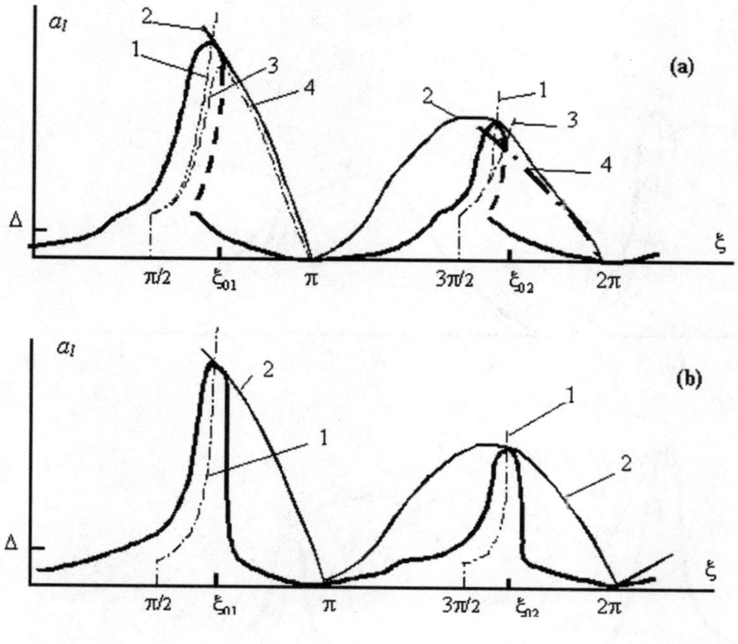

Fig. 3.12.

$$a_l = \frac{2\Delta}{C(\xi)} = \frac{2\Delta}{\xi \cot \xi + 2}, \tag{3.109}$$

which follows from (3.108) when $\delta = 0$ and $\psi = 0$.

The backbone curve's definition domains are described by (3.102), in which the value $\xi_n = \pi(2n-1)/2$ should be assigned. $\xi_{Hn} = \pi n$ and ξ_{0n} are solutions of equation $\xi \cot \xi = -2$, the first four roots being $\xi_{01} = 2,289$, $\xi_{02} = 5,087$, $\xi_{03} = 8,096$ and $\xi_{04} = 11,173$.

From relations (3.108) and (3.109), equation (3.104), giving the limiting amplitude line 2, is obtained. Its intersections with the backbone curves give the peak points in the amplitude-frequency characteristics.

At $\Delta = 0$ equation (3.108) takes the form,

$$a_l = \delta(\xi)/\sqrt{C^2(\xi) + \Psi^2(\xi)},$$

and the resonant curves are single-valued throughout the whole frequency range (Fig.3.11).

By applying a vibro-impact regime condition $a_l > 0$ to (3.108), it is found that for systems with a clearance (Fig.3.12, a), or an interference (Fig.3.13, a), the resonant curves in the backbone curve definition domains have zones of ambiguity. This happens when $\delta(\xi) < 2|\Delta|$. The upper (continuous line) and lower (dashed line) branches of the resonant curve meet at the point at which

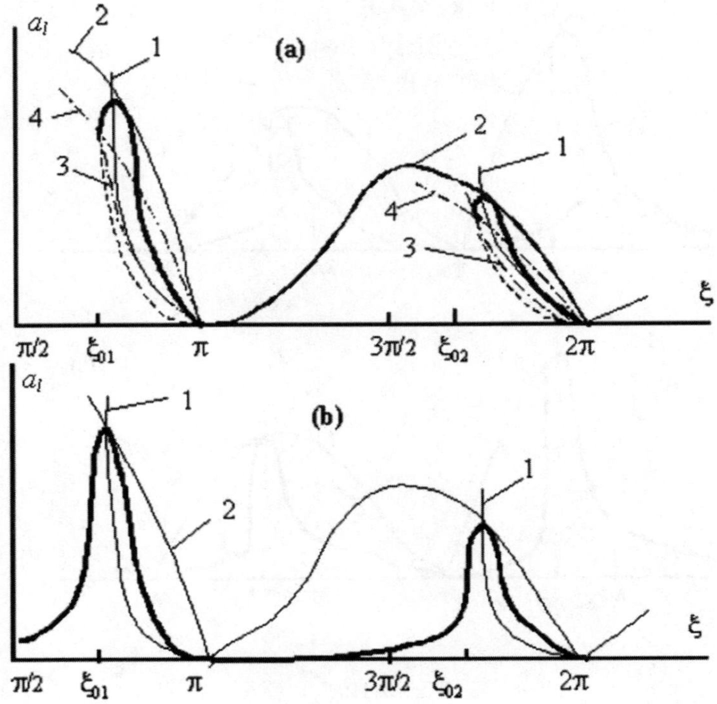

Fig. 3.13.

the radicand in (3.108) turns to zero, i.e. if the equality,

$$C(\xi) = \Psi(\xi)\sqrt{\left(\frac{2\Delta}{\delta(\xi)}\right)^2 - 1} \tag{3.110}$$

holds true.

In order to find the boundary of the resonant curve's ambiguous region, from (3.108) and (3.110), the following system of equations are obtained,

$$a_l = \frac{2\Delta C(\xi)}{C^2(\xi) + \Psi^2(\xi)}, \tag{3.111}$$

$$a_l = \frac{\delta(\xi)}{\Psi(\xi)}\sqrt{1 - \left(\frac{\delta(\xi)}{2\Delta}\right)^2}. \tag{3.112}$$

In Fig.3.12, a and Fig.3.13, a dashed-dotted curves 3 and 4, are plotted in accordance with equations (3.111) and (3.112). From the comparison between expressions (3.111) and (3.112), it is seen that the separating curve 3 passes below the backbone curve and is situated between the branches of the amplitude-frequency characteristic within the ambiguous zone. According to (3.112) and (3.104), curve 4 passes below the limiting amplitude line 2. The point of its intersection with the separating line 3, coincides with the meeting point of the upper and lower branches of the resonant curve.

When $\delta(\xi) \geq 2|\Delta|$, the resonant curves for systems with both a clearance (Fig.3.12, b), and an interference (Fig.3.13, b) are single-valued. They are described by equation (3.108), in which a plus sign is present in front of the radical. The resonant curves shown in Fig.3.12, a, b, are completed by the linear system's realizable branches $a_l \leq \Delta$.

It is noted that, in the absence of damping ($\Psi = 0$), equation (3.108), by virtue of (3.97) can be converted into the following simple expression:

$$a_l = \frac{2\Delta \sin \xi \pm \delta \sin \chi \cdot \xi}{\xi \cos \xi + 2 \sin \xi}. \tag{3.113}$$

The separating curve (3.111) coincides with backbone curve (3.109).

5. Using condition (3.93), the stability of the obtained solutions shall be studied. Firstly, a rigid limiter ($k_0 \to \infty$) shall be considered.

Taking expressions (3.95), (3.96) and (3.106) into account for the derivatives in (3.93) the following expressions exist: $\partial V/\partial a_l = 0$ and $\partial U/\partial a_l = \partial k(a_l)/\partial a_l = 2\Delta/a_l^2 L_{ll}(0)$. After substituting these quantities, and the corresponding expressions from (3.95), (3.96) and (3.106), condition (3.93), by virtue of the notation used in (3.108), takes the form,

$$\left[C(\xi) - \frac{2\Delta}{a_l}\right] C(\xi) + \Psi^2(\xi) > 0.$$

By solving this inequality with respect to amplitude a_l, the following is finally obtained,

$$a_l > \frac{2\Delta C(\xi)}{C^2(\xi) + \Psi^2(\xi)}. \tag{3.114}$$

The right-hand side of stability condition (3.114) coincides with separating curve equation (3.111). Therefore, the resonant curves' lower branches, as shown in Fig.3.12, a and Fig.3.13, a by dashed lines, transpire to be unstable. As follows from (3.114), and the resonant curve ambiguity conditions, the regimes corresponding to single-valued amplitude-frequency characteristics (Fig.3.11, Fig.3.12, b and Fig.3.13, b) are stable in the whole defined domain.

It is now shown that, in general, for an elastic limiter of finite stiffness, only regimes corresponding to the lower branches of a resonant curve's ambiguous regions are unstable. The stability condition (3.93) by virtue of expressions (3.85), (3.95) and (3.96) takes the form,

$$C_1^2(\xi, \alpha) + \Psi^2(\xi) + a_l \frac{\partial q[\alpha(a_l)]}{\partial a_l} > 0, \tag{3.115}$$

where $C_1(\xi, \alpha) = \xi \cot \xi + q(\alpha)$.

The derivative can be calculated,

$$\frac{\partial q}{\partial a_l} = \frac{\partial q}{\partial \alpha} \cdot \frac{\partial \alpha}{\partial a_l} \tag{3.116}$$

using relations (3.100) and (3.99),

$$\begin{aligned} \frac{\partial q}{\partial \alpha} &= -\frac{2q_0}{\pi}\sqrt{1-\alpha^2}, \\ \frac{\partial \alpha}{\partial a_l} &= -\frac{\Delta}{a^2}\left(1 + \frac{q_0}{\pi}\arccos\alpha\right)^{-1}. \end{aligned} \tag{3.117}$$

After the substitution of (3.116) and (3.117), stability condition (3.115) takes the form,

$$a > -\frac{2\Delta C_1(\xi,\alpha)\sqrt{1-\alpha^2}}{[C_1^2(\xi,\alpha) + \Psi^2(\xi)](\pi/q_0 + \arccos\alpha)}. \tag{3.118}$$

Inequality (3.118) is definitely fulfilled for all regimes when $\Delta = 0$ and for those regimes that correspond to resonant curve branches situated above the backbone curve when $\Delta \neq 0$ for which,

$$\Delta C_1(\xi,\alpha) = \Delta[\xi\cot\xi + q(\alpha)] > 0.$$

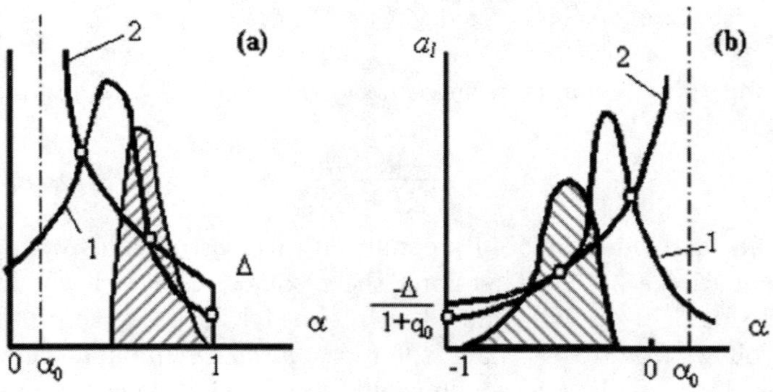

Fig. 3.14.

The graphical solution of equations (3.98) (curve 1) and (3.99) (curve 2) for systems with both a clearance ($\Delta > 0$), and an interference ($\Delta < 0$) are shown in Fig.3.14, a, b. For a given frequency ξ, there are two solutions. The equation on the right-hand side of inequality (3.118) allows the stability region's boundary to be found. The instability region is shown by hatching. It contains the solution that occurs due to smaller vibration amplitude.

The graphical constructions described show that, at the meeting points between the upper and lower resonant curve branches, not only should the quantities in (3.98) and (3.99) be equal, but their derivatives also. Differentiating (3.98) and taking account of (3.99) and (3.117) gives,

$$\frac{\partial a_l}{\partial \alpha} = \frac{\partial a_l}{\partial q} \cdot \frac{\partial q}{\partial \alpha} = \frac{2a_l C_1(\xi,\alpha) q_0 \sqrt{1-\alpha^2}}{\pi[C_1^2(\xi,\alpha) + \Psi^2(\xi)]}. \tag{3.119}$$

From (3.99) the following derivative is found,

$$\frac{\partial a_l}{\partial \alpha} = -\frac{a_l^2}{\Delta}\left(1 + \frac{q_0}{\pi} \arccos \alpha\right). \tag{3.120}$$

By equating (3.119) and (3.120), the following equality is obtained, and should be satisfied at the meeting point of the upper and lower branches:

$$a_l = -\frac{2\Delta C_1(\xi,\alpha)\sqrt{1-\alpha^2}}{[C_1^2(\xi,\alpha) + \Psi^2(\xi)](\pi/q_0 + \arccos \alpha)}. \tag{3.121}$$

This equality coincides with the right-hand side of stability criterion (3.118). Thus, the meeting point between the resonant curve's upper and lower branches belong to the stability region's boundary. All the regimes that correspond to the ambiguous zones' lower branches transpire to be unstable.

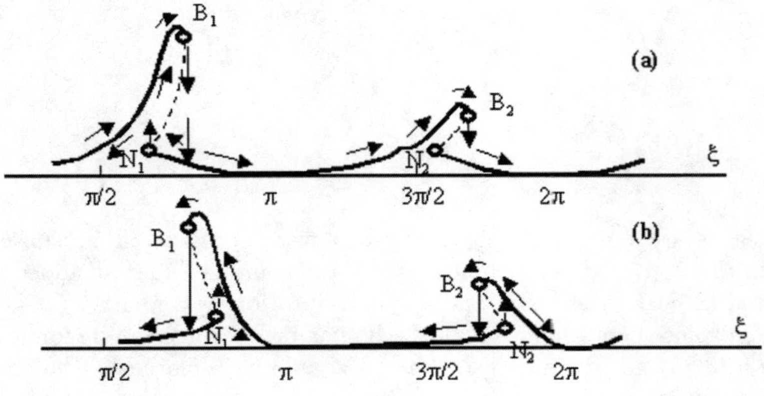

Fig. 3.15.

Resonant curves for systems with an initial clearance ($\Delta > 0$) and interference ($\Delta < 0$) are shown in Fig.3.15, a, b. The continuous lines correspond to stable branches and dashed lines to unstable branches. If the excitation frequency slowly varies in such a way that every possible value within the regime of forced vibration occurs, it is seen that the branch of the resonant curve followed depends on the direction of the change in frequency. This is shown by the arrows. As the frequency is altered, the vibration amplitude is seen to hop: the first occurrences of this take place at the meeting points B_i $(i = 1, 2, \ldots)$ between the nonlinear branches, the second, at the meeting points between the unstable and linear regime branches N_i within the clearance (Fig.3.15, a) and interference (Fig.3.15, b). Unstable branches are physically unrealizable.

6. Ultrasonic bar systems that perform bending vibration and interact with nonlinear loads demonstrate the same behavioural peculiarities as found in longitudinal vibration systems. In order to demonstrate this, consider a cantilever bar with a body fixed at its end, under the influence of bending vibration, and interacting with a single-sided limiter (Fig.3.16). The system vibrates due to a harmonic driving force $f_c(t) = F_c e^{j\omega t}$ which is applied in the added body's mass centre C. The way in which the bar's cross sections displace relative to their undistorted state is described by the function $u_x(t)$. The way in which the added body displaces is determined by its mass centre coordinate u_c, and rotation angle θ.

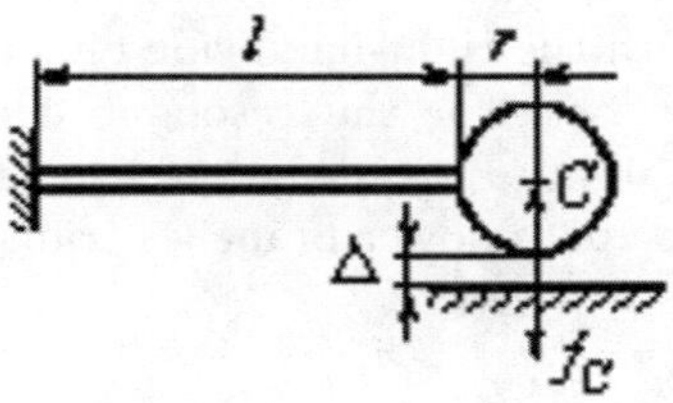

Fig. 3.16.

Linear system dynamic characteristics were constructed in subsection 2.3.3. In this section, the dynamic stiffness and compliance for an added body, subjected to both progressive and angular vibration, was found.

The displacement amplitude of the body's mass centre can be found from equation (3.88), in which indexes l and s should be replaced by the index c, and the reduced dynamic stiffness is given by,

$$W(j\omega, m_c, a_c) = W_c(j\omega) + k(m_c, a_c) + j\omega b(m_c, a_c), \qquad (3.122)$$

where, W_c is the linear system's dynamic stiffness at point C, as determined from relation (2.220).

The angular vibration amplitude θ_c is described by equality (2.223).

Consideration is now restricted to a system in which a rigid limiter is positioned with a clearance Δ from the body. By using an expression of the form (3.105) in order to determine the harmonic linearization coefficient, equation (3.88), describing the system's resonant curves in the absence of dissipation

$(V(a_l, \omega) = 0)$, is converted to the form,

$$a_c = \frac{2\Delta \pm \delta}{2 + L_c(0)U_c(\omega)}, \tag{3.123}$$

where $U_c(\omega) = (\mathrm{Re})|\mathrm{W_c(j\omega)}|$, $L_c(0) = U_c^{-1}(0)$ and $\delta = F_c L_c(0)$.

The domain in which solutions exist can be found from the conditions $a \geq \Delta$ for $\Delta > 0$ and $a > 0$ for $\Delta \leq 0$. By letting $\delta = 0$ $(F_c = 0)$ in (3.123), backbone curve equations are determined in the ranges,

$$\begin{aligned} &\omega_n \leq \omega < \omega_{0n} \qquad \text{for } \Delta > 0, \\ &\omega = \omega_{0n} \qquad \text{for } \Delta = 0, \\ &\omega_{0n} < \omega \leq \omega_n^* \qquad \text{for } \Delta < 0. \end{aligned} \tag{3.124}$$

ω_n and ω_n^* are the resonant and antiresonant frequencies of the linear system. They are determined by relations (2.225) – (2.228) and from the graphs in Fig.2.25; ω_{0n} are the natural frequencies of the vibro-impact system when $\Delta = 0$. They are obtained from the solutions of the equation $2U_c(0) + U_c(\omega) = 0$ and depend only on the linear system's properties.

As follows from (3.123), when $|\Delta| \leq \delta/2$ the resonant curves are single-valued and exist in the whole frequency domain. When $|\Delta| > \delta/2$,the resonant curves are determined in the domain $\omega_{0n} < \omega \leq \omega_n^*$ when $\Delta < 0$, and in the adjacent domains when $\Delta > 0$. In this circumstance, vibro-impact regimes can exist in pairs.

The amplitude-frequency characteristics of a body's transverse vibration for the first three vibration modes are plotted in Fig.3.17, a for a system with an initial clearance $(\Delta > 0)$, and Fig.3.18, a for a system with an initial interference $(\Delta < 0)$. The resonant curve branches envelope the backbone curves 1, and in the absence of dissipation, they continue along them infinitely without intersection. The amplitude is limited due to the energy losses of overcoming internal friction in the bar's material, and dissipation caused by impacts with the limiter.

According to (3.88), the maximum amplitudes realizable are determined by the intersection points between the backbone curves and the limiting amplitude line. The latter is described by the equation,

$$a_c = F_c/V(a_\omega, \omega) = F_c/[V_c(\omega) + \omega b(a_c, \omega)], \tag{3.125}$$

where, $V_c(\omega) = \mathrm{Im}W_c(j\omega)$, and is given by formula (2.222).

Assuming that the added body's impact with the limiter is instantaneous, and by estimating its effect using the restorability factor R, the harmonic linearization coefficient $b(a, \omega)$ for the dissipative part of the impact's characteristic is described by expression (3.44). Taking into account that the backbone curve is determined by the expression,

$$U(\omega) = U_c(a_c, \omega) + k(a) = 0, \tag{3.126}$$

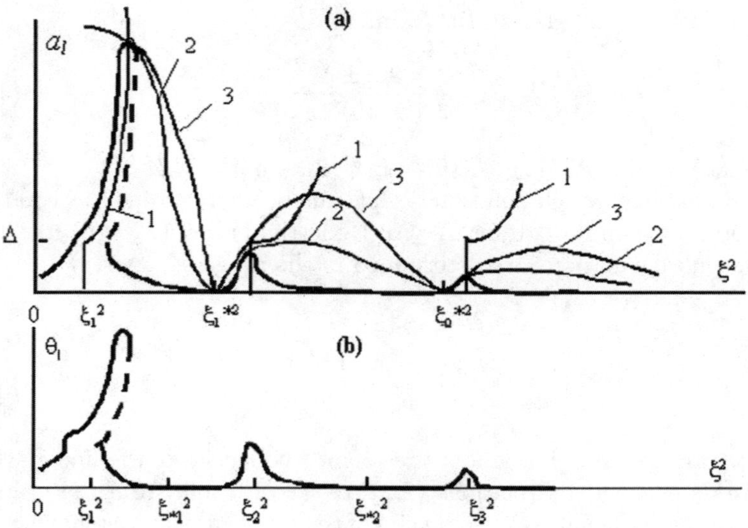

Fig. 3.17.

equation (3.125), giving the limiting amplitude line, can be converted to the form,

$$a_c = F_c \left[V_c(\omega) + \frac{\pi}{2} \cdot \frac{1-R}{1+R} \cdot \frac{U_c^2(\omega)}{M\omega^3} \right]^{-1} . \tag{3.127}$$

In Fig.3.17, a and Fig.3.18, a, the limiting amplitude lines, plotted in accordance with equation (3.127), are marked by the index 2. For comparison, the limiting amplitude lines 3 for a perfectly elastic collision ($R = 1$), are plotted as well. It can be seen that the influence of the impact's dissipation becomes more pronounced with an increase in vibration mode number. The dashed lines correspond to the resonant curves' unstable, nonlinear branches. Resonant curves for systems with a clearance are shown in Fig.3.17, a. They are complemented by the realizable branches of a linear system in which the amplitude $a \leq \Delta$.

Fig.3.17, b and Fig.3.18, b show resonant curves for a body's angular vibration. These are plotted in accordance with equations (2.216) and (2.220). The nonlinear branches of the angular vibration characteristic have the same frequency domains as the transverse vibration resonant curves. At angular vibration antiresonant frequencies ξ_{*n}, the body performs progressive vibration. At linear vibration antiresonant frequencies ξ_n^*, the body performs purely angular vibration. It must be noted that the vibro-impact system's angular vibration amplitude can be significantly larger than the angular vibration amplitudes of the linear system. This is very pronounced in branches situated

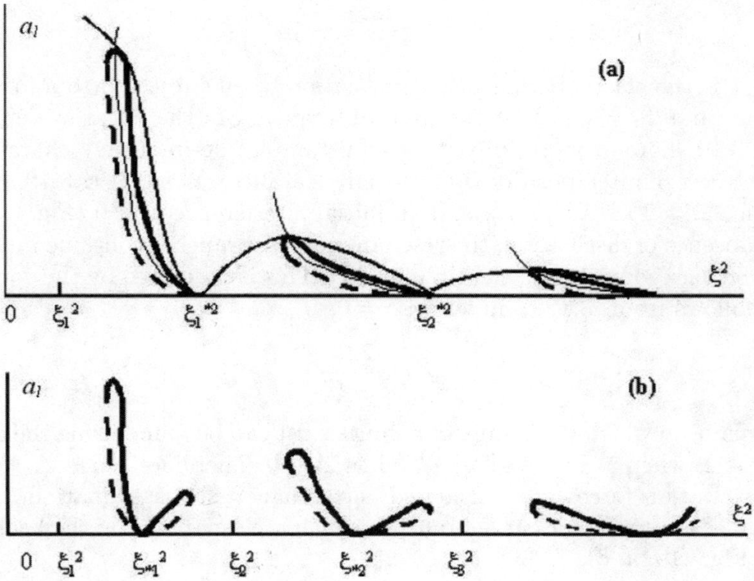

Fig. 3.18.

after antiresonant frequencies ξ_{*n} in systems with an initial interference (Fig.3.18, b).

7. Using the results from the previous subsection, a very simple vibro-impact system is considered. The bar's mass is assumed to be negligible and the added body is modelled as a point mass M.

The existence and stability of an oscillator's periodic motion, in which a fixed limiter is present, was studied in a number of papers using exact methods of stitching and point mapping. The amplitude-frequency characteristic of an oscillator colliding with a limiter is plotted in Astashev (1971) using exact solutions. The approximate solutions for the same system are obtained in Babitsky (1998).

By setting values $J = 0$ and $r = 0$ in (2.220), and subjecting the obtained expression to the limiting case $\rho \to 0$, the dynamic stiffness of the lumped parameter system is found,

$$W_c(j\omega) = c + j\omega b - M\omega^2, \tag{3.128}$$

where, $c = 3EI/l$ is the cantilever bar's static stiffness and $b = c\psi/2\pi\omega$ is the linearized internal friction force coefficient.

By substituting (3.128) into (3.123) an equation describing the system's amplitude-frequency characteristics, in which no dissipation is present, is obtained,

$$a_c = \frac{2\Delta \pm \delta}{3 - \xi^2}. \tag{3.129}$$

$\delta = F_c/c$ is the elastic bar's static deformation when subjected to a force F_c; $\xi = \omega/\omega_0$ and $\omega_0 = \sqrt{c/M}$ is the natural frequency of the linear system.

Fig.3.19 shows a vibro-impact system's amplitude-frequency characteristics for three limiter positions: an initial clearance $\Delta > 0$ (Fig.3.19, a), no clearance $\Delta = 0$ (Fig.3.19, b), and an initial interference $\Delta < 0$ (Fig.3.19, c). In the absence of dissipation, the resonant curves' branches continue infinitely along the backbone curves (the thin lines 1). This is described by the equation which follows from (3.129), in which $\delta = 0$,

$$a_c = \frac{2\Delta}{3 - \xi^2}. \tag{3.130}$$

Domains in which vibro-impact regimes exist can be found using the condition $a_c \geq \Delta$ when $\Delta > 0$, and $a_l > 0$ when $\Delta \leq 0$. Therefore, when $\Delta \neq 0$ and small excitation forces are considered, such that $\delta < 2|\Delta|$, equation (3.129) determines two resonant curve branches which are separated by the backbone curve (Fig.3.19, a, c).

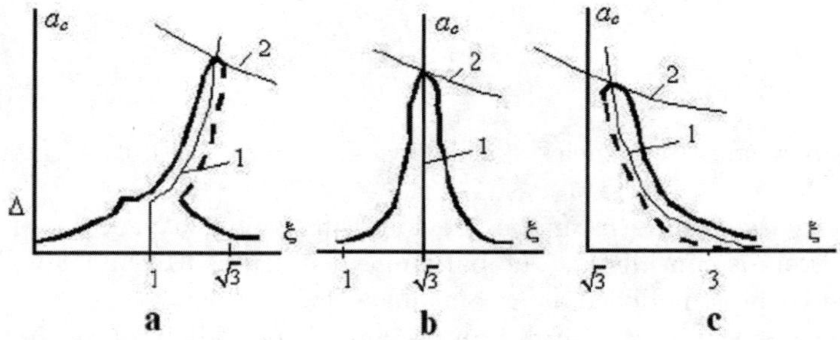

Fig. 3.19.

The lower branches, shown by dashed lines, transpire to be unstable. A system with no clearance ($\Delta = 0$) has an isochronous resonant curve for the whole domain in which a vibro-impact regime exists. The system's resonant curve, in which $\Delta \neq 0$ and large excitation forces ($\delta > 2\Delta$) are experienced, has a similar form.

Linear harmonic vibration with an amplitude $a_c \leq \Delta$ can also be realized in a system with an initial clearance ($\Delta > 0$). It is easy to show that resonant curves (3.129) originate from the points at which the linear oscillator's amplitude-frequency characteristic impinges on the limiter $a_c = \Delta$ (Fig.3.19,a).

The resonant curves have a finite nature due to energy dissipation. The maximum value occurs at the intersecting point between the backbone curve

1 and limiting amplitude line 2, as described by equation (3.127). After substituting quantities $V_c = \omega b$ and $U_c = c(1 - \xi^2)$ from (3.128) into equation (3.127), it takes the form,

$$a_c = \frac{\delta}{\xi n + B\frac{(1-\xi^2)^2}{\xi^2}}, \tag{3.131}$$

where, $n = b/\sqrt{cM}$ and $B = \frac{\pi}{2} \cdot \frac{1-R}{1+R}$.

According to (3.129) and (3.130), the resonant curve's maximum point, for a system with no clearance ($\Delta = 0$), is attained at a frequency $\xi = \sqrt{3}$. The amplitude at resonance (3.131), is equal to,

$$a_c = \frac{\delta}{\sqrt{3}n + 4B/3}. \tag{3.132}$$

In the absence of viscous friction $b_0 = 0$ ($n = 0$), and taking account of the notation used above,

$$a_c = \frac{3}{2\pi} \cdot \frac{F_c}{c} \cdot \frac{1+R}{1-R}. \tag{3.133}$$

It is noted that, in Astashev (1971), the following resonant amplitude formula was found, using exact methods, when considering the same problem:

$$a_c = \frac{1}{3} \cdot \frac{F_c}{c} \cdot \frac{1+R}{1-R}. \tag{3.134}$$

From the comparison between (3.133) and (3.134), they are seen not only have a similar structure, but give quantitatively close values. The resonant curves shown in Fig.3.19 possess all of the amplitude-frequency characteristic peculiarities of those constructed in Astashev (1971) resulting from exact solutions.

It should be noted that in the studies of vibro-impact systems considered at the beginning of this section, only the energy dissipation due to impact interaction was accounted for. The approximate solutions obtained allow the influence of any kind of energy dissipation to be estimated. For example, the analysis shows that the influence of viscous and impact damping on resonant regimes when working at maximal amplitude varies for systems with differing structures. By considering the instances when $n = 0$, and $B = 0$ in (3.131), equating the results to expression (3.130), and solving the equations obtained, the frequencies for the regimes giving the maximum amplitude are found: for a system with impact dissipation ($n = 0$),

$$\xi = \sqrt{\frac{3\delta + 4\Delta B + \sqrt{(9\delta + 16\Delta B)\delta}}{2(\delta + 2\Delta B)}}; \tag{3.135}$$

and for a system with viscous dissipation ($B = 0$).

$$\xi = \frac{-\Delta n + \sqrt{(\Delta n)^2 + 3\delta^2}}{\delta}. \tag{3.136}$$

The dependence of the resonant vibration's frequency ξ and amplitude a_c on the amount of clearance (interference) Δ, for systems with impact (curves 1) and viscous (curves 2) dissipation, is shown in Fig.3.20, a, b. It can be seen that viscous damping limits an increase in the clearance's value $\Delta = \delta/n$ in order for the linear system's natural vibration frequency $\xi = 1$ to be achieved, it does not however restrict the magnitude of interference $\Delta \to -\infty$ as $\xi \to \infty$. On the contrary, impact damping does not restrict the clearance value $\Delta \to \infty$ as $\xi \to 1$, but restricts the magnitude of interference to,

$$\Delta = -\frac{\delta}{2B} = -\frac{\delta}{\pi} \cdot \frac{1-R}{1+R}, \tag{3.137}$$

when the frequency $\xi \to \infty$.

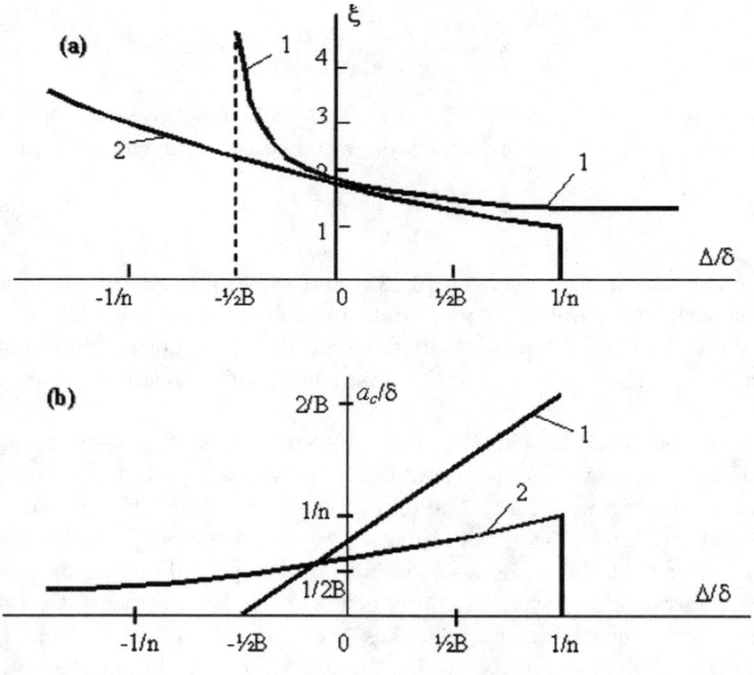

Fig. 3.20.

The limiting value of interference (3.137), as found from (3.135), coincides with value from book Babitsky (1998) in which it was obtained using exact solutions.

Finally, it is noted that the approximate solutions obtained above allow the impulse's magnitude between the colliding bodies of the vibro-impact system to be estimated. This is an important factor when considering systems with a vibro-impact nature. In order to do this, relation (3.40) should be used, in which (3.105) and (3.129) are substituted, leading to the oscillator's impulse-frequency characteristic equation. The limiter's influence is incorporated into the derived equation, and no energy dissipation is present ($b_0 = 0$ and $R = 1$):

$$J = \frac{2\pi}{\xi}\sqrt{cM}\frac{\Delta(1+\xi^2) \pm \delta}{3-\xi^2}. \tag{3.138}$$

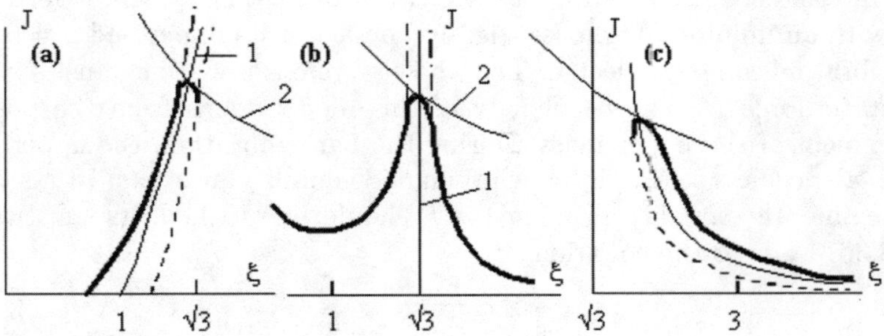

Fig. 3.21.

The impulse-frequency characteristics for systems with an initial clearance $\Delta > 0$ (Fig.3.21, a), no clearance $\Delta = 0$ (Fig.3.21,b) and an initial interference $\Delta < 0$ (Fig.3.21, c) are plotted in Fig.3.21. In the absence of dissipation, the characteristics continue to follow the backbone curves 1 with no restriction (dashed-dotted lines). Backbone curve 1's equation follows from (3.138) when $\delta = 0$. The energy dissipation restricts the characteristic at the intersection point between the backbone curve and the limiting momentum line 2. The limiting momentum line's equation can be obtained from (3.40), (3.105) and (3.131):

$$J = \frac{2\pi}{\xi}\sqrt{cM}\left[\frac{\delta}{\xi n + B\frac{(1-\xi^2)^2}{\xi^2}} - \Delta\right]. \tag{3.139}$$

As an example, in a system in which no initial clearance is present ($\Delta = 0$), the maximum value of momentum is achieved when $\xi = \sqrt{3}$. The value at this point, when $n = 0$, is equal to,

$$J = \delta\sqrt{3cM}\frac{1+R}{1-R}. \tag{3.140}$$

The following expression, as obtained in Astashev (1971), Babitsky (1998), gives the maximum momentum value for the same system, but is calculated from exact solutions:

$$J = \frac{4}{3}\delta\sqrt{cM}\frac{1+R}{1-R}.$$

This expression has the same structure as (3.140) and gives close quantitative values.

As such, the approximate solutions obtained using the harmonic linearization method are seen to provide clear qualitative and sufficiently correct quantitative information about resonant phenomena, even when considering strongly nonlinear systems such as those of vibro-impact (Babitsky & Krupenin (2001)).

8. To conclude this section, consider the vibration of a system in which an initial interference is created by a force (Fig.3.6, b). The system is formed of a bar with an infinitely big mass, rigidly bonded at its upper end $x = 0$. The mass vibration can be neglected. The whole system is pressed against a limiter by a static force P. The peculiarity of this model stems from the possible displacement of the upper mass. During the bar's vibration, the upper mass reaches a specific position in which dynamic equilibrium exists. In a steady-state regime, the equality $P_l(a_l, m_l) = P$ should be satisfied. Its substitution into (3.30) leads to the equation,

$$a_l = \pi\delta_p/q_0(\sqrt{1-\alpha^2} - \alpha \arccos\alpha), \tag{3.141}$$

where, $\delta_p = Pl/ES$. The notation is the same as that used in (3.99) and (3.100).

The problem is now reduced to a system of equations (3.98), (3.100) and (3.141) and can be solved using the methods described in subsection 2. Fig.3.22, a demonstrates the graphical solution of these equations, while Fig.3.22, b shows the resonant curve's configuration. It can be seen that, in this instance, three solutions are possible. The first, when $\alpha = -1$, corresponds to linear vibration inside the initial deformation of the limiter ($a_l \le \delta_p/q_0$), the second and third correspond to nonlinear vibration accompanied by a separation from the limiter. Fig.3.22, b also shows backbone curve 1 and limiting amplitude line 2, as described by equation (3.104). Their intersection corresponds to the resonant curve's maximum point.

This problem can be solved using an alternative method by which expression (3.34) is used to determine the linearization coefficient. Taking account of the notation used above, this expression can be written in the form,

$$q(a_l) = \frac{2\delta_p}{a_l + \delta_p/q_0}. \tag{3.142}$$

By substituting (3.142) into equality (3.101), the following backbone curve equation is obtained:

$$a_l = -\delta_p\left(\frac{2}{\xi_0 \cot\xi_0} + \frac{1}{q_0}\right). \tag{3.143}$$

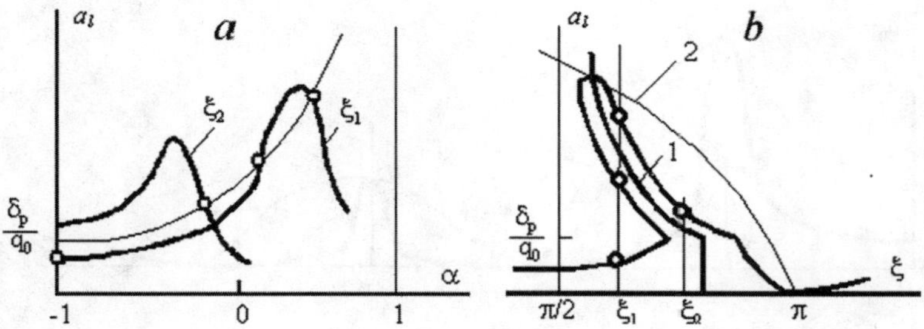

Fig. 3.22.

The intersection of backbone curve 1 (Fig.3.23, a) with limiting amplitude line 2 gives the resonant curve's maximum point. The rest of the curve can be found by solving the equation system (3.98) and (3.142) graphically, as shown in Fig.3.23, b. By choosing q as an independent parameter in these equations, the dependency graphs $a_l(q)$ can be plotted. Their intersection points determine the required solutions. The resonant curves are complemented by branches that correspond to vibration occurring within the limiter's initial deformation ($a_l \leq \delta_p/q_0$).

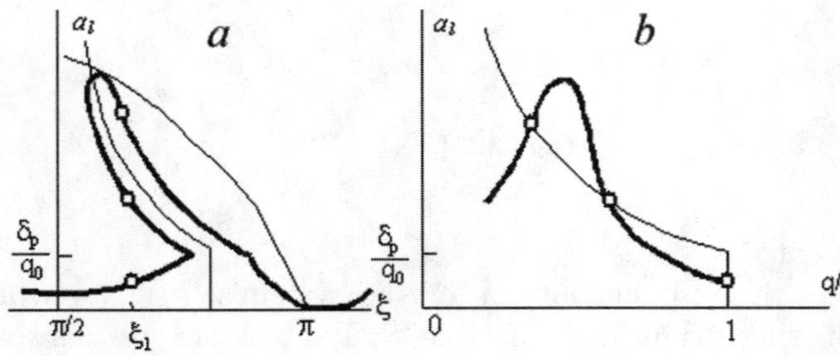

Fig. 3.23.

The system's interaction with a rigid limiter ($q_0 \to \infty$) may now be considered in detail. In this situation, the linearization coefficient is described by expression (3.38) or,

$$q(a_l) = 2\delta_p/a_l. \tag{3.144}$$

By substituting coefficient (3.144) into equality (3.98), the amplitude-frequency characteristic equation is obtained. It is similar to (3.108):

$$a_l = \frac{-2\delta_p C_l(\xi) \pm \sqrt{[\delta(\xi) C_l(\xi)]^2 + \Psi^2(\xi)[\delta^2(\xi) - 4\delta_p^2]}}{C_l^2(\xi) + \Psi^2(\xi)}, \tag{3.145}$$

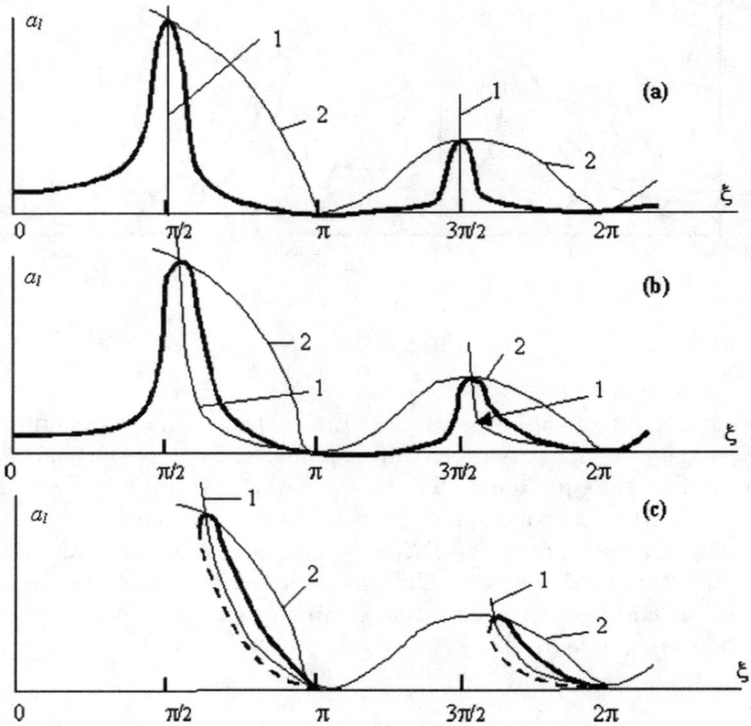

Fig. 3.24.

where, $C_l(\xi) = \xi \cot \xi$.

Fig.3.24 shows the resonant curves for the system's first two vibration modes under differing magnitudes of force P. Index 1 denotes a backbone curve and is described by the equation,

$$a_l = -2\delta_p / C_l(\xi) = -2\delta_p / \xi \cot \xi \tag{3.146}$$

which is obtained from (3.145) when $\delta = 0$ and $\psi = 0$.

Limiting amplitude line 2 is described by equation (3.104), which follows from (3.145) and (3.146). An intersection between the backbone curve and limiting amplitude line, gives the resonant curve's maximum point. In the absence of a pressing force ($\delta_p = 0$), the system is linear, and its resonant curves (Fig.3.24, a) are described by equation (3.98) in which $q = 0$. The system's resonant frequencies are given by $\xi_n = \frac{\pi}{2}(2n - 1)$, $n = 1, 2, \ldots$

The condition by which vibro-impact regimes exist, $a_l > 0$, yields that when $\delta_p \leq \delta(\xi)/2$, the resonant curves (Fig.3.24, b) are single-valued throughout the whole frequency domain. With an increase in pressing force, the resonant frequencies also increase, and they drift from those of the linear system's natural frequencies. When $\delta_p > \delta(\xi)/2$ the resonant curves (Fig.3.24, c) have ambiguous zones. The upper (continuous lines), and lower (dashed lines) branches meet at the point at which the radicand in (3.145) becomes zero, i.e.

$$C_l(\xi) = \Psi(\xi)\sqrt{2\delta_p/\delta(\xi) - 1}.$$

Taking account of this, from (3.145), an equation which determines the boundaries of the resonant curve's ambiguous zones is found:

$$a_l = -\frac{2\delta_p C_l(\xi)}{C_l^2(\xi) + \Psi^2(\xi)}, \tag{3.147}$$

$$a_l = \frac{\delta(\xi)}{\Psi(\xi)}\sqrt{1 - \left(\frac{\delta(\xi)}{2\delta_p}\right)^2}. \tag{3.148}$$

Curve (3.147) separates the resonant curve's branches, and is situated below the backbone curve (3.146). In the absence of damping ($\psi = 0$), resonant curve equation (3.145) takes the form,

$$a_l = \frac{-2\delta_p \sin\xi \pm \delta \sin\chi\xi}{\xi\cos\xi}, \tag{3.149}$$

and separating curve (3.147) coincides with backbone curve (3.146).

The stability of the solutions found above can be investigated using condition (3.93). By substituting the derivatives $\partial V/\partial a_l = 0$ and $\partial U/\partial a_l = \partial q/\partial a_l = -2\delta_p/a_l^2$, the stability condition can be converted into the form,

$$a_l > \frac{-2\delta_p C_l(\xi)}{C_l^2(\xi) + \Psi^2(\xi)}. \tag{3.150}$$

As follows from (3.147) and (3.150), the boundary of the stability domain coincides with the separating curve. Therefore, the lower branches of the resonant curve (Fig.3.24, c), shown by dashed lines, are unstable. Regimes corresponding to single-valued amplitude-frequency characteristics (Fig.3.24, b) are stable throughout the whole domain of their existence.

The situation considered above, in which a bar system is subjected to a compressive force, is very important for understanding the dynamic properties of a number of ultrasonic machines in which the tool is subjected to a similar force during its working process (Fig.2.1–2.4).

3.3 Vibro-impact interactions in ultrasonic bar systems

1. Subsection 2.1.2 is devoted to ultrasonic machine arrangements in which a tool interacts with a passive waveguide. In the systems shown in Fig.2.3 and Fig.1.2, a passive waveguide-reflector serves as an acoustic decoupler between

the base and the load's dynamic component. In a number of cases (Fig.1.4 and Fig.2.5), the workpiece acts as a passive waveguide. This is the case if its length is comparable to, or exceeds, the acoustic wavelength within the waveguide's material.

In this section, effects due to the vibro-impact interaction of bar systems are considered. Similar effects take place, not only in ultrasonic systems, but in various vibro-conductive constructions that incorporate gaps in their connections or joints. These gaps open under peak loading conditions and give rise to vibro-impact effects (Babitsky (1998)). Some aspects of vibration theory, such as that of construction, were studied in Katkovnik (1968).

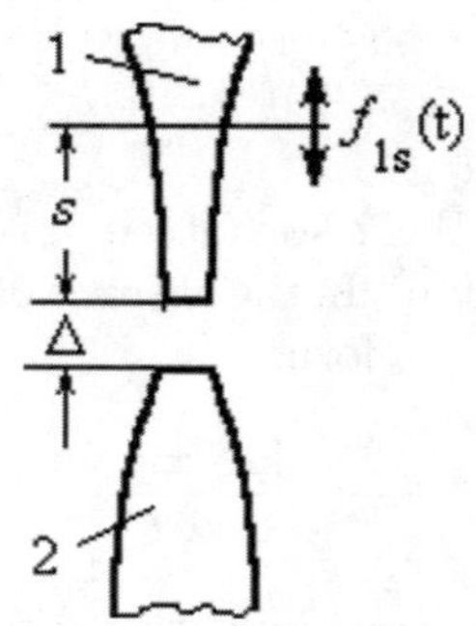

Fig. 3.25.

Consider two bar systems (Fig.3.25) performing one-dimensional vibration. The adjacent system elements, in which their closest faces are assigned to be at point zero, are positioned slightly apart, with a clearance of Δ (a negative value of Δ corresponds to an initial compression). They can contact each other with no interpenetration. The displacement of the i-th system's ($i = 1, 2$) arbitrary element x, from its non-deformed state, is described by the function $u_{ix}(t) = u_i(x, t)$. System 1's vibration is set in motion by a periodic force $f_{1s}(t) = F_{1s}e^{j\omega t}$ which acts at a specific element s. The secondary system's vibration occurs as a result of its interaction with the primary system, during the contact of elements at $x = 0$.

The relative displacement of the system's contacting elements are denoted by,

$$v(t) = u_{10}(t) - u_{20}(t). \tag{3.151}$$

Their interaction forces are represented by a nonlinear static characteristic of the form (3.29):

$$f(v) = k_0(v - \Delta)\eta(v - \Delta) \tag{3.152}$$

The limit $k_0 \to \infty$ is considered, in which there is an absence of interpenetration between the contacting elements. This approach reproduces the main interaction effects. It also allows specific problem ambiguities to be removed, and some simple design relations to be obtained.

As before, the i-th system's dynamic compliance operator is denoted $L_{sx}^{(i)}(p)$. This operator relates element x's displacement to an arbitrary force applied at element s. Equations describing the system's vibration can be written in the form,

$$\begin{aligned} u_{1x} &= L_{sx}^{(i)}(p) f_{1s}(t) - L_{0x}^{(i)}(p) f(v), \\ u_{2x} &= L_{0x}^{(2)}(p) f(v). \end{aligned} \tag{3.153}$$

By substituting equation (3.153), in which $x = 0$, into (3.151), an equation describing the contacting elements' relative motion is obtained,

$$v(t) = L_{sx}^{(i)}(p) f_{1s}(t) - L(p) f[v(t)], \tag{3.154}$$

where, $L(p) = L_{00}^{(1)}(p) + L_{00}^{(2)}(p)$.

In order to look for an approximate periodic solution of equation (3.154) in the form,

$$v(t) = m + v^0(t), v^0(t) = ae^{j(\omega t - \phi)}, \tag{3.155}$$

harmonic linearization of function (3.152)is performed,

$$f(v) = P_0(m, a) + k(m, a)v^0. \tag{3.156}$$

By substituting (3.155) and (3.156) into (3.154) and separating the constant and periodic components, when $p = j\omega$, it is found that,

$$m = -L(0)P_0, \tag{3.157}$$

$$[1 + kL(j\omega)]v^0(t) = L_{s0}^{(1)}(j\omega)F_{1s}e^{j\omega t}. \tag{3.158}$$

Taking account of (3.155), and from (3.158), expressions for the amplitude and phase of the relative motion's periodic components are obtained,

$$a = F_{1s}|W(j\omega, k)|^{-1} = F_{1s}(U^2 + V^2)^{1/2}, \tag{3.159}$$

$$\cos\varphi = aU/F_{1s}, \sin\varphi = aV/F_{1s}, \tag{3.160}$$

where,

$$W(j\omega, k) = [1 + kL(j\omega)]/L_{s0}^{(1)}(j\omega), U = ReW, V = ImW. \tag{3.161}$$

Expressions (3.157), (3.159) and (3.162) along with the harmonic linearization coefficient $k = k(m, a)$ expressions, form an equation system. These allow the amplitude a, the relative motion's constant component m, and the interaction force's constant component P_0 to be found. Using (3.160), the phase φ can be found, and from (3.153) the periodic motion parameters of bar systems 1 and 2 can be determined.

As shown in section 3.1, by subjecting the limit $k \to \infty$ to the expressions for nonlinear function's (3.152) harmonic linearization coefficients, (3.30) and (3.31), the following relations are found (see subsection 3.1.4):

$$a = \Delta - m, \tag{3.162}$$

$$k(a) = 2P_0/a \tag{3.163}$$

From (3.157), (3.162) and (3.163) the linearization coefficients for bar systems with an initial clearance (and interference) Δ are found,

$$k(a) = \frac{2}{L(0)}\left(1 - \frac{\Delta}{a}\right) \tag{3.164}$$

An approximate periodic solution of equation (3.153) which describes the absolute motion of an arbitrary cross section within the i-th bar, is sought. It is required in the form,

$$u_{ix}(t) = m_{ix} + u_{ix}^0(t), u_{ix}^0(t) = a_{ix}\exp[j(\omega t - \varphi_{ix})], (i = 1, 2) \tag{3.165}$$

By substituting (3.165) and (3.156) into (3.153), and separating the constant and periodic components, by virtue of (3.158), the following is obtained,

$$m_{ix} = (-1)^i L_{0x}^{(i)}(0)P_0, \tag{3.166}$$

$$u_{ix}^0(t) = \frac{F_{1s}}{W_x^{(i)}[j\omega, k(a)]}e^{j\omega t}, \tag{3.167}$$

where,

$$W_x^{(1)}(j\omega, k) = \frac{1 + kL(j\omega)}{[1 + kL(j\omega)]L_{sx}^{(1)} + kL_{0x}^{(2)}(j\omega)L_{s0}^{(1)}(j\omega)}, \tag{3.168}$$

$$W_x^{(2)}(j\omega, k) = \frac{1 + kL(j\omega)}{kL_{0x}^{(2)}(j\omega)L_{s0}^{(1)}(j\omega)}. \tag{3.169}$$

Using (3.167), the amplitudes and phases of the motion's periodic components for an arbitrary element x within the i-th bar system are found,

$$a_{ix} = F_{1x}|W_{ix}(j\omega, k)|^{-1} = F_{1s}(U_{ix}^2 + V_{ix}^2)^{1/2}, \tag{3.170}$$

$$\cos\varphi = aU_{ix}/F_{1s}, \sin\varphi = aV_{ix}/F_{1s}, \tag{3.171}$$

where, $U_{ix} = ReW_{ix}$ and $V_{ix} = ImW_{ix}$.

The relations obtained allow all the parameters which characterize vibration between two impacting systems of arbitrary structure to be found. This method is now applied to some typical examples.

2. Consider the vibration of two identical impacting bars with lengths $l_i = l$ $(i = 1, 2)$. In order to simplify the final relations, it is assumed that a driving force $f_{10} = f$ is applied to cross section $s = 0$ of the first bar. In this

situation, $L_{sx}^{(1)}(j\omega) = L_{sx}^{(2)}(j\omega) = L(j\omega)$, and expressions (3.161), (3.168) and (3.169) take the form,

$$W(j\omega, k) = W_{00}(j\omega) + 2k(a), \tag{3.172}$$

$$W_x^{(1)}(j\omega, k) = W_{0x}(j\omega)\frac{W_{00}(j\omega) + 2k(a)}{W_{00}(j\omega) + k(a)}, \tag{3.173}$$

$$W_x^{(2)}(j\omega, k) = W_{0x}(j\omega)\frac{W_{00}(j\omega) + 2k(a)}{k(a)}. \tag{3.174}$$

The dynamic stiffness of a homogeneous bar with one end fixed, and its free end subjected to a driving force, was found in the previous section. This is given by expression (3.96), and has the form,

$$W_{00}(j\omega) = \frac{ES}{l}\left[\xi \cdot \cot\xi + j\frac{\psi\xi}{4\pi} \cdot \frac{(\xi + 0{,}5\sin 2\xi)}{\sin^2\xi}\right], \tag{3.175}$$

where, $\xi = \omega l/c$, $c = \sqrt{E/\rho}$ is the speed of sound within the bar's material, E and ρ are the elastic modulus and density of the bar's material, S and l are the bar's cross sectional area and length, and ψ is the energy absorption coefficient in the bar's material.

By substituting (3.172) and (3.175) into (3.159), the following expression for the relative motion's amplitude a is obtained,

$$a = \delta\left|\left[\xi\cot\xi + 2q + j\frac{\psi\xi}{4\pi} \cdot \frac{(\xi + 0{,}5\sin 2\xi)}{\sin^2\xi}\right]^{-1}\right|, \tag{3.176}$$

where, $\delta = Fl/ES$ and $q = q(a) = k(a)l/ES$.

Similarly, from (3.167), taking account of (3.173) – (3.175) in which $x = 0$, the absolute motion amplitudes a_{i0} of the bars' contacting cross sections are found,

$$a_{10} = \delta\left|\frac{\xi\cot\xi + q + j\Psi}{\xi\cot\xi(\xi\cot\xi + 2q) + 2j\Psi(\xi\cot\xi + q)}\right|, \tag{3.177}$$

$$a_{20} = \delta q|[\xi\cot\xi(\xi\cot\xi + 2q) + 2j\Psi(\xi\cot\xi + q)]^{-1}|, \tag{3.178}$$

where, $\Psi = \Psi(\xi) = \frac{\psi\xi}{4\pi} \cdot \frac{(\xi+0.5\sin\xi)}{\sin^2\xi}$.

Equations (3.176) – (3.178) along with the harmonic linearization coefficient expressions, (3.163) and (3.164), allow the amplitude-frequency characteristics of the relative and absolute vibration of the bars' linked cross sections to be constructed.

3. Suppose that the bars are set apart with an initial clearance (or interference) Δ (Fig.3.26). Using the notation introduced previously, the harmonic linearization coefficient expression (3.164) can be written in the form,

$$q = q(a) = 1 - \frac{\Delta}{a} \tag{3.179}$$

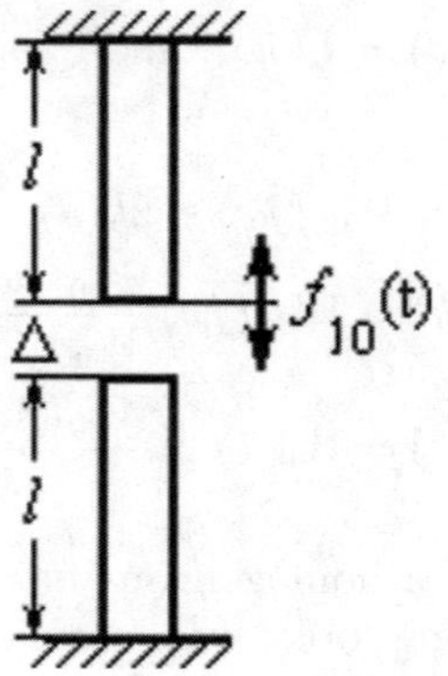

Fig. 3.26.

The resonant curve equation for the impacting cross sections' relative vibration can be found from (3.176),

$$a = \delta \left\{ (\xi \cot \xi + 2q)^2 + \left[\frac{\psi \xi (\xi + 0{,}5 \sin 2\xi)}{4\pi \sin^2 \xi} \right]^2 \right\}^{-2} . \quad (3.180)$$

It must be noted that in the situation under consideration, the relative motion of the bars' adjacent faces is described by equations equivalent to the equations describing the motion of a bar end when impacting against a rigid limiter (see subsection 3.1.4). Therefore, all of the expressions obtained in subsection 3.2.4 are valid for analysing the relative motion between elastic bars. Only the key points of such an analysis are discussed here.

According to (3.180), the backbone curves that relate to the amplitude-frequency characteristics of the bars' impacting cross sections relative motions are described by the relation,

$$\xi \cot \xi + 2q = 0. \quad (3.181)$$

After the substitution of (3.179), this yields,

$$a = -2\Delta / (\xi \cot \xi + 2). \quad (3.182)$$

From (3.180), and using (3.182), an equation describing the limiting amplitude line is found,

$$a = \frac{4\pi\delta \sin^2 \xi}{\psi \xi (\xi + 0.5 \sin 2\xi)} . \quad (3.183)$$

The intersection between the backbone curves and limiting amplitude line on plane (ξ, a) determines, according to (3.180) and (3.181), the resonant curve's maximum value. All of the other points can be found by using the expressions discussed in the previous section. In the absence of internal friction energy losses in the bar's material ($\psi = 0$), (3.180) and (3.179) yield the following equation determining the relative motion's amplitude-frequency characteristic,

$$a = \frac{2\Delta \pm \delta}{\xi \cot \xi + 2}. \tag{3.184}$$

Fig.3.27 a, d shows the relative motion's amplitude-frequency characteristic for two identical impacting bars for situations in which the contacting cross sections are set to an initial clearance ($\Delta > 0$), and an initial interference ($\Delta < 0$).

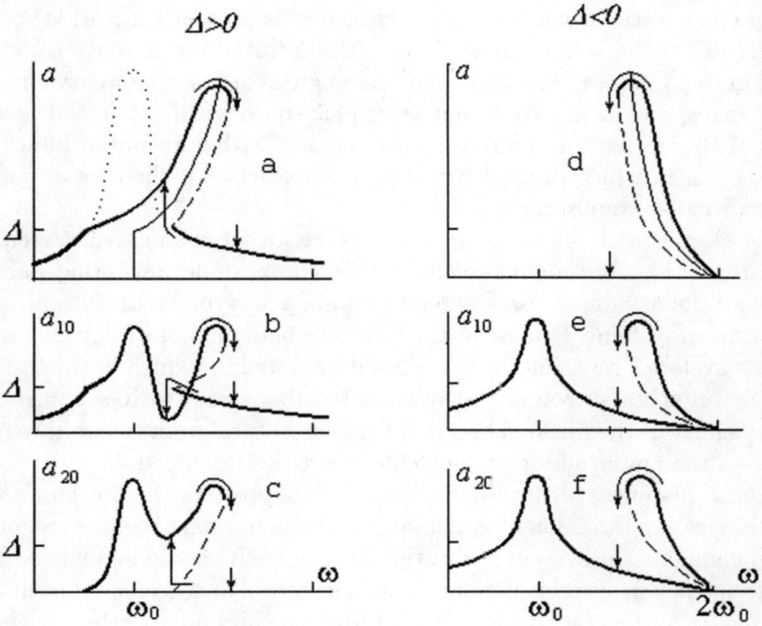

Fig. 3.27.

Fig.3.27 a, d shows that the type of nonlinearity present is dependant on the sign of Δ: if the bars are set with an initial clearance then the nonlinearity experienced is of the hard type, whilst when an initial interference is present, the nonlinearity is of the soft type. In these figures and those subsequent, thin lines correspond to backbone curves, while dashed lines show resonant curve branches that corresponding to unstable solutions. The dotted lines shown in Fig.3.27, a show resonant curve branches $a > \Delta$ which are realizable only in the absence of a limiter. The curve sections that are always realizable $a < \Delta$ are shown by thick lines.

If the relative motion amplitude a is known, then, using expressions (3.177) – (3.179), the absolute motion amplitude of the bar's contacting cross sections can be obtained. In the absence of dissipation ($\psi = 0$), equations for the contacting cross sections' amplitude-frequency characteristics can be written in the form,

$$a_{10} = \left| \frac{2\Delta\xi \cot\xi \pm \delta(\xi \cot\xi + 1)}{\xi \cot\xi(\xi \cot\xi + 2)} \right|;$$
$$a_{20} = \left| \frac{-\Delta\xi \cot\xi \pm \delta}{\xi \cot\xi(\xi \cot\xi + 2)} \right|. \tag{3.185}$$

The absolute motion's amplitude-frequency characteristic of the bars' contacting cross sections when an initial clearance is present ($\Delta > 0$) is shown in Fig.3.27, b, c. The same, but for bars with an initial interference ($\Delta < 0$), is shown in Fig.3.27, e, f. Also shown in these figures are resonant curve branches which correspond to vibration that takes place both within an initial clearance (Fig.3.27, b) i.e. with an amplitude $a < \Delta$, and within an initial interference (Fig.3, e, f), in which mutual vibration occurs between the contacting cross sections with an amplitude $a < |\Delta|/2$.

The characteristics discussed reveal behavioural analogies between complex structured vibro-impact systems that contain one impacting pair with both traditional lumped parameter vibro-impact systems, and usual nonlinear vibration systems. Primarily, the existence of ambiguous frequency regions within a system's resonant curves, should be noted, in which both stable and unstable branches surround the systems backbone curves. Resonance curves in the vicinity of the linear subsystem's higher natural frequencies have a form similar to the amplitude-frequency characteristics discussed.

In both instances of the bars' relative initial position (i.e. initial clearance or interference), transition to maximal amplitude resonant regimes can be realized by changing the system's vibration frequency. It should be increased from a low frequency in systems with an initial clearance, and decreased from a high frequency in systems with an initial interference and force acting downwards on the system. This is shown by the arrows in Fig.3.27. Changing the frequency further, once maximal vibration amplitudes have been realized, leads to disruption of the system's vibration, and a termination of the vibro-impact process. If this happens, the active bar 1, in a system with an initial clearance, will perform linear vibration within the clearance present (Fig.3.27, a, b). In a system in which an initial interference is present, the bars perform mutual vibration and act as a single bar of combined length (Fig.3.27, d, e).

Note the appearance of additional resonances in the bars' absolute motion. These occur at the subsystems' natural frequencies. In systems in which an initial clearance is present, these resonances (Fig.3.27, b, c) occur during the cophasal motion of the subsystems, in which light impacts occur, leading to motion within the passive bar. In systems with an initial interference, this resonance (Fig.3.27, e, f) occurs in the bars due to their effective combined length. In both instances, the resonances take place at the natural frequencies $\xi = \pi 2(r-1)/2$ ($r = 1, 2, \ldots$). The appearance of nonlinear antiresonance should be noted in systems that have an initial clearance (Fig.3.27, b, c). This is present in the frequency range between the two resonances. In this situation, the passive bar acts as a vibro-impact dynamic absorber to the active bar. The active bar's vibration amplitude transpires to be small, and tends to zero when the energy dissipation in the bar's material is reduced.

If vibration is generated at a fixed frequency belonging to the resonance curve's ambiguous regions, a vibro-impact regime can be instigated by hard excitation. This happens when the system is forced to receive enough additional energy for the upper stable branch of the characteristic to be realized. Under such a condition, the system allows an increase in vibration amplitude simply due to the level of vibration present. This is realized by an increase in clearance, interference or compressive force, until the amplitude reaches the limiting value as determined by the energy boundary of the vibro-impact domain.

4. The amplitude-frequency characteristic of two identical bars' impacting cross sections, which are pressed together under the action of a longitudinal force P, (Fig.3.28) may be plotted. It is assumed that the passive bar 2 is fixed, while the active bar 1 is coupled to a sufficiently large mass whose vibration can be neglected.

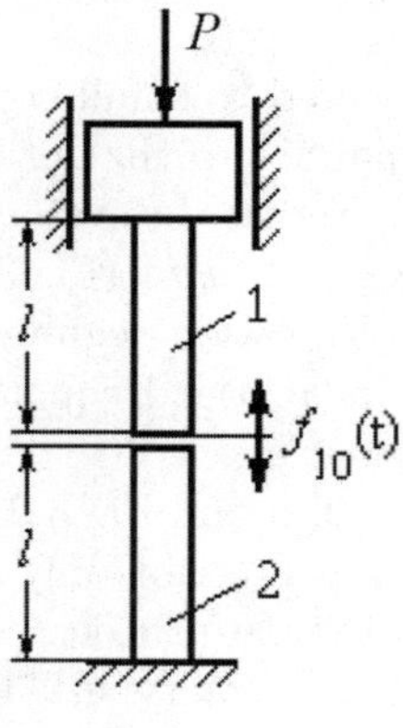

Fig. 3.28.

In this situation, harmonic linearization coefficient expression (3.163) takes the form,

$$q = q(a) = 2\delta_p/a, \tag{3.186}$$

where, $\delta_p = Pl/ES$, and the equation determining the resonant curve for the bars' relative vibration can be found from (3.180),

$$a = \delta\left\{\left(\xi\cot\xi + 4\frac{\delta_p}{a}\right)^2 + \left[\frac{\psi\xi(\xi + 0{,}5\sin 2\xi)}{4\pi\sin^2\xi}\right]^2\right\}^{-2}. \qquad (3.187)$$

From (3.187) the explicit equation for the amplitude-frequency characteristic can easily be obtained,

$$a = \frac{-4\delta_p\xi\cot\xi \pm \sqrt{\delta^2\xi^2\cot^2\xi - \Psi^2(\xi)(16\delta_p^2 - \delta^2)}}{\xi^2\cot^2\xi + \Psi^2(\xi)}, \qquad (3.188)$$

where, $\Psi(\xi) = \frac{\psi\xi}{4\pi}\cdot\frac{\xi+0.5\sin 2\xi}{\sin^2\xi}$.

According to (3.187) (or (3.188) in which $\delta = 0$ and $\psi = 0$), the amplitude-frequency characteristic's backbone curve for the relative motion of the bar's impacting cross sections, can be described by the relation,

$$a = -4\delta_p/\xi\cot\xi. \qquad (3.189)$$

The equation for the limiting amplitude line is still determined by equality (3.183).

Fig.3.29, a, d shows the relative motion's amplitude-frequency characteristic for the contacting cross sections of two identical impacting bars which are subjected to a constant force compressing them together. The characteristic is shown in the vicinity of the first vibration mode and for different ratios of excitation and compressive forces.

As can be seen, when subjected to small compressive forces (Fig.3.29, a) the system acts in a similar manner to that of a linear system. However, an increase in compressive force increases the system's resonant frequency. When under static compressive forces $P > F/4$ (Fig.3.29, d), the resonant curve's character changes drastically. Vibration regimes that are accompanied with impacts can only exist in pairs in the frequency region between the bar's natural and antiresonant frequencies.

Comparison between Fig.3.29, d and Fig.3.27, a, d demonstrates that the type of nonlinearity present, depends not only on Δ's sign, but also on the model's structure. In systems that are subjected to a compressive force, the nonlinearity present is only of the soft type. The thin lines in Fig.3.29, a, d show backbone curves while the dashed lines show resonant curve branches that correspond to unstable solutions.

If the relative motion's amplitude a is known, then from expressions (3.178), (3.179) and (3.186), the amplitude of the bars' contacting cross sections' absolute motion can be found. In the absence of dissipation ($\psi = 0$), equations for the contacting cross sections' amplitude-frequency characteristics can be written in the form,

$$a_{10} = \left|\frac{2\delta_p \pm \delta}{\xi\cot\xi}\right|; a_{20} = \left|\frac{2\delta_p}{\xi\cot\xi}\right|. \qquad (3.190)$$

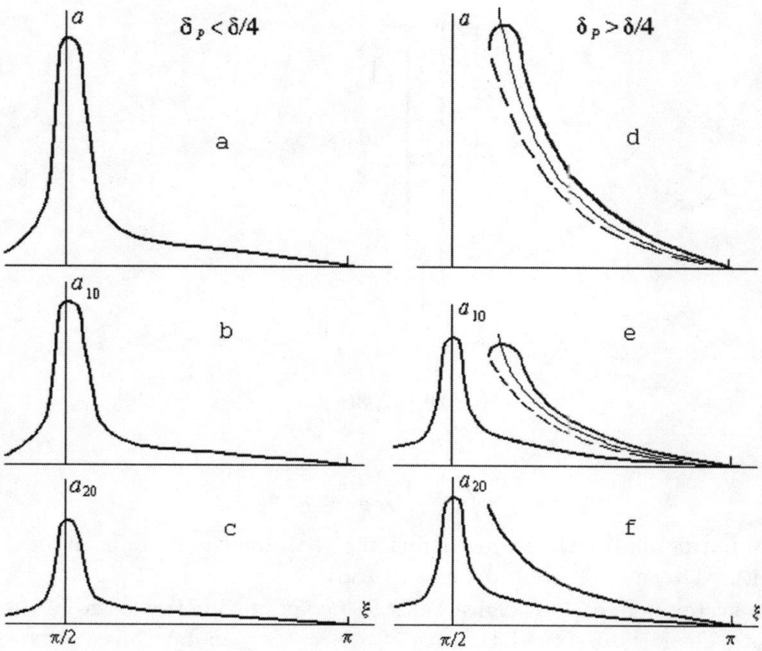

Fig. 3.29.

Fig.3.29 b, c, e, f show the absolute motion's amplitude-frequency characteristic for the bars' contacting cross sections when under the influence of a static compressive force. It should be noted, that when a compressive force is present, passive bar 2's vibration amplitude when in regimes of vibro-impact, is not dependant on the excitation force. It is therefore not dependant on the backbone curve either, and the stable and unstable branches merge (Fig.3.29, e). Moreover, notice the appearance of additional resonances in the bar's absolute motion. These occur at the natural frequencies of the subsystems. In systems with an initial but minimal compression (Fig.3.27, d, e), the resonance encountered is that of the bars' combined length.

5. Consider the vibration of the system shown in Fig.3.30. Bar 1, with length l, collides with a semi-infinite bar 2. The system's compression is maintained by a static force P. The dynamic compliance of a semi-infinite bar which is subjected to an excitation in cross section $x = 0$ was found in subsection 2.2.2. It has the form,

$$L_{0x}^{(2)}(x, j\omega) = -j\frac{c_2}{\omega E_2 S_2} \exp\left[-\left(j + \frac{\psi_2}{4\pi}\right)\frac{\omega x}{c_2}\right]. \tag{3.191}$$

When $x = 0$, the following formula for bar 2's contacting cross section is obtained,

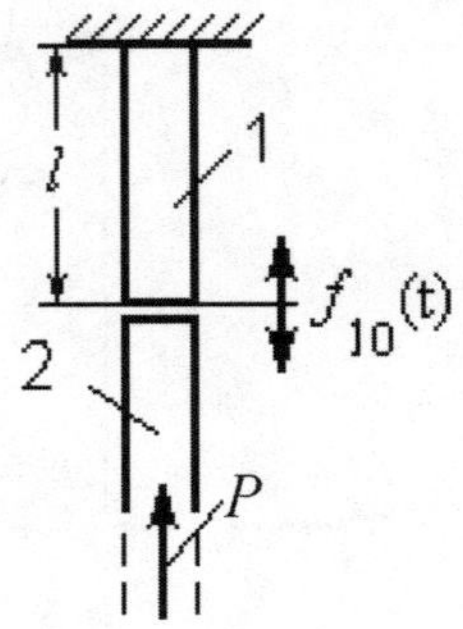

Fig. 3.30.

$$L_{00}^{(2)}(0, j\omega) = -j\frac{c_2}{\omega E_2 S_2}. \tag{3.192}$$

The harmonic linearization coefficient, $q(a)$, for the system under consideration, is determined from formula (3.186).

The system's frequency characteristic, neglecting energy losses in the bar's material, i.e. it is assumed that $\psi_1 = \psi_2 = 0$, can be constructed. From (3.159), (3.161), (3.175), (3.186) and (3.192), and after some transformations, the following expression for the contacting cross sections' relative motion's amplitude, is obtained,

$$a = -\frac{2\delta_p \pm \sqrt{\delta^2 - 4\delta_p^2 \frac{\cot^2 \xi}{w^2}}}{\xi \cot \xi}, \tag{3.193}$$

where, $\delta = Fl_1/E_1S_1$, $\delta_p = Pl_1/E_1S_1$ and $w = w_2/w_1$ is the ratio of wave resistance $w_i = S_i\sqrt{E_i\rho_i}$ $(i = 1, 2)$ within bars 1 and 2.

Fig.3.31, a shows the relative vibration's amplitude-frequency characteristic of the bars' contacting cross sections in the vicinity of active bar 1's first natural frequency.

Analogously, from (3.168) – (3.170), and after the necessary substitutions and transformations, the vibration amplitudes of the bars' contacting faces are found,

$$\begin{aligned} a_{10} &= \left|\frac{\delta}{\xi \cot \xi}\right| \sqrt{1 + \left(\frac{2\delta_p}{\delta}\right)^2 \pm \frac{4\delta_p}{\delta}\sqrt{1 - \left(\frac{2\delta_p \cot \xi}{\delta w}\right)^2}}; \\ a_{20} &= \frac{2\delta_p}{\xi w} \end{aligned} \tag{3.194}$$

Fig.3.31, b, c show the contacting cross sections' amplitude-frequency characteristics for bars 1 and 2. Curves 1 correspond to vibration without separation. This happens at an amplitude which can be determined by applying the limit $k \to \infty$ to expressions (3.168) and (3.169), and when $x = s = 0$ and $\psi = 0$,

$$a_{10} = a_{20} = \delta / \sqrt{(\xi \cot \xi)^2 + (\xi w)^2}. \tag{3.195}$$

As follows from Fig.3.31, two vibration regimes are possible within a specific frequency range. The bars may vibrate as one, with no separation, or the active bar 1 may exhibit resonant vibration, characterized by large amplitude and impacts in the joint between the bars.

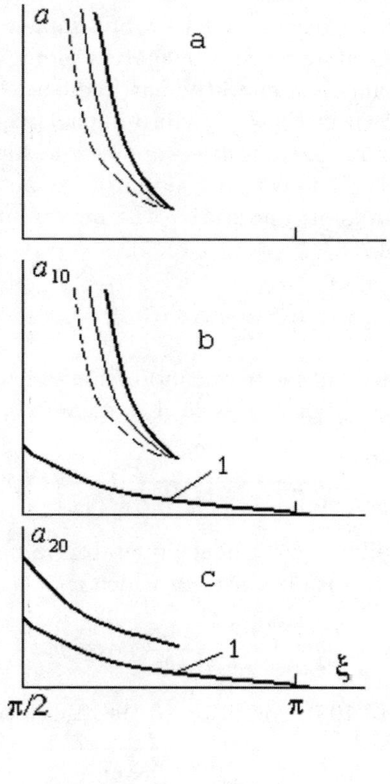

Fig. 3.31.

According to (3.193), vibro-impact regimes exist when the condition $\delta/\delta_p > 2|\cot\xi|/w$ is satisfied. Ambiguity within the amplitude-frequency characteristic occurs in the frequency range $(2k-1)\pi/2 < \xi < k\pi$ ($k = 1, 2, \cdots$ is the active bar's natural vibration mode number) when $\delta < 2\delta_p$. Note that the domains in which vibro-impact regimes exist broaden with an increase in the limiter's relative wave resistance w. If the limit $w \to \infty$ is applied, the situation considered in the previous section is obtained, i.e. collisions occur between a finite bar and a rigid limiter under the action of a constant compressive force.

As in the situation when a bar of finite length experiences a vibro-impact regime under the action of a compressive force, the passive bar's vibration amplitude, and therefore the amount of energy transferred through the joint, is not dependant on the excitation amplitude. This is related to the fact, that the interaction force between the bars during the impacts,

$$f[v(t)] \approx P + 2P \exp j(\omega t - \varphi)$$

is independent of their relative motion. Therefore, the amplitude-frequency characteristic of the passive bar 2 (Fig.3.31, c) is single-valued, and has a single-valued branch corresponding to vibro-impact regimes. The energy transfer due to the travelling wave propagating along waveguide 2 does not limit the active bar's vibration amplitude at frequencies $\xi \to (2k-1)\pi/2$.

In the example shown in Fig.3.31, vibro-impact regimes can only appear as a result of hard excitation. Conditions are sought, for which a soft opening of the joint occurs. It is obvious that the joint does not open until the stress present in bar 2, at the joint, and during the bars' mutual motion, does not exceed the value P/S_2, i.e.,

$$E_2 \frac{\partial u_{2x}^0}{\partial x}\bigg|_{x=0} > \frac{P}{S_2} \tag{3.196}$$

A function u_{2x}^0 describing the bars' motion, for which no separation occurs, can be found from (3.167) when $k \to \infty$. Letting $s = 0$,

$$u_{2x}^0(t) = F_{10} \frac{L_{0x}^{(2)}(j\omega)}{1 + L_{00}^{(2)}(j\omega)/L_{00}^{(1)}(j\omega)} \exp(j\omega t). \tag{3.197}$$

By substituting dynamic compliance operator values (3.175) and (3.192) into (3.197), the following is obtained, in which $\psi_1 = \psi_2 = 0$,

$$u_{2x}^0(t) = -F_{10} \frac{c_2}{E_2 S_2 \omega} \left(1 - j\frac{\cot \xi}{w}\right) \exp(j\omega t - \frac{\omega x}{c_2}). \tag{3.198}$$

Taking account of (3.198), condition (3.196) takes the form,

$$F_{10} > P\sqrt{1 + \left(\frac{\cot \xi}{w}\right)^2}.$$

It is worth noting that the results obtained can be applied, not only to ultrasonic systems, but to numerous structures which have joints and are subjected to vibration. In such structures, the initial compression present in its joints is usually specified in order to ensure that the joint does not open under the influence of the vibration present. However, it is shown in the above calculations that a vibro-impact process can occur even when large compressive forces are present. This may happen as a result of hard excitation stemming from a single sporadic overload. This process can have serious negative, or even catastrophic consequences, and may cause the destruction of a structure's parts or the loss of a seal etc.

3.4 Vibratory displacement of a viscoelastic bar in a media with dry friction

Ultrasonic machine arrangements that are suitable for deep drilling and boring were described in subsection 2.1.2. Significant frictional forces act on the sides of their vibrating tools, which progress forward as the machining processes take place. In order to study these systems, a vibrating object's oscillatory displacement, within a media exhibiting dry frictional characteristics must be considered. An effective reduction in a medium's resistive force due to the tool's vibration, i.e. a vibratory smoothing of nonlinearities (Blekhman (1999), Krasovsky (1948)), arises in such processes. This effect is employed in the building industry for vibratory pile driving, in ultrasonic cutting in which vibration is superimposed on to a cutting tool's motion, in vibro-conveyance, etc. Analysis of the vibration smoothing effect has been carried out, but only for systems with lumped parameters: Andronov (1967*b*, 1970, 1975), Blekhman (1999).

This section is devoted to the study of a viscoelastic bar's motion within a medium which exhibits dry friction. It is under the action of both a constant and a harmonic force which vibrates at the bar's natural frequency. The analysis is conducted using the harmonic linearization method and by separating its fast, vibrational motion, and its slow motion for which it acts as a rigid body. Their influence on each other is studied for steady-state and transient regimes.

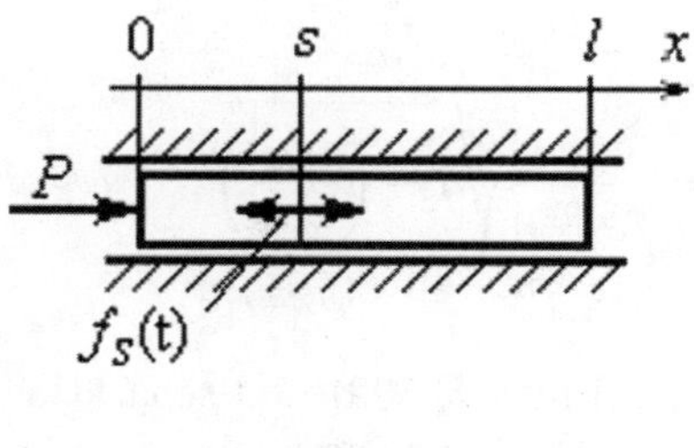

Fig. 3.32.

A viscoelastic bar with length l (Fig.3.32) is subjected to a harmonic force $f_s(t) = F_s e^{j\omega t}$, with a frequency ω, acting at cross section $x = s$. It is also under the influence of a static force P which acts at cross section $x = 0$. The dry frictional force ϑl, with a density of ϑ, is uniformly distributed along the whole of the bar's length, and $P < \vartheta l$. The total motion at cross section x can be determined by summing the translational motion y and the relative displacement $u_x(t)$. The determined motion of the bar is relative to the location of the cross section under consideration, in its undistorted state. The equation of motion for a viscoelastic bar has the form,

$$\rho S\frac{\partial^2 u_x}{\partial t^2} - ES\left(\frac{\psi}{2\pi\omega}\frac{\partial^3 u_x}{\partial x^2 \partial t} + \frac{\partial^2 u_x}{\partial x^2}\right) - \rho S\frac{d^2 y}{dt^2} -$$
$$-\vartheta \operatorname{sgn}\left(\frac{dy}{dt} + \frac{\partial u_x}{\partial t}\right) + F_s e^{j\omega t}\delta(x-s) + P\delta(x), \qquad (3.199)$$

where, E and ρ are the bar material's elastic modulus and density, S is the bar's cross sectional area, ψ is the bar material's energy absorption coefficient and $\delta(x)$ is the Dirac function.

If the relative motion's speed does not change significantly over a vibrational period, it can be assumed that the derivative of the relative motion becomes $dy/dt = \dot{y} = const$.

An approximate solution is sought for cross section x's relative motion. It is required in the form,

$$u_x \approx a_x \exp j(\omega t - \varphi_x), \qquad (3.200)$$

where, a_x and φ_x are the slowly varying amplitude and phase of vibration. Harmonic linearization is performed in order to determine the dry frictional force,

$$\vartheta \operatorname{sgn}\left(\dot{y} + \frac{\partial u_x}{\partial t}\right) \approx \vartheta\left(m_x + b_x\frac{\partial u_x}{\partial t}\right). \qquad (3.201)$$

In the above equation, the slow component m_x, and the coefficient in front of the fast speed b_x, are given by the following relations,

$$m_x = \frac{2}{\pi}\arcsin\frac{\dot{y}}{\omega|a_x|},$$
$$b_x = \frac{4}{\pi\omega|a_x|}\sqrt{1-\left(\frac{\dot{y}}{\omega a_x}\right)^2} \quad (\dot{y} \le \omega a_x), \qquad (3.202)$$
$$m_x = 1, \quad b_x = 0 \quad (\dot{y} \ge \omega a_x). \qquad (3.203)$$

By substituting (3.201) into (3.199) and separating the relative motion's slow and fast components, the following is obtained,

$$\rho S\frac{\partial^2 u_x}{\partial t^2} - ES\left(\frac{\psi}{2\pi\omega}\frac{\partial^3 u_x}{\partial^2 x \partial t} + \frac{\partial^2 u_x}{\partial t^2}\right) =$$
$$= f_s(t)\delta(x-S)e^{j\omega t} - \vartheta b_x\frac{\partial u_x}{\partial t}. \qquad (3.204)$$

By integrating (3.199) over the bar's length and using (3.201) and (3.204), an equation is obtained that describes the bar's slow component of motion as if it were a rigid body,

$$M\ddot{y} = P - \vartheta\int_0^l m_x dx, \qquad (3.205)$$

where $M = \rho l S$ is the bar's mass.

When $\dot{y} \geq \omega a_x$, equation (3.205) by virtue of (3.203) takes the form $M\ddot{y} = P - \vartheta l$. As such, the fast (3.204) and slow components of motion transpire to be independent. The frictional force ϑl only influences the bar's translational (or slow) motion.

By substituting an expression for m_x from (3.202) into equation (3.205), the following is obtained, in which $\dot{y} \leq \omega a_x$,

$$M\ddot{y} = P - \frac{2\vartheta}{\pi}\int_0^l \arcsin\frac{\dot{y}}{\omega|a_x|}dx. \tag{3.206}$$

In this circumstance, the fast and slow motions' influence on each other can be qualitatively estimated. It is assumed that $\dot{y} \ll \omega a_x$, and by writing (3.206) to an approximation of its first order terms,

$$M\ddot{y} + \beta\dot{y} = P, \vartheta = \frac{2\beta}{\pi\omega}\int_0^l \frac{dx}{|a_x|}.$$

Thus, at small speeds of translational motion, and when under the influence of vibration, the dry friction present is transformed. On average, it has the same effect as viscous friction with a coefficient β. This is dependant on the amplitude distribution along the bar. Conversely, the last term in (3.204) demonstrates that the slow motion has an opposite effect on the bar's vibration.

The periodic solution of equation (3.204) is required in the form,

$$u_x = \int_0^l \left[f_s(t)\delta(z-s) - \vartheta b_z\frac{\partial u_z}{\partial t}\right]L_{zx}(j\omega)dz, \tag{3.207}$$

where, $L_{zx}(j\omega)$ is the dynamic compliance operator which relates the displacement of cross section x with the harmonic force applied at cross section z.

Taking expressions (3.200) and (3.202) into account, from (3.207) an equation governed by the vibration's amplitude and phase is obtained,

$$a_x e^{-j\varphi_x} = F_s L_{sx}(j\omega) - j\frac{4\vartheta}{\pi}\int_0^l e^{-j\varphi_z}\sqrt{1-\left(\frac{\dot{y}}{\omega a_z}\right)^2}L_{zx}(j\omega)dz. \tag{3.208}$$

The dynamic compliance operator $L_{sx}(j\omega)$ was found previously in section 2.2.12 (formula (2.164)) and has the form,

$$L_{sx}(j\omega) = \left\{\begin{array}{ll} -\frac{\lambda}{\rho S\omega^2}\cdot\frac{\cosh\lambda x\cosh\lambda(l-s)}{\sinh\lambda l} & 0 \leq x \leq s \\ -\frac{\lambda}{\rho S\omega^2}\cdot\frac{\cosh\lambda(1-x)\cosh\lambda s}{\sinh\lambda l} & s \leq x \leq l \end{array}\right\}, \tag{3.209}$$

where, $\lambda = \left(j + \frac{\psi}{4\pi}\right)\frac{\omega}{c}$ (see (2.49)) and c is the speed of sound within the bar's material.

From now on, the bar's resonant vibration is considered. This occurs at frequencies $\omega_n = n\pi c/l$, in which $n = 1, 2, \dots$ is the vibration's natural mode number.

By expanding the hyperbolic functions in (3.209) into a series of the small parameter ψ, and retaining only the linear terms, the following expression is obtained for the dynamic compliance operator under resonant conditions:

$$L_{sx}(j\omega_n) = -j\frac{4l}{\psi\pi n^2 ES}\cos\frac{\pi ns}{l}\cos\frac{\pi nx}{l}. \tag{3.210}$$

Before solving equation (3.208), it must be noted that the influence of the dissipative forces on the distribution of the system's natural frequencies is weak. Therefore, by considering only resonant motion, it is assumed that in the presence of dry friction, the vibration modes and phases are close to the resonant modes and phases for a viscous-elastic bar's forced vibration without Coulomb friction. This is given by relation (3.210), i.e.,

$$a_x \approx a_0 \cos\frac{\omega x}{c} = a_0 \cos\frac{\pi nx}{l}, \tag{3.211}$$

$$\varphi_x \approx \arcsin\left[\text{sgn}\left(\cos\frac{\pi nx}{l},\right)\right] \tag{3.212}$$

where, a_0 is the unknown variable.

Substituting (3.210) – (3.212) into equation (3.208) the following expression for amplitude a_0 is obtained,

$$a_0 = \frac{4l}{\psi\pi n^2 ES}[F_s\cos\frac{\pi ns}{l} - \frac{4\vartheta}{\pi}\int_0^l \text{sign}\left(\cos\frac{\pi nz}{l}\right) \times$$

$$\times\sqrt{1-\left(\frac{\dot y}{\omega_n a_z}\right)^2}\cos\frac{\pi nz}{l}dz]. \tag{3.213}$$

In order to calculate the definite integral in equation (3.213), a power expansion is used for the radicand (Bronshtein & Semendyayev (1998)),

$$\sqrt{1-\nu_z^2} = 1 - 1/2\nu_z^2 - 1/8\nu_z^4 - \dots, \tag{3.214}$$

where, $\nu_z = \dot y/\omega_n a_z$.

The inequality,

$$\dot y \le \omega_n a_z, \tag{3.215}$$

should be taken into account when calculating the integral in (3.213). This inequality follows from the conditions applied to the harmonic linearization coefficients (3.202) and (3.203). It means that the absolute speeds of the cross sections do not change their sign in the vicinity of the modes' nodes, and the dry friction present in these parts of the bar has no effect on the vibration parameters. By using (3.211), the length of the segment (Fig.3.33) in which condition (3.214) is satisfied, can be found,

$$l_1 = \frac{l}{\pi n} \arccos \frac{\dot{y}}{\omega_n a_0}. \tag{3.216}$$

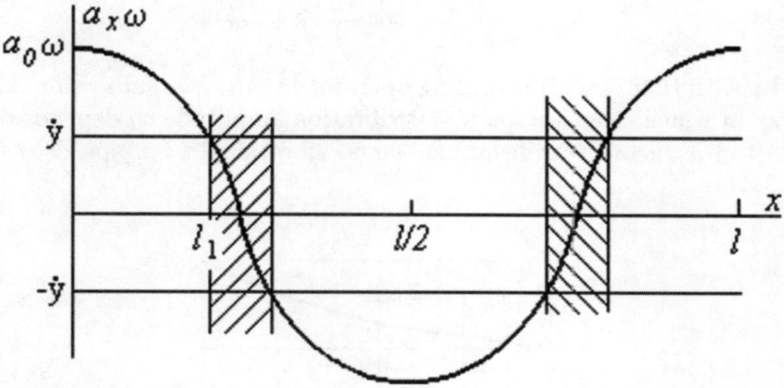

Fig. 3.33.

Due to the symmetry of vibration modes at resonance, the following expression exists for the integral in equation (3.213),

$$\int_0^l \operatorname{sgn}\left(\cos\frac{\pi n z}{l}\right)\sqrt{1-\left(\frac{\dot{y}}{\omega_n a_z}\right)^2}\cos\frac{\pi n z}{l}dz =$$

$$= 2n\int_0^{l_1}\sqrt{1-\left(\frac{\dot{y}}{\omega_n a_z}\right)^2}\cos\frac{\pi n z}{l}dz. \tag{3.217}$$

By substituting (3.214) – (3.217) into the right-hand side of equation (3.213), truncating expansion (3.214) after the first two terms, and with integration and transformations, the following is obtained,

$$a_0 = \frac{4l}{\psi\pi n^2 ES}(F_s\cos\frac{\pi n s}{l} - \frac{8\vartheta l}{\pi^2}[\sqrt{1-(\frac{\dot{y}}{\omega_n a_0})^2} -$$

$$-\frac{1}{2}(\frac{\dot{y}}{\omega_n a_0})^2 \ln\frac{\omega_n a_0}{\dot{y}}(1+\sqrt{1-(\frac{\dot{y}}{\omega_n a_0})^2})]). \tag{3.218}$$

As follows from (3.218), in the absence of slow motion ($\dot{y}=0$), vibration can occur subject to the condition,

$$F_s\cos\frac{\pi n s}{l} > \frac{8}{\pi^2}\vartheta l = 0.81\vartheta l. \tag{3.219}$$

An increase in the slow motion's speed improves vibration conditions and reduces the damping due to dry friction. If the absolute speed's sign does not change for all the bar's cross sections, i.e. speed $\dot{y} > \omega_n a_0$, condition (3.215) is not satisfied along the whole length of the bar and the dry friction present has no effect on vibration. The vibration amplitude,

$$a_x^* = \frac{4F_s l}{\psi \pi n^2 ES} \cos \frac{\pi n s}{l} \cos \frac{\pi n x}{l}$$

coincides with that determined using operator (3.210) (see subsection 2.2.12). The way in which cross section $x = 0$ vibration amplitude a_0 depends on the ratio $\nu_0 = \dot{y}/\omega_n a_0$ of the translational $\dot{y}$ and vibrational $\omega_m a_0$ speeds is shown in Fig.3.34, a.

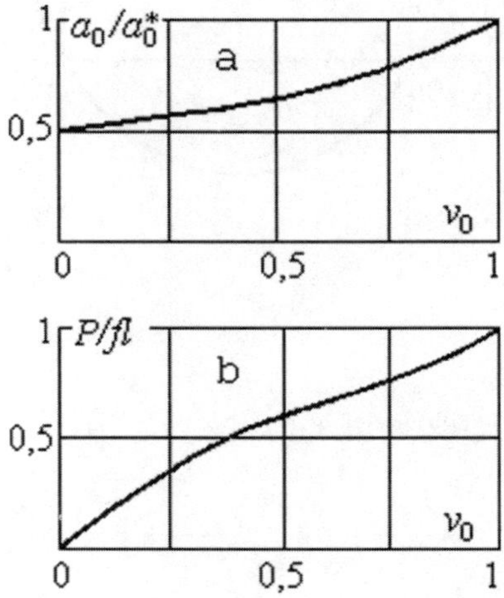

Fig. 3.34.

The slow speed $\dot{y}$ in equation (3.218) can be determined from the general solution of equation (3.206). Let it be denoted that,

$$G = \frac{2\vartheta}{\pi} \int_0^l \arcsin \frac{\dot{y}}{\omega |a_x|} dx. \tag{3.220}$$

Using the expansion (Bronshtein & Semendyayev (1998)),

$$\arcsin \nu = \nu + 1/6\nu^3 + \ldots \tag{3.221}$$

for the integrand in (3.220), taking account of (3.203), (3.211) and (3.216), and after integration, the following is obtained,

$$G = \frac{2}{\pi}\vartheta l \left[\frac{2}{\pi} \frac{\dot{y}}{\omega_n a_0} \ln \frac{\omega_n a_0}{\dot{y}} \left(1 + \sqrt{1 - \left(\frac{\dot{y}}{\omega_n a_0} \right)^2} \right) + \arcsin \frac{\dot{y}}{\omega_n a_0} \right] \tag{3.222}$$

When integrating (3.222), only the first term of series (3.221) was taken into consideration since the second term improves the result by less than an order. Note that the second term in (3.222) corresponds to parts of the bar with length $l_2 = l/n - 2l_1$. These are situated near the vibration nodes, where due to a small vibrational speed, condition (3.215) is not satisfied and the dry frictional force remains at a constant density ϑ.

When $G = P$ in (3.222) and in accordance with (3.206), a relation connecting the parameters of fast and slow motion within a steady-state regime $\dot{y} = const$ is found. This dependence is shown in Fig.3.34, b. It allows the ratio $\nu_0 = \dot{y}/\omega_n a_0$ between translational $\dot{y}$ and vibrational $\omega_m a_0$ speeds, in a steady-state regime, for a given value of $P/\vartheta l$, to be determined. Using (3.218), the vibration amplitude a_0 at the bar's end may then be found. Consequently, the amplitude distribution (3.211) along the bar's length may also be found.

Thus, the vibration of a distributed parameter system within a medium in which dry friction is present, is accompanied by a smoothing of the non-linear frictional characteristic. Using this fact, transient processes may be constructed, in which relation (3.222) is used to account for the slow motion's simplified frictional characteristic. Since the curve shown in Fig.3.34, b is close to being linear, equation (3.205) can be written in the form,

$$M\ddot{y} + \frac{fl}{\omega_n a_0}\dot{y} = P. \tag{3.223}$$

In this equation, according to relation (3.218) (Fig.3.33), variable a_0 is a function of the form $a_0 = a_0(\dot{y})$. By solving the equation system (3.218) and (3.223), the required slow transient processes $\dot{y}(t)$ and $a_0(t)$ can be obtained. Fig.3.35 shows the transient process $\dot{y}(t)$ and the vibration envelope $a_0(t)$ for the following system parameters: $f = 2\pi/\omega = 20.4kHz$, $l = 0.507m$, $n = 4$, $S = 3.14 \times 10^{-4}m^2$, $\psi = 0.05$, $F_0 = 39kg$, $P = 20kg$, $\vartheta l = 24kg$ and $a_0^* = 5\mu m$.

The transient process limiting values can be obtained by substituting constant value for amplitude a_0 (corresponding to the initial $\dot{y} = 0$ and steady-state values of translational speed $\dot{y}$) into (3.223). The solution of (3.223) takes the form,

$$\dot{y} = \frac{P}{\vartheta l}\omega_n a_a(1 - \exp bt), \tag{3.224}$$

where, $b = \frac{\vartheta l}{M\omega_n a_0}$ is the medium's resistance coefficient to the bar's motion (the bar is considered as a rigid body). The resistance coefficient is transformed due to the vibration present.

Expression (3.224), determining the amplitude a_0 values given above, describes the boundary curves surrounding the real transient process.

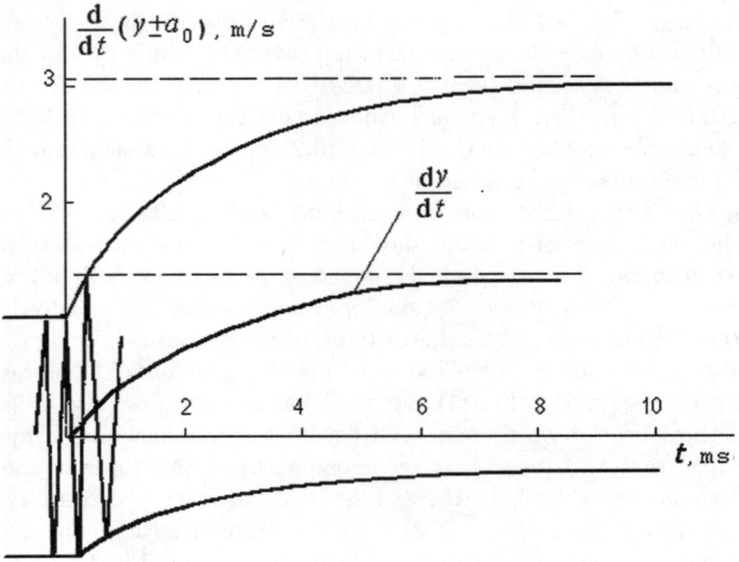

Fig. 3.35.

4

The resonant tuning of ultrasonic machines

Take but degree away, untune that string,
And hark what discord follows ! Each thing meets
In mere oppugnancy.

William Shakespeare (1564-1616): Troilus and Cressida

4.1 Methods of increasing the efficiency of ultrasonic machines when under load

1. Various arrangements of ultrasonic machining processes were discussed in detail in section 1.1. The same section contains descriptions of the process' main features as have been found experimentally. Experiments have established (Averianova et al. (1962), Averianova & Milovidov (1964), Rosenberg & Kazantsev (1959*a*,*b*), Rosenberg et al. (1964)) that ultrasonic machining takes place as the result of a tool's impact against abrasive particles contained within a slurry. These abrasive particles are forced onto the workpiece's surface and effectively erode it. Extensive experimental evidence is reviewed in Kazantsev et al. (1966), Markov (1980), Neppiras (1956), Rosenberg & Kazantsev (1959*a*). This allows the influence of both the drive's static force and the main vibration parameters on the cutting speed to be estimated. It was found that in order to achieve a high output capacity, a large vibration amplitude is required at the tool, and a considerable driving force must be provided.

In order to excite vibration at maximal amplitude, the vibrating system should be tuned to resonance. This is usually undertaken when the system is under idling conditions, and the assumption is usually made that the tool-workpiece interaction does not significantly influence the system's operation. Conversely, investigations of an ultrasonic machine's characteristics when under load (Astashev (1972), Astashev & Babitsky (1972, 1981, 1982*b*)) show that by increasing the driving force at the idling motion's resonant frequency, during the tool's working process, leads to a decrease in the tool's vibration amplitude. This is due to nonlinear distortions and is caused by the tool-workpiece interaction during the cutting process. The corresponding nonlinear effects are discussed in section 3.2.

In this section, the machine's output capacity is related to its design, and dynamic characteristics. A rheological model is used to describe the ultrasonic cutting process. Optimum methods of tuning the vibrating system are considered and experimental results are discussed. The results obtained are compared with experimental data.

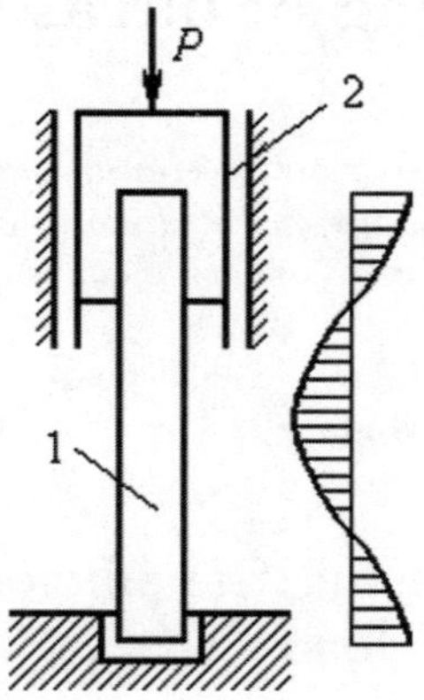

Fig. 4.1.

The cutting process main features can be obtained by considering the vibration system shown in Fig. 4.1. The system is forced against the workpiece by a static force P. Consider the tool's motion in the form,

$$u(t) = vt + u^0(t) = vt + a \exp j(\omega t - \varphi)$$

where, v is a small constant speed that describes the system's motion as a result of penetration, a, ω and φ are the amplitude, frequency and phase of the tool motion's periodic component $u^0(t)$.

The interaction force between the tool and workpiece is described by a rigid-plastic material characteristic (Fig.4.2),

$$f(u) = 1/2D\eta(u - \Delta)(1 + \operatorname{sgn} \dot{u}), \tag{4.1}$$

which, after harmonic linearization, is approximately equal to,

$$f(u) \approx f_0(a) + [k(a) + j\omega b(a)]u^0. \tag{4.2}$$

The value of D in expressions (4.1) and (4.2) is dependant on the workpiece's material, the tool's cross sectional area, the type of abrasive slurry used, and the conditions of the abrasive slurry in the cutting zone; Δ is the value of coordinate $u(t)$ at the beginning of interaction, $\eta(u)$ is the unit step-function, $f_0(a), k(a)$ and $b(a)$ are harmonic linearization coefficients. These coefficients were calculated in section 3.1. In order to simplify the analysis, the following arguments are used: since, in the process under consideration,

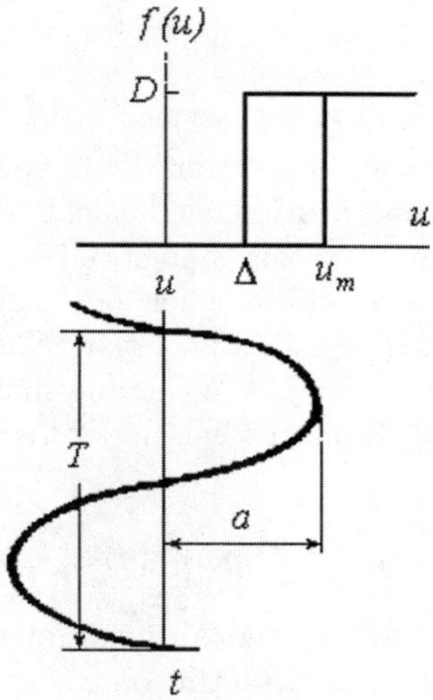

Fig. 4.2.

the tool motion's constant speed v is much smaller than its vibration speed $a\omega$, i.e. $v \ll a\omega$, the system's slow motion during one cycle of vibration can be neglected, and the harmonic linearization coefficients can be found from (3.24)–(3.26) in which $v = 0$. As a result,

$$f_0(a) = \frac{D}{2\pi} \arccos \frac{\Delta}{a} \tag{4.3}$$

$$k(a) = \frac{D}{\pi a}\sqrt{1 - \left(\frac{\Delta}{a}\right)^2}, \quad b(a) = \frac{D}{\pi a\omega}\left(1 - \frac{\Delta}{a}\right) \tag{4.4}$$

Taking account of the relation $f_0(a) = P$, from (4.3), it is found that,

$$\Delta / a = \cos(2\pi P/D) \tag{4.5}$$

and after substituting (4.5) into (4.4),

$$k(a) = \frac{D}{\pi a} \sin \frac{2\pi P}{D}, b(a) = \frac{2D}{\pi a\omega} \sin^2 \frac{\pi P}{D} \tag{4.6}$$

Coefficients (4.6) determine the elastic and dissipative components of the load applied to the vibrating system during the cutting process. Their dependence on amplitude a and static force P is in good agreement with that obtained experimentally in Ganeva et al. (1981).

The difference $a - \Delta$ represents the penetration over a cycle (Fig.4.2). Therefore, the cutting speed is given by, $v = (a - \Delta)\omega/2\pi$. By using (4.5), it is found that,

$$v = \frac{a\omega}{\pi} \sin^2 \frac{\pi P}{D} \tag{4.7}$$

It is noted that relations (4.6) and (4.7) coincide with formulae (3.70) and (3.71) which were obtained in subsection 3.1.6 from the complete dynamic characteristic's harmonic linearization coefficients. They were obtained under the assumption that the tool's progression speed was small, $v \ll a\omega$.

The tool's vibration amplitude a when performing work is significantly dependant on the load acting on it. Suppose that when idling ($P = 0$), the tool performs vibration $u_0(t) = a_0 \exp j\omega t$. Assuming that the machine's vibrating system is linear, the tool's vibration when under load can be described by the following equation:

$$u(t) = u_0(t) - W^{-1}(j\omega)[k(a) + j\omega b(a)]u(t) \tag{4.8}$$

where, $W(j\omega)$ is the vibrating system's dynamic stiffness which relates the displacement of the tool to the forces acting on it. Let the following be denoted,

$$W(j\omega) = U(\omega) + jV(\omega) \tag{4.9}$$

During idling motion, the tool's vibration amplitude a achieves its maximum value $a_0 = a_0^*$ at resonant frequency $\omega = \omega_0$, $U(\omega_0) = 0$. The quantity $a_0^* V_0$, in which $V_0 = V(\omega_0)$, characterizes the system's excitation level. The idling motion's amplitude at frequencies in the vicinity of resonance is given by,

$$a_0 = a_0^* V_0 / |W(j\omega)| \tag{4.10}$$

Equation (4.8), together with (4.9) and (4.10), yields the following vibration amplitude expression for the system when under load,

$$a = \frac{a_0^* V_0}{\sqrt{[U(\omega) + k(a)]^2 + [V(\omega) + \omega b(a)]^2}} \tag{4.11}$$

Some ultrasonic machine tuning methods are now considered in which the system is under load.

2. The majority of modern ultrasonic machines are tuned at their idling frequencies $\omega = \omega_0$. This type of tuning restricts their use to small compressive forces $P \ll D$. According to (4.6), an initial approximation can be made in which $k(a) = 2P/a$. It is assumed that the load's dissipative component can be neglected due to its small value in comparison to the losses experienced in the vibration system ($\omega_0 b \ll V_0$). Under these assumptions, and from (4.11), the tool's vibration amplitude, when under load, is found,

$$a = a_0^* \sqrt{1 - \left(\frac{2P}{a_0^* V_0}\right)^2}. \tag{4.12}$$

Expression (4.7) is expanded in a small parameter (P/D) power series and truncated to its first significant terms. By using (4.12), an expression is obtained which determines the tool's cutting speed under load at the idle motion's resonant frequency, $\omega = \omega_0$,

$$v = \pi a_0^* \omega_0 \left(\frac{P}{D}\right)^2 \sqrt{1 - \left(\frac{2P}{a_0^* V_0}\right)^2}. \tag{4.13}$$

Formula (4.13) proves the existence (as is known from experimentation) of an optimal driving force $P = P_0$. This driving force corresponds to maximum cutting speed (Markov (1962), Neppiras (1956), Rosenberg et al. (1964)), and is found by applying the condition $dv/dP = 0$. The following is obtained,

$$P_0 = 0.4 a_0^* V_0. \tag{4.14}$$

The maximum cutting speed which corresponds to force (4.14) is,

$$v_0 = 0.3 a_0^{*3} \omega_0 (V_0/D)^2. \tag{4.15}$$

It is noted that the proportionality between the optimal driving force P_0, and idle motion resonant amplitude a_0^*, is in good agreement with the experimental data reported in Markov (1962), Rosenberg et al. (1964). Paper Kazantsev et al. (1966) validates experimentally that force P_0 is independent of D. D is proportional to the tool's area. An empirical formula obtained in Kazantsev et al. (1966), Markov (1980), Goliamina (1979) yields, by virtue of (4.14), that the cutting speed is dependant on the amplitude a_0^* and frequency ω_0. This is similar to (4.15).

The cutting speed's dependence on compressive force is shown in Fig.4.3, a for different values of amplitude a_0^* (1-8, 2-11, 3-16, 4-20.5 μm). This is calculated from formula (4.3). The experimental results (shown by dots), and initial data for use in calculations, are taken from paper Neppiras (1956). Values $V_0 = 1800 N/mm$ and $D = 3700 N$ are calculated using (4.14), (4.15) and an experimental point which corresponds to the optimal compressive force. The graph in Fig.4.3, b is plotted in accordance with (4.12). It shows the decrease in vibration amplitude as the compressive force increases. It is noted that at the optimal value of compressive force P_0, the vibration amplitude is given by,

$$a_0 = 0.6 a_0^*. \tag{4.16}$$

Consider formula (4.15) which determines a machine's maximum output capacity. When designing a vibrating system, an excitation frequency is chosen that lies in the low ultrasonic range ($f = \omega/2\pi = 18 - 22 kHz$). The limiting amplitude value a_0^* is determined by the waveguide's strength. Therefore, ω_0 and a_0^* are considered as initial values. Parameter V_0 is determined by the vibrating system's dissipative properties. During idling motion, the dissipated power is equal to $N_x = \frac{1}{2} a_0^{*2} \omega_0 V_0$. Thus, for the tuning method under consideration, an increase in cutting speed can be achieved by increasing the vibrating system's dissipative loses as characterized by value V_0. This leads to an increase in power, and is confirmed by industrial ultrasonic machine data.

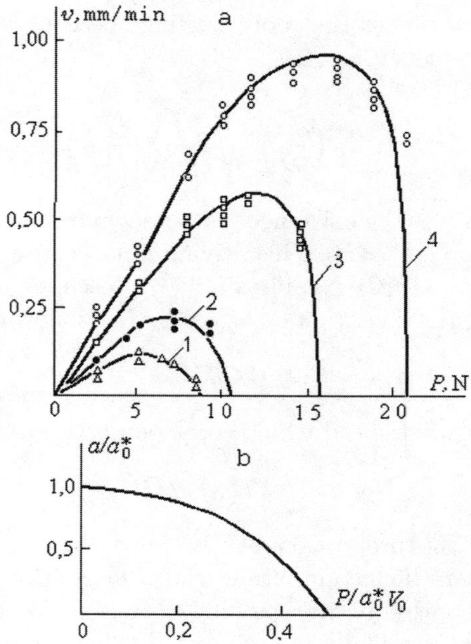

Fig. 4.3.

Some technical specifications of two industrial ultrasonic machines, models 4770 and 4772, are given in Table 4.1.

Table 4.1.

Machine model		4770	4772	A R
Power of the generator		0.25	1.5	0.15
Maximal driving force P, N		45	150	600
Treated area, S, mm^2	Driving force, P, N	Output capacity, mm^3/min		
80	30	280	430	480
80	100	-	2800	3900
80	200	-	-	9000
800	30	28	48	48
800	100	-	280	470
800	150	-	720	1200
800	300	-	-	3800
800	600	-	-	9000

The machines' output capacities are given for various values of area treated S and driving force P.

An ultrasonic machine's Q-factor is now estimated for the vibrating system tuning method under consideration. The Q-factor is determined by the formula $\eta = N_p/(N_p + N_k)$, where, $N_p = Dv$ is the useful power supplied to the cutting process, and $N_k = 1/2a^2\omega_0 V_0$ is the power dissipated in the vibrating system. At the optimal driving force P_0, and with the tuning method chosen, by using (4.15) and (4.16), it is found that $\eta = \frac{1}{1+(0.6D/a_0^* V_0)}$. By using data corresponding to the experiment described above (see Fig.4.3, a), an example of a machines Q-factor may be given. At an amplitude $a_0^* = 20.5\mu m$, it is found that $\eta \approx N_p/N_x = 0.017 = 1.7\%$. Thus, for usual tuning methods, power is mostly dissipated within the vibrating system as a result of its material's internal friction.

In order to estimate the dependence of V_0 on a vibrating system's construction, size and material, the model shown in Fig.4.1 is considered. The vibrating system consists of a homogeneous elastic bar 1, excited in one of its natural modes, and fixed in a housing 2 at one of its nodal points. (Fig.4.1 shows a displacement diagram of the bar's cross-sections). The length of a bar that has a resonant frequency ω_0 is determined by the relation $l = \pi nc/2\omega_0$. $c = \sqrt{E/\rho}$ is the longitudinal wave propagation speed, E and ρ are the elastic modulus and density of the bar's material, and n is the number of quarter-waves in the length of the bar in which odd values of n correspond to constructions where the bar is joined to its housing by its upper end.

The dynamic stiffness (4.9) of such a system can be found by the method discussed in section 2.2, and can be written in the form,

$$W(j\omega) = w\omega[\tan \tfrac{\pi n}{2} \left(1 - \tfrac{\omega}{\omega_0}\right) + j\tfrac{\psi}{4\pi} \tfrac{\pi n\omega/\omega_0 - (-1)^n \sin(\pi n\omega/\omega_0)}{1+(-1)^n \cos(\pi n\omega/\omega_0)}] \tag{4.17}$$

where, $w = S\sqrt{E\rho}$ is the wave resistance, S is the bar's cross sectional area, and ψ is the absorption coefficient.

From (4.17), when $\omega = \omega_0$, it is found that,

$$V_0 = |W(j\omega_0)| = 1/8\psi w\omega_0 n \tag{4.18}$$

By substituting (4.18) into formula (4.15), the maximum achievable cutting speed is obtained for a bar tuned under idling motion conditions,

$$v_0 = 3.75 \cdot 10^{-2}(a_0^*\omega_0)^3(\psi nw/D)^2$$

The last relation shows that an increase in the processes output capacity can be achieved in several ways. The absorption coefficient ψ can be increased i.e. the vibrating system's Q-factor may be reduced. The number of quarter-waves n in the bar may be increased i.e. the vibrating system's length may be increased thereby using higher natural modes of oscillation. Finally, the wave resistance w may be increased by enlarging the bar system's cross sectional area, increasing its elastic modulus, or increasing the material's density. All of the above is validated by ultrasonic machine design experience.

3. It is noted, in connection with expressions (4.7) and (4.12), that a decrease in cutting speed, when driving forces exceed value (4.14), is caused by an abrupt decrease in the tool's vibration amplitude (Fig.4.3, b). This can be avoided by tuning the vibrating system to resonance when under load. According to (4.11), the maximum tool vibration amplitude, when under load, is achieved by satisfying the following relations:

$$U(\omega) + k(a) = 0, a = a_0^* V_0/[V(\omega) + \omega\beta(a)]$$

By the substitution of coefficients (4.6), the following equation system is obtained,

$$\begin{aligned} a &= -\frac{D}{\pi U(\omega)} \sin \frac{2\pi P}{D} \\ a &= a_0^* \frac{V_0}{V(\omega)} \left(1 - \frac{2D}{\pi a_0^* V_0} \sin^2 \frac{\pi P}{D}\right) \end{aligned} \tag{4.19}$$

The solutions of these equations determine the resonant frequency ω^* and corresponding amplitude a^* when the system is under the action of the static compressive force P. The first equation determines the amplitude-frequency characteristic's backbone curve, the second determines the limiting amplitude line.

The advantages of this tuning method can be estimated by considering the system shown in Fig.4.1. Using (4.9) and (4.17), equations (4.19) take the form,

$$\begin{aligned} a &= -\frac{D}{\pi w \omega} \sin \frac{2\pi P}{D} \cot \frac{\pi n}{2} \left(1 - \frac{\omega}{\omega_0}\right), \\ a &= \frac{a_0^*}{2} \frac{\pi n \omega_0}{\omega} \left(1 - \frac{2D}{\pi a_0^* V_0} \sin^2 \frac{\pi P}{D}\right) \times \\ &\quad \frac{1+(-1)^n \cos(\pi n\omega/\omega_0)}{(\pi n\omega/\omega_0)-(-1)^n \sin(\pi n\omega/\omega_0)} \end{aligned} \tag{4.20}$$

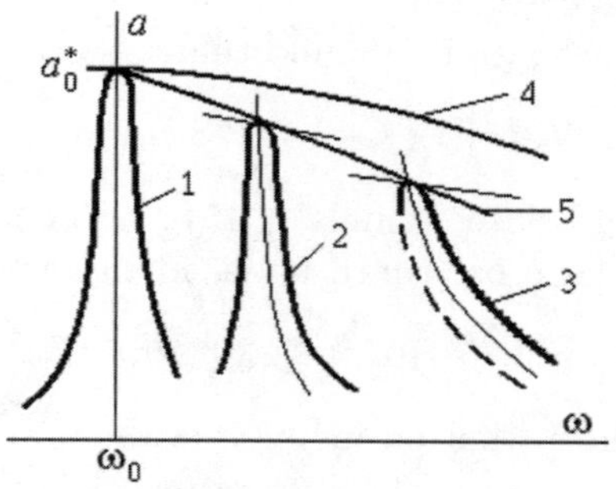

Fig. 4.4.

The graphical solution of equation system (4.20), for different values of P, is shown in Fig.4.4. The curves corresponding to equations (4.20) are shown by thin lines. The resonant curve sections also included in this figure are analogous to those described in section 3.2. When under the action of relatively small compressive forces, $P \leq a_0^* V_0/2$, the shapes of the resonant curves when under load (curve 2) or during idling motion (curve 1) are the same. It is noted however that their maximum values are shifted to higher frequencies as P is increased. A further increase in compressive force ($P > a_0^* V_0/2$) drastically changes the resonant curve's character (curve 3) and leads to the appearance of an unstable branch. This is shown by a dashed line.

Curve 4 is the limiting amplitude line for a system with a rigid limiter (see section 3.2); its characteristics can be obtained from the second equation in (4.20) by applying the limit $D \to \infty$. Line 5, which is formed from the solutions of equation system (4.20), is the resonant curve's envelope curve. It determines all of the resonant regimes possible when the system is subjected to differing driving force values P.

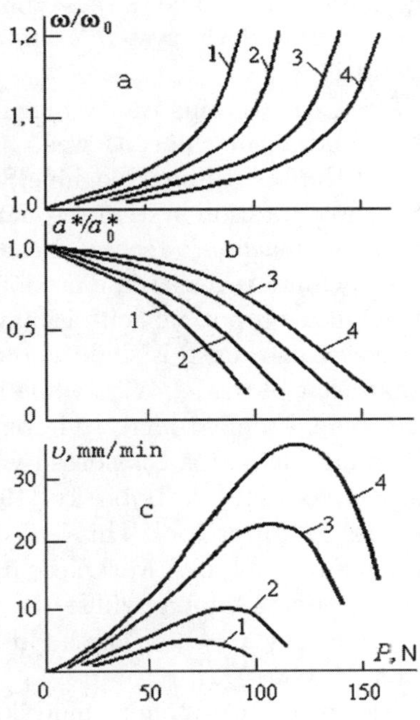

Fig. 4.5.

Fig.4.5 shows how the resonant frequency ω^*, amplitude a^*, and cutting speed v depend on the driving force P and amplitude a_0^*. The results refer to a system in which resonant tuning is performed when the system is under load and the notation is the same as that used in Fig.4.3, a. The dependencies shown in Fig.4.5, a, b are plotted from the results obtained by solving equation system (4.20). The same values of $V_0 = 1800N/mm$ and $D = 3700N$ were used, as were used in Fig.4.3, a, and the absorption coefficient is assigned the value $\psi = 0.05$. Fig.4.5, b shows graphs that are plotted using formula (4.7). Fig.4.5 shows that the maximum cutting speed occurs at a specific driving force value. In this regime the vibration amplitude is equal to $a = a_0^*/2$. A further increase in driving force leads to an abrupt decrease in vibration amplitude and a reduced cutting speed.

Comparisons between Fig.4.3, a and Fig.4.5, b illustrate how an increase in efficiency is attainable when the machine's resonant tuning is performed when under load. Calculations show that the machine's efficiency, when these tuning methods are employed can reach values of up to 30%. This is very beneficial, although serious difficulties are encountered in the practical realization of this forced vibration regime due to the resonant curve's distortion (Fig.4.3) as caused by the load's nonlinearity

It must be noted, albeit a little in advance, that the automatic tuning system, considered in detail in section 4.3, allows these difficulties to be avoided. When this tuning system is employed, the system's only stable state transpires to be that of resonant vibration.

The last column in Table 4.1 contains the technical characteristics of an experimental prototype ultrasonic machine. It was designed in the laboratory of vibration systems of the Institute of Machine Studies (Moscow, Russia). The machine implements vibration system resonant tuning when under load. Comparison between this machine's technical characteristics and those of commercially available machines reveals some pronounced advantages. For example, the Institute machine's output capacity is 30 times larger than that of model 4770, while the power consumed is slightly lower. The power it consumes is 10 times less than that of model 4772, while its capabilities are far in advance, allowing more than a tenfold increase in output capacity.

4. Finally, one more tuning method is considered when the system is under load. It was proposed in Astashev & Babitsky (1982*a*) and is based on attaching an additional mass M to the tool. This leads to a compensation of the detuning experienced due to the tool – workpiece interaction. In order to achieve this compensation, resonant tuning whilst the system is under load, is required to occur at a frequency $\omega = \omega_0$. The additional mass required is determined by the relation $M = k(a)/\omega_0^2$.

The vibration amplitude when the system is under load can be found from the second equation in (4.19),

$$a^* = a_0^* \left(1 - \frac{2D}{\pi a_0^* V_0} \sin^2 \frac{\pi P}{D}\right) \tag{4.21}$$

By substituting (4.21) into (4.7), the cutting speed is obtained,

$$v = \frac{a_0^* \omega}{\pi} \sin^2 \frac{\pi P}{D} \left(1 - \frac{2D}{\pi a_0^*} \sin^2 \frac{\pi P}{D}\right) \tag{4.22}$$

From the condition $\partial v/\partial P = 0$, the compressive force P^* which corresponds to the maximum cutting speed can be found,

$$P^* = \frac{D}{\pi} \arcsin \sqrt{\frac{\pi a_0^* V_0}{4D}} \tag{4.23}$$

Usually, in real conditions, $a_0^* V_0 \ll D$, and formula (4.22) takes the form,

$$P^* = \sqrt{a_0^* V_0 D/4\pi}$$

According to (4.22) and (4.23), the maximum cutting speed is given by,

$$v^* = \frac{1}{8} a_0^{*2} \omega_0 V_0/D \tag{4.24}$$

and the tool's vibration amplitude (4.21) in this regime is equal to,

$$a^* = a_0^*/2 \tag{4.25}$$

For the system considered above (Fig.4.3, a), (4.24), in which $a_0^* = 20.5\mu m$, gives $v^* = 49mm/min$. This cutting speed is attained for a compressive force of $P^* = 104N$.

It is interesting to note that when the machine works at maximum output capacity, the power N_p supplied for material fracture is equal to the power N_k dissipated in the vibration system, i.e.

$$N_p = N_k = 1/8 a_0^{*2} \omega_0 V_0 \tag{4.26}$$

This can easily be shown. Indeed, using (4.6) and (4.7) it is found that $N_p = \frac{1}{2}(a\omega)^2 \beta(a) = Dv$. Taking into account that $N_k = \frac{1}{2} a^{*2} \omega_0 V_0$, and using (4.24) and (4.25), equality (4.26) can be found. Thus, the tuning method under consideration provides the best matching between the vibration system and both the elastic and dissipative components of load. It allows the ultrasonic machine's optimal capabilities to be realized. Its efficiency is equal to $\eta = 50\%$. Relations (4.24) and (4.26) are convenient for designing a tool's vibration system, for which an output capacity is prescribed.

An increase in driving force leads to an increase in Q-factor (4.23) but a decrease in cutting speed. As such, the Q-factor can not be taken as an indicator of how efficiently a machine is tuned. It is convenient to estimate the tuning efficiency using the efficiency coefficient $\phi = v/v^*$ (Astashev et al. (2000)). This is given by the ratio between the attained cutting speed v and its limiting value v^* as given by equality (4.24). For example, according to formulae (4.15) and (4.24), the tuning efficiency coefficient during idling motion is given by,

$$\phi = v_0/v^* = 2.4 a_0^* V_0/D$$

The quantity $a_0^* V_0$ characterizes the system's excitation level. It has the same order of magnitude as the small dissipative forces. In ultrasonic vibration systems within the power range $0.1 - 10kW$, this quantity has a value of the order $a_0^* V_0 = 10 - 10^4 N$. The values of D are in the region of $D = 10^3 - 10^6 N$. Thus, idle motion tuning is very inefficient: in the case under consideration $\phi = 0.024$. This means that in this instance, only $\approx 2.5\%$ of the machine's potential capability is being used. At the same time, tuning the machine to resonance whilst under load allows an efficiency coefficient of $\phi \approx 0.7$ to be attained. Moreover, the efficiency coefficient can be increased to a value closer to $\phi = 1$ by the introduction of the compensation detailed above.

By comparing dependencies (4.24) and (4.15) it can be seen that tuning machines when under load can be especially efficient for the machining of stiff workpieces whilst using low power machines with high Q-factor vibration systems (the workpiece is referred to as stiff if it either has a large treatment area, or is manufactured from a material which is hardly suitable for cutting). A similar conclusion was made in paper Astashev & Sakaian (1967) based on experimental evidence.

Note that the results obtained above can be used also to estimate the efficiency and to tune various ultrasonic machines (Goliamina (1979)) for the surface hardening of components, the welding of plastics and synthetic fabrics, the excitation of a metal-cutting tool's high-frequency vibration as used in machines for vibration cutting (see Chapter 5), etc.

5. The results gained from an experimental study of an ultrasonic machine's vibration system dynamics are now described. The system was analysed when under load. The experimental aim was to find the main dynamic characteristics of an ultrasonic machine's vibration system. It was also required in order to validate assumptions made in the previous theoretical analysis, and to estimate the influence of the tool – workpiece interaction's nonlinearities on the vibration system.

The experiments were performed on the ultrasonic machine detailed in Fig.4.6. The main component of the machine is its acoustic head. This consists of a housing 1, a magnetostrictive transducer 2, a step-type concentrator 3 which is soldered to the face of the transducer, and a tool 4. The magnetostrictive transducer's winding is being fed by a generator which supplies a power of up to $1kW$. The generator used during the experiments allows a wide range of control over both the excitation frequency and the transducer core's vibration amplitude. Control of the systems amplitude was maintained by adjusting the current in the transducer's winding.

Vibration is transmitted between the transducer and the tool via a waveguide-concentrator. This has the effect of increasing the vibration amplitude. The workpiece is a glass plate 5 and is situated under the tool. It has a diameter of $12mm$, a thickness of $5mm$, and is attached to the stiff massive base 6. The plate 5 and tool 4 are placed in a tray 7 with a suspension slurry.

The cutting process' continuity is provided by pressing the tool against the workpiece with a static driving force. This force is generated due to the weight difference between the head and the bobs 8 connected to it. In order to

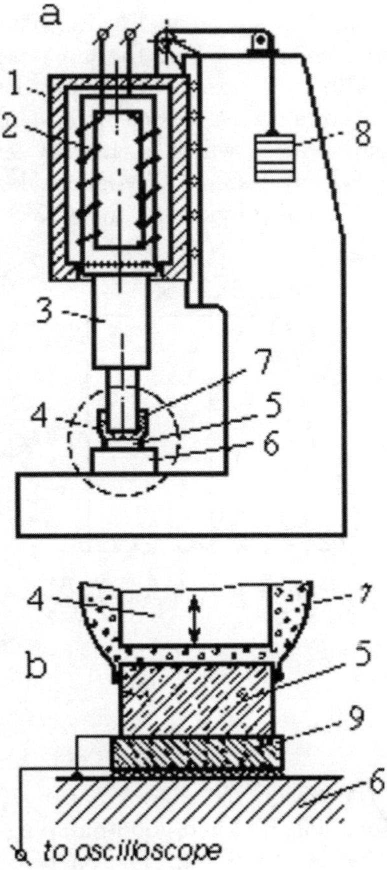

Fig. 4.6.

ensure that the head has a smooth motion, it is mounted within ball bearing slides.

The interaction force between the tool and workpiece was recorded during the cutting process. The force sensor's arrangement is shown in Fig.4.6, b. The sensor's piezoelectric element 9 is glued between the base 6 and glass plate 5. The bottom surface of the piezoelectric element is insulated from the base by a thin caprone net which also acts as a damper. The sensor's top surface is earthed, and the sensor's signal is taken from the bottom surface. The signal is inputted into an oscilloscope. Such a construction provides good damping of the sensor's natural vibration and reliable protection from electromagnetic noise and capacitance induced by the vibrating tool.

The sensor's natural properties were obtained by taking readings of balls impacting on the sensor's plate. They were dropped from various heights.

One of the graphs obtained is shown in Fig.4.7. The graph's upper curve represents the signal that emerges during the ball's impact and the sensor's subsequent natural vibration. The graph was taken with a delayed sweep and was triggered by the sensor's signal. As such, the curve doesn't start from zero, but from a nonzero value u which is the first level recorded after the triggering delay. The lower curve shows the signal taken from a generator. It has a frequency of $200kHz$ and provides a time scale.

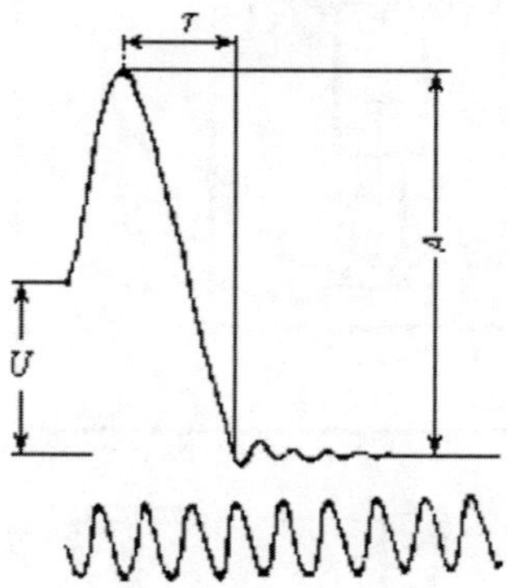

Fig. 4.7.

The tests show that the sensor (Astashev & Sakaian (1967)) has a high natural frequency (exceeding $300kHz$) and good damping properties. Its decrement in natural vibration is equal to 2. The analysis of many oscilloscope readings has proven that the sensor has a linear nature and a sufficient accuracy. For example, the impact duration is not dependant on the height h from which the ball is dropped, and the amplitude of the signal recorded during the impact is proportional to the ball's speed at the beginning of the impact. This means that the sensor behaves as a stiff spring with a linear elastic characteristic. The sensor's calibration is therefore easily facilitated.

In order to calibrate the sensor, the momentum theorem is used. For an impacting body dropped from a height H without initial velocity,

$$m\sqrt{2gH}(1+R) = \int_0^{2\tau} F(t)dt \tag{4.27}$$

where, m is the ball's mass, $R = \sqrt{h/H}$ is the speed restitution factor during the impact, h is the height which the ball bounces to after the impact, $F(t)$ is the impact force's time-dependence, and 2τ is the duration of impact. It must be noted that in the experiments conducted, the restitution coefficient was equal to $R \approx 1$.

Assuming that the sensor has a linear characteristic, it is accepted that,

$$F(t) = k_F A \sin \frac{\pi}{2\tau} t$$

where, A is the amplitude of the sensor's signal during the impact, k_F is the force's scale factor, and from (4.27) it is found that,

$$A\tau = \frac{\pi m \sqrt{2gH}}{4k_F}(1 + R)$$

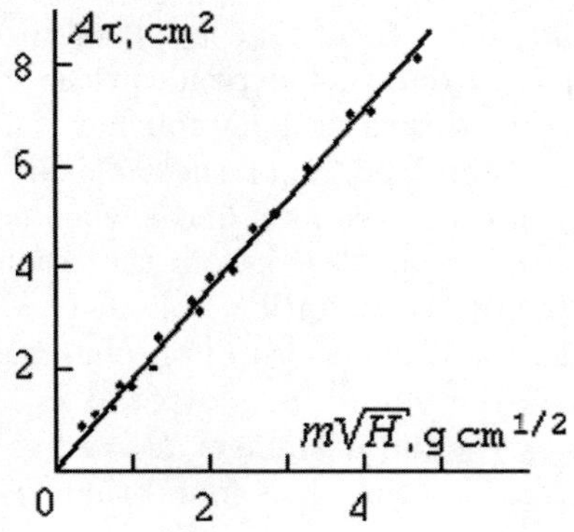

Fig. 4.8.

As a consequence, the product $A\tau$ is a linear function of the variable $m\sqrt{H}$. The graph of this function is plotted in Fig.4.8. It results from the analysis of many oscilloscope readings taken for balls of various masses impacting on the sensor when dropped from various heights. It can be seen from the graph that all the experimental points fall on a straight line with sufficient accuracy. This proves that the sensor has indeed got a linear characteristic, and the scale factor is a constant given by the formula,

$$k_F = \frac{\pi m \sqrt{2gH}}{4A\tau}(1 + R)$$

6. The vibrating system's dynamic characteristics were studied experimentally both under load and during idling motion.

During idling motion, the vibrating system's amplitude-frequency characteristic was obtained using a piezoelectric element glued to the tool's face and used as an accelerometer. The signal was taken from the piezoelectric element's inner surface which was insulated from the tool by a caprone net. The outer surface was earthed. This ensured that the sensor was shielded from the magnetostrictive transducer's external electromagnetic field. The sensor demonstrates a linear behaviour with a high accuracy. Its calibration was performed by measuring the tool's vibration amplitude a using a microscope as it oscillates at a frequency ω in one of its vibration modes. This was then compared with the sensor's signal amplitude A. The sensor's scale factor is calculated using the formula $k_a = a/A$.

The idling motion's resonant amplitude a^* is dependant on the current present in the magnetostrictive transducer's winding. In the following analysis, variable a^* is used to characterize the system's excitation level, i.e. it is assumed that it is independent of the vibration mode under consideration and that the excitation corresponds to the systems initial tuning as it oscillates under idling conditions. The vibrating system's amplitude-frequency characteristic when idling has a form that is typical for a linear system. The system's resonant frequency does not depend on the excitation level and is equal to $f_p = 18kHz$. When the excitation frequency is increased or decreased, the experimental readings obtained fall on the same curve. In the following, it is important that the vibrating system is linear when considered in itself. Its dissipative properties can be calculated from the resonant curves obtained. First of all, the Q-factor of the vibrating system $Q = f_p/(f_2 - f_1)$ should be found. f_2 and f_1 are the frequencies of the intersecting points between the resonant curve and a horizontal line plotted at $a/2$. The vibration's logarithmic decrement $d = \pi/Q$ and decrement $D = e^d$ can then be calculated. For the system under consideration, it is found that $Q = 97, d = 0.0324$ and $D = 1.0329$.

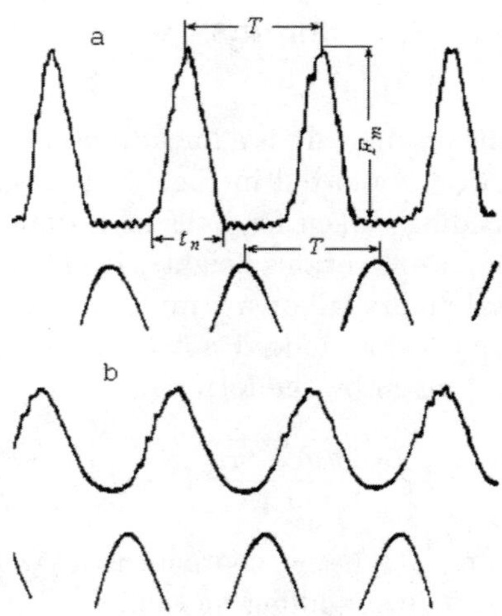

Fig. 4.9.

7. Regimes accompanied by a tool-workpiece interaction will, as of now, be referred to as working regimes. Such an interaction will be characterized by the impact force's amplitude F_m. This is understood to be the force's maximum deviation from zero. F_m was measured by obtaining a large number of oscilloscope readings at different values of tuning amplitude a^*, excitation frequency f, and static driving force P.

Two typical oscilloscope readings for a working regime are shown in Fig.4.9. They are obtained for when $f = 18kHz, a^* = 15\mu m, P = 100N$ (Fig.4.9, a) and $P = 160N$ (Fig.4.9, b). The upper curves of these readings show the interaction force between the tool and workpiece, the lower curves show the signal from the magnetostrictive transducer's generator. As follows from these readings, the working regimes are seen to be periodic with a frequency equal to the systems excitation frequency.

There are two vibrational regimes possible. The regime shown in Fig.4.9, a has an interaction force with an impulsive character. The force acts during the interval t_n, with an amplitude F_m. For the remainder of the period $T - t_n$ the tool performs vibration whilst being separated from the workpiece, i.e. the interaction force is equal to 0. Such regimes will be designated as impact regimes. The theoretical analysis of the vibrating system's work when under load, as carried out above, was devoted to finding this type of regime.

In the second type of regime (shown in Fig.4.9, b), the tool is continuously forced against the workpiece. It performs vibration with a small amplitude inside the limiter's elastic zone. This type of regime shall be designated as being bumpless. By comparing oscilloscope readings obtained at constant values of f and a^*, it has been shown that an increase in driving force leads to a gradual transition from impact regimes to bumpless regimes.

Consider the main dynamic characteristics of the system, obtained as a result of analysing working regime oscilloscope readings. Fig.4.10 shows the dependence of the impact force's amplitude (Fig.4.10, a) and impact duration (Fig.4.10, b) on the static driving force. These dependencies are obtained for a constant excitation frequency $f = 18kHz$ (the linear system's natural frequency) and different values of amplitude a^*. As can be seen from these plots, an optimal driving force exists. It corresponds to the impact force's maximum value. The optimal driving force's magnitude is proportional to the amplitude a^*. A further increase in driving force leads to a decrease in the interaction force accompanied by an increase in impact duration. This then gradually transforms the vibration present into a bumpless regime. It must be noted that in regimes with maximum impact force, the impact duration is equal to $t_n = 0.6T$ (the dashed line in Fig.4.10, b).

These dependencies are important characteristics of an ultrasonic machine since the impact force's amplitude is the main quantity which characterizes the ultrasonic machining efficiency. This follows from comparisons between the graphs shown above, and the ultrasonic machining speed's dependency on the driving force. The latter was obtained experimentally in Neppiras (1956) in which the theoretical reasoning in subsection 2 (see Fig.4.3) was also found.

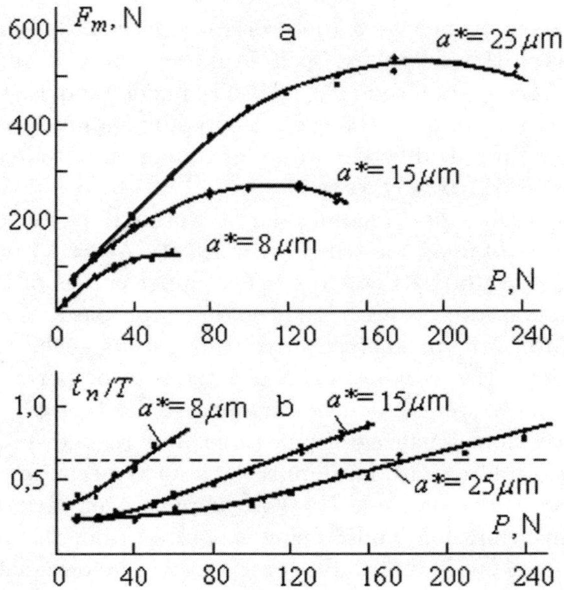

Fig. 4.10.

The characteristics of these curves are the same. Therefore, the machine's maximum output capacity, when working at constant frequency and a given amplitude, will occur at the optimal driving force in which the impact force reaches its maximum value.

8. In order to increase the machining output capacity, ways should be found of intensifying the machine's working regime. At the present time, machining is carried out at a constant frequency which is equal to the resonant frequency of the system's idling motion. As shown above, the only way to increase the output capacity under these conditions is to increase the driving force which requires an increase in power in the vibrating system. Therefore this method of ultrasonic machine design, as is commonly employed, is very inefficient.

As shown theoretically in section 4.1, and confirmed by experiment, the best way of increasing the efficiency of an ultrasonic machine is to tune it to resonance when under load.

A typical frequency characteristic of an ultrasonic system is shown in Fig.4.11, a. The dependence of the impact force's amplitude $F_{m}.$ on the excitation frequency f is obtained at a constant amplitude $a^* = 15\mu m$ and compressive force $P = 200N$. The dependence of the impact duration on the excitation frequency is shown in Fig.4.11, b for the same experimental conditions. The system's frequency characteristic has a distinctive resonant character and the frequency f_p that corresponds to the impact force's maximum value is considered to be the system's resonant frequency.

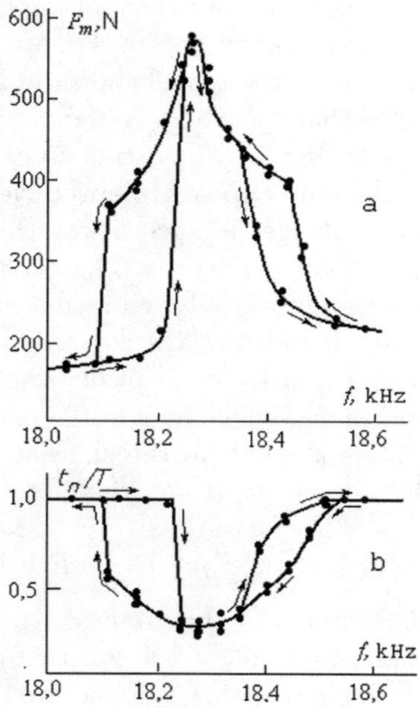

Fig. 4.11.

The existence of ambiguous domains within the frequency characteristic is an interesting feature of the system. These domains are situated on both the left and the right of the resonance peak. The branch taken by the system as it oscillates is dependant on the direction of frequency change and is shown by the arrows. In the left hand side ambiguous domain, the characteristic's lower branch corresponds to a bumpless working regime, while the upper branch to that of an impact regime. The transition between the branches occurs as a jump. Here, the distinctive effect of vibro-impact regime pulling is observed. Such effects are predicted by theoretical consideration within the framework of the accepted models.

In the right hand side ambiguous domain, both branches correspond to impact working regimes. The transition to the lower branch can be realized by increasing the frequency from that of resonance. This is accompanied by a decrease in force amplitude and the regime gradually becomes bumpless. The transition to the upper branch can be achieved by changing the excitation frequency in the opposite direction.

9. It should be noted that the appearance of a second ambiguous zone within the experiment's amplitude-frequency characteristic was completely

unexpected. This effect did not appear in the framework of accepted models. The same effect, however, has also been observed in similar experiments conducted later, by researches at the Vitebsk institute of solid-state physics. The explanation of this effect is therefore of principal importance since otherwise the adequacy of the models considered becomes questionable.

In order to explain this effect, the model shown in Fig.4.1 is again considered. The reader is reminded that the whole system, including the carriage M, is being forced against the limiter by a constant force. In this situation, the system reaches an equilibrium and appears to hover over the limiter. The level at which the system hovers changes in such a way that the impulse's value, which is known $J = PT$, remains constant. The carriage rises if the amplitude is increased, and lowers if the tool's vibration amplitude is decreased. It is now noted that although the carriage is retained within the frame by ball bearing slides, it is subjected to forces of dry friction $P_\tau \text{sgn}\, \dot{a}$. As such, the constant force component F_0, which acts in the contact zone between the tool and the workpiece, is dependant on the character by which changes in the tool's vibration amplitude occur. It can therefore take values within the range,

$$P_1 = (P - P_\tau) \leq F_0 \leq P_2 = (P + P_\tau) \tag{4.28}$$

The limiting values are determined by the carriage's motion upwards or downwards. If the carriage remains motionless, the constant component F_0 can take any value from the range given above. Until the force F_0 reaches a limiting value, the system behaves as a system with an initial interference as created by the previous interactions within the process.

In Fig.4.12, lines 1 and 2 show the system's amplitude-frequency characteristics for the constant force components P_1 and P_2 respectively. The lines are plotted using the method described in section 3.2. The thin lines $1'$ and $2'$ show the process' backbone curves. The backbone curves' vertical sections correspond to the linear system's natural frequencies during its vibration inside the limiter's elastic zone when no separation occurs. Also shown are linear vibration branches. These occur inside the workpiece's elastic zone at amplitudes $a \leq P_i/k_0$, where k_0 is the workpiece's static stiffness, and $i = 1, 2$. The realizable sections of these characteristics are shown by thick lines, while thin lines show the unrealizable sections. Dashed lines show unstable branches. Arrows show the direction in which the excitation frequency changes.

The system's behaviour is now followed during a change in excitation frequency. It is supposed that, initially, the system is forced against the workpiece and is excited at a frequency corresponding to point A. The constant force component is equal to $F_0 = P_1$. As the frequency is increased to point B, where the system starts to bounce, the amplitude follows the linear branch of characteristic 1 and the vibration encountered is bumpless. After the jump to point C, a further increase in frequency is accompanied by a decrease in amplitude, i.e. the system follows characteristic 1. The vibration present in

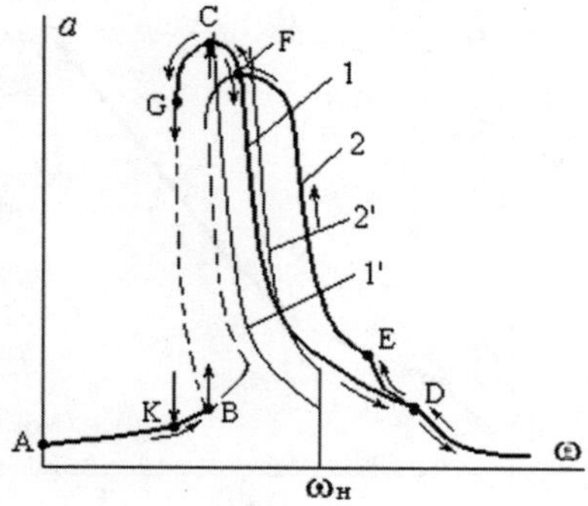

Fig. 4.12.

the system makes the transition to the bumpless regime present at point D and continues to follow the linear branch of the characteristic.

When the vibration's frequency changes in the opposite direction, the amplitude increases along the linear branch of characteristic 1 up until point D. This point corresponds to the appearance of an impact regime. Due to the increase in amplitude and a change in the direction of the frictional force present in the machines guides, force F_0 increases. This happens until equality $F_0 = P_2$ holds true and the system follows section DE which belongs to the resonant curve of a system in which an initial interference is present. After point E the system follows characteristic 2. The rest is clear from the picture.

Thus, all of the effects found experimentally can be explained by the approach developed in this book.

10. Experiments have shown that the system's working resonant frequency is shifted when compared to the resonant frequency of the system's idling motion. The magnitude of the shift can be estimated from Fig.4.13.

The above figure shows how the vibrating system's resonant frequency f_p is dependant on the static driving force. This dependence was obtained when the system was tuned to an amplitude of $a^* = 15\mu m$. As can be seen from the figure, an increase in driving force results in the system having a higher natural frequency. It is worth noting that all points on the graph correspond to impacting working regimes. The impact force's amplitude in these regimes is 2.5 to 3 times larger than the driving force.

The effects caused by adjusting the excitation frequency can now be estimated. For example, when the system is working at the idle motion's resonant frequency (Fig.4.10), the maximum impact force occurs when $a^* = 25\mu m$ and $P = 200N$. As can be seen from Fig.4.11, a and Fig.4.13, much larger impact forces and smaller levels of initial tuning amplitude can be attained by increasing the driving force and frequency control in accordance with Fig.4.13.

It is recalled at this point, that a machine's output capacity is determined by the impact forces present: the volume of the material fractured is

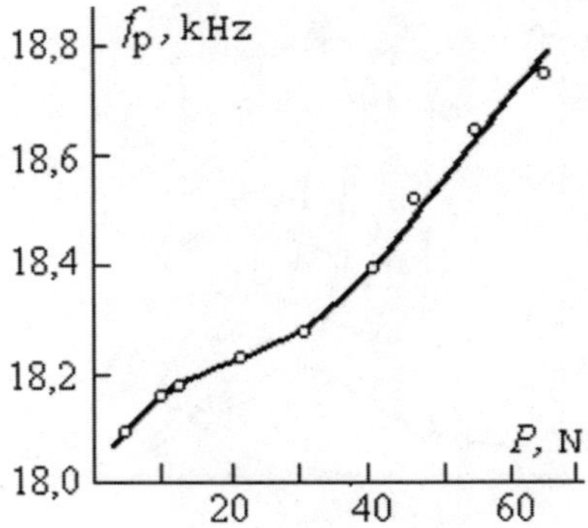

Fig. 4.13.

proportional to the cube of the impacting force's amplitude (Kazantsev et al. (1966)). In addition, the reader is reminded that increasing the machine's output capacity by increasing its power is extremely inefficient. This all goes to show the obvious necessity of tuning a machine to resonance whilst under load. As noted above and validated by the experimental results discussed, such a tuning method can only be realized in an automatic regime. This is due to the amplitude-frequency characteristic's strongly nonlinear distortion. The most efficient method of excitation and stabilization of an ultrasonic machine's resonant vibration when under load will be considered in section 4.3.

4.2 Nonlinear theory of ultrasonic concentrators

1. In the previous section a resonant tuning method was considered in which the process' detuning effect was compensated for by selecting an appropriate mass for the tool attached to the system. In a similar way, the system's resonant tuning, when under load, can be achieved by selecting the correct geometrical parameters for its concentrator. This is considered in this section along with the theory behind an ultrasonic concentrator which is subject to an elastic-dissipative nonlinear load. With respect to ultrasonic machining processes, the way in which the cutting speed, and both the load's elastic and dissipative components are dependant on the tool's vibrational parameters are found. Calculations are made for both exponential and stepped concentrators.

Ultrasonic concentrator theory which determines its ideal tuning during idling motion has been developed in papers Merkulov (1957), Merkulov & Kharitonov (1959), Teumin (1959). A bar concentrator with internal losses is now considered.

Concentrator 1's (Fig. 4.14) $x = 0$ cross section is connected with vibration actuator 2. Tool 3, which interacts with workpiece 3 or medium 4, is fixed in

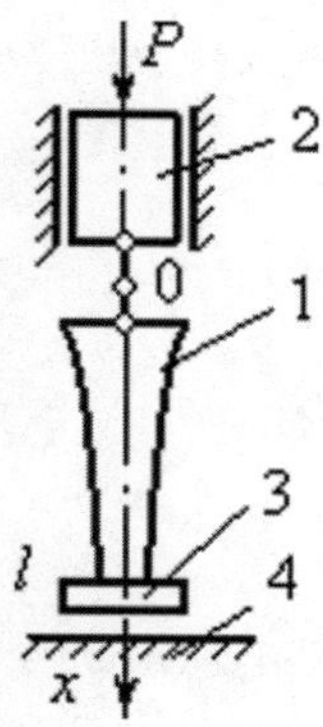

Fig. 4.14.

cross section $x = l$. The vibration present in cross section x of the concentrator is described by the function,

$$u_x(t) = a_x \exp j(\omega t - \varphi_x) \tag{4.29}$$

where, a_x, ω and φ_x are the amplitude, angular frequency and initial vibration phase.

As shown in section 2.2, the periodic load acting on the vibrating system as a result of its interaction with the workpiece has a nonlinear character and can be represented by the force reaction,

$$f(u_l) = [k(a_l) + j\omega\beta(a_l)]u_l \tag{4.30}$$

The load's nonlinearity manifests itself in the way that the elastic $k(a_l)$ and dissipative $b(a_l)$ load components are dependant on the tool's vibration amplitude. Such a dependence was repeatedly observed in experiments (Rosenberg & Kazantsev (1959*a*)).

The interaction force between the concentrator and vibration actuator shall be estimated from the force acting in their connecting cross section,

$$f_0(t) = F_0 \exp j\omega t \tag{4.31}$$

As before, the concentrator's motion is described at cross sections $s = 0, l$, using its dynamic compliance $L_s(x, j\omega)$. It relates the displacement u_x of cross section x with reaction forces (4.30) and (4.31), and the attached tool's inertial force $-M\omega^2 u_l$ which results from its mass M,

$$u_x(t) = L_{ox}(j\omega)F_0 \exp j\omega t - L_{0x}(j\omega)[k(a_l) + b(a_l) - M\omega^2]u_l(t) \tag{4.32}$$

Using (4.29), the following expressions are obtained:

$$\exp j\omega t = \frac{u_l}{a_l} \exp j\varphi_l, u_x = u_l \frac{a_x}{a_l} \exp j(\varphi_l - \varphi_x)$$

By substituting them into equation (4.32), the following is obtained when $x = 0$ and $x = l$,

$$a_0 \left[W_{l0}(j\omega) + \frac{a_l}{a_0} [k(a_l) - M\omega^2 + j\omega\beta(a_l)] e^{j(\varphi_0 - \varphi_l)} \right] =$$

$$= \frac{W_{l0}(j\omega)}{W_{00}(j\omega)} F_0 e^{j\varphi_0} \quad (4.33)$$

$$a_l W(a_l, j\omega) = F_0 e^{j\varphi_l} \quad (4.34)$$

where, $W(a_l, j\omega) = [W_{ll}(j\omega) + k(a_l) - M\omega^2 + j\omega b(a_l)] \frac{W_{0l}(j\omega)}{W_{ll}(j\omega)}$ and $W_{sx}(j\omega)$ is the concentrator's dynamic stiffness $(s, x = 0, l)$.

By separating the real and imaginary parts of formula (4.34), expressions are obtained that determine the concentrator motion's amplitude a_l and phase φ_l at cross section $x = l$:

$$a_l = \frac{F_0}{|W(a_l, j\omega)|} =$$

$$= \frac{F_0}{\sqrt{[\text{Re } W(a_l, j\omega)]^2 + [\text{Im } W(a_l, j\omega)]^2}} \quad (4.35)$$

$$\cos\varphi_l = \frac{a_l}{F_0} \text{Re } W(a_l, j\omega), \sin\varphi_l = \frac{a_l}{F_0} \text{Im } W(a_l, j\omega)$$

Using (2.79) for the dynamic stiffness gives,

$$W_{sx}(j\omega) = U_{sx}(\omega) + jV_{sx}(\omega) \quad (4.36)$$

In the following it is assumed that $V_{sx}(\omega)$ and $b(a_l)$, which characterize the dissipative losses in the concentrator and load are small, and all calculations are restricted to first order small terms. As shown in section 3.2, equation (4.35) determines the concentrator's amplitude-frequency characteristic when it is subjected to a harmonic force (4.31) acting on its end ($x = 0$). The maximum tool amplitude a_l is realized when the condition Re $W(a_l, j\omega) = 0$ holds true. By using (4.36), this condition can be converted into the form,

$$U_{ll}(\omega) + k(a_l) - M\omega^2 = 0 \quad (4.37)$$

Clearly, a minimal force F_0 may be used to realize any required amplitude a_l. As such, relation (4.37) may be used to determine the optimal conditions by which the vibration actuator excites vibration within the concentrator which is subjected to the tool's mass M and the nonlinear load.

Excluding F_0 from equations (4.33) and (4.34), and taking account of relations (4.36) and (4.37), the following is obtained,

$$K = \frac{a_l}{a_0} = \left| \frac{U_{00}(\omega)}{U_{0l}(\omega)} \right| \quad (4.38)$$

Similar to in Merkulov (1957), Merkulov & Kharitonov (1959) and subsection 2.2.5, the variable $K = a_l/a_0$ may be defined as the concentrator's amplification coefficient. It characterizes how well the vibration is guided and amplified as it travels between the actuator and tool. It is noted that expression (4.38) has the same form as equality (2.85) which is used to determine the amplification coefficient of an unloaded concentrator. For a given value of actuator amplitude a_0, relation (4.38) may be used to determine the amplitude present at the concentrator's working end. It satisfies condition (4.37), in which optimal levels of vibration are assumed for a system under load.

2. By attaching the concentrator, a load is experienced by the actuator. When the actuator is not loaded by the concentrator, it performs vibration which may be described as $u_0^*(t) = a_0^* \exp j(\omega t - \varphi)$. At the point where the concentrator and the actuator are connected, the actuator is subjected to the force $-f_0(t)$. Its motion, when under load, can be described by the following equation:

$$u_0(t) = u_0^*(t) - \frac{f_0(t)}{W_n(j\omega)} \tag{4.39}$$

where, $W_n(j\omega) = U_n(\omega) + jV_n(\omega)$ is the actuator's dynamic stiffness at the point where it is connected to the concentrator.

By using relations (4.34),(4.37) and (4.38), reaction (4.31) can be represented in the form,

$$f_0(t) = \dot{u}_0(t) K^2 [V_{ll}(\omega) + \omega b(a_l)] \tag{4.40}$$

As shown in section 2.4, the most common type of actuators that are used in ultrasonic machines are resonant electroacoustic transducers. Their resonant tuning condition is of the form $U_n(\omega) = 0$. From (4.39) and (4.40), for a loaded transducer, it is found that:

$$u_0(t) = u_0^*(t) \left[1 + \frac{V_{ll}(\omega) + \omega b(a_l)}{V_n(\omega)} K^2\right]^{-1} \tag{4.41}$$

Thus, by attaching a correctly matched concentrator, a transducer's resonant tuning remains unchanged and the change in vibration level is determined by the ratio of dissipative parameters in the transducer, concentrator and load. From (4.38) and using (4.41), the tool's vibration amplitude when operating in a working regime is found:

$$a_l = a_0^* K \left[1 + \frac{V_{ll}(\omega) + \omega b(a_l)}{V_n(\omega)} K^2\right]^{-1} \tag{4.42}$$

3. Consider the situation that emerges when ultrasonically processing brittle materials. In the previous subsection it was shown that treatment occurs as a result of high-frequency vibro-impact interactions between the tool and workpiece when subjected to a static compressive force P. By using a rheological model of the cutting process, the following expressions for the load's (4.30) elastic and dissipative components were obtained:

$$k(a_l) = \frac{D}{\pi a_l} \sin \frac{2\pi P}{D}; b(a_l) = \frac{2D}{\pi \omega a_l} \sin^2 \frac{\pi P}{D} \tag{4.43}$$

where, D is a quantity which is proportional to the tool's area. It is also dependant on both the workpiece's material and type of abrasive slurry used.

By substituting (4.43) into (4.37) and (4.42), and performing some specific transformations ($P \ll D$), the following equation system is obtained:

$$a_l = \frac{2P}{M\omega^2 - U_{ll}(\omega)} \tag{4.44}$$

$$a_l = a_0^* K \frac{1 - K\frac{\pi P^2}{DV_n(\omega)}}{1 + K^2 \frac{V_{ll}(\omega)}{V_n(\omega)}} \tag{4.45}$$

Its solution determines the matched concentrator's vibration parameters and amplitude when the system is under load.

Example calculations for the ultrasonic concentrator types most frequently used are given below. Concentrator 1's end ($x = 0$) (Fig. 4.14) is attached to the driving surface of vibration actuator 2. The dynamic compliances for various concentrator types were determined in section 2.2 by solving the equations governing longitudinal vibration within a bar of variable cross section.

Consider an exponential concentrator (see subsection 2.2.5) whose cross sectional area x is given by the equation,

$$S_x = S_0 \exp(2\gamma x/l)$$

where, $\gamma = 1/2 \ln(S_l/S_0)$.

Expressions for an exponential waveguide's dynamic compliance were obtained in subsection 2.2.6. The operator $L_{lx}(j\omega)$ has the following form:

$$L_{lx}(j\omega) = \frac{\gamma \sinh \lambda x - \lambda l \cosh \lambda x}{\omega^2 \rho S_0 l \sinh \lambda} e^{\gamma\left(1+\frac{x}{l}\right)} \tag{4.46}$$

where, $\lambda = \frac{\omega}{c}\left(j\frac{c^2}{c_\phi^2} + \frac{\psi}{2\pi}\right)^{1/2}$, $c_\phi = \frac{c}{\sqrt{1-(\gamma c/l\omega)^2}}$ is the vibration's phase velocity within the exponential waveguide, $c = \sqrt{E/\rho}$ is the speed of sound in the concentrator's material, and E and ρ are the concentrator's elastic modulus and material density.

The absorption coefficient ψ is supposed to be small. By performing a small parameter expansion of it in (4.46), and retaining only linear terms, expressions are found that determine the exponential concentrator's dynamic stiffness components (4.36):

$$U_{lx}(\omega) = -\frac{ES_0}{l} \frac{(\xi^2 + \gamma^2)\exp[\gamma(1 + x/l)]}{\xi \cos \xi x/l + \gamma \sin \xi x/l} \tag{4.47}$$

$$V_{lx}(\omega) = \frac{\psi E S_0}{4\pi l \xi} \frac{(\xi^2 + \gamma^2)^2 \exp[\gamma(1 + x/l)]}{\xi \cos \xi x/l + \gamma \sin \xi x/l} \times$$

$$\times \left[\cos \xi - \frac{(1 + \gamma x/l)\cos \xi x/l - (\xi x/l)\sin \xi x/l}{\xi \cos \xi x/l + \gamma \sin \xi x/l} \sin \xi\right] \tag{4.48}$$

where, $\xi = \omega l/c_\phi$.

Taking (4.47) into account, expression (4.38) that determines the loaded concentrator's amplification coefficient takes the form,

$$K = \frac{a_l}{a_0} = \left|\frac{\gamma \cos\xi + \xi \sin\xi}{\xi}\right| \sqrt{\frac{S_0}{S_l}} \tag{4.49}$$

For a free unloaded concentrator ($M = 0, P = 0$), formula (4.44) gives $U_{ll} = 0$. This is achieved when $\xi = \pi n$ ($n = 1, 2, \ldots$). Expression (4.49) gives the amplification coefficient $K = \sqrt{S_0/S_l}$ which coincides with that found in subsection 2.2.6. In the presence of a load, the concentrator is matched to the actuator at a different value of ξ. In order to find this value, (4.47) (in which $x = l$) is substituted into expression (4.44),

$$a_l = \frac{2P}{\omega w_l}\left(\mu + \frac{\sqrt{1 + (\gamma/\xi)^2}}{\cos\xi + (\gamma/\xi)\sin\xi}\right)^{-1} \tag{4.50}$$

where, $\mu = M\omega/w_l$ and $w_l = S_l\sqrt{E\rho}$ is the wave resistance at the concentrator's end.

For a given amplitude at the waveguide's driven end, the solution of equation system (4.49) and (4.50) determines the unknown exponential concentrator values ξ and a_l. In this instance, the concentrator is matched to the actuator when the system is under load. In order to find the solution of this equation system, it is convenient to use the graphical method shown in Fig. 4.15.

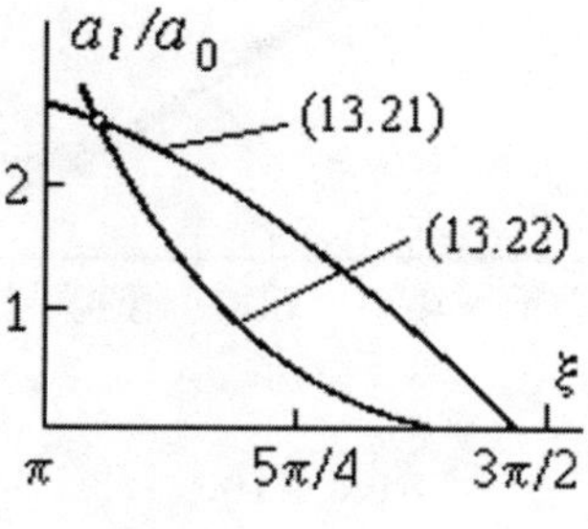

Fig. 4.15.

Subsequently, the concentrator's length can be calculated from the formula,

$$l = \frac{c}{\omega}\sqrt{\xi^2 + \gamma^2} \tag{4.51}$$

which, in the absence of load ($\xi = \pi n$), coincides with that obtained in Teumin (1959) and subsection 2.2.6.

Fig. 4.16, a shows the corresponding dependence of the concentrator's length l, when matched under load, on the compressive force P for different values of the attached tool's mass M. In this instance, l_0 is the free unloaded concentrator's length which is determined in accordance with formula (4.51) when $\xi = \pi n$. The calculations were performed for the following system parameters: $f = \omega/2\pi = 20kHz, K_0 = 4, w_l = 3.14 \cdot 10^3 N \cdot s/m, \psi = 0.01$ and $D = 5 \cdot 10^4 N$.

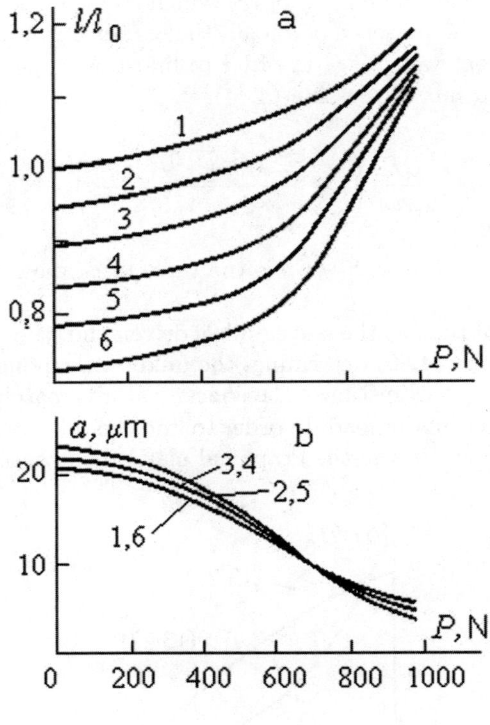

Fig. 4.16.

The curves shown in Fig.4.16, a are plotted for the following values of μ: $1 - \mu = 0, 2 - \mu = 0.2, 3 - \mu = 0.4, 4 - \mu = 0.6, 5 - \mu = 0.8$ and $6 - \mu = 1$.

This would be the complete solution of the problem if the electroacoustic transducer's vibration amplitude was not influenced by the attachment of the concentrator. However, as follows from (4.45), the vibration amplitude at the concentrator's end is dependant on the ratio of the dissipative losses in the transducer and concentrator.

Supposing, for the sake of definiteness, that the transducer is a homogeneous bar with a length l_n and cross sectional area S_n. Its dynamic stiffness, when tuned to the resonant frequency of its first natural vibration mode, can be found from formula (4.48) when $\gamma = 0, \xi = \pi$ and $x = l$:

$$V_n = \psi_n \omega w_n / 4 \tag{4.52}$$

where, $w_n = S_g \sqrt{E_n \rho_n}$ is the transducer's wave resistance at the connecting point between it and the concentrator, and E_n, ρ_n and ψ_n are the elastic modulus, density and absorption coefficient of the transducer's material.

Even the attachment of an unloaded ($P = 0, M = 0$) concentrator to a resonant actuator changes its amplitude. This can be found from formulae (4.45), (4.48) and (4.52),

$$a_l^* = a_0^* K_0 \left[1 + \frac{\psi w_l K_0^2 (1 + \gamma^2 / n^2 \pi^2)}{\psi_n w_n} \right]^{-1} \tag{4.53}$$

As follows from the last expression, the vibration amplitude of a concentrator with internal losses is decreased when compared to that of an ideal concentrator. The bigger the losses, the more pronounced the difference. The change in amplitude at the concentrator's end is dependant on its amplification coefficient K_0.

The dependence of the exponential concentrator's free end vibration amplitude (curve 1) on the amplification coefficient K_0 is calculated from formula (4.53) in which $\psi_n w_0 / \psi w_l = 65$. This is shown in Fig. 4.17. It can be seen that a specific value of the amplification coefficient K_0 corresponds to a maximum vibration amplitude at the concentrator's end.

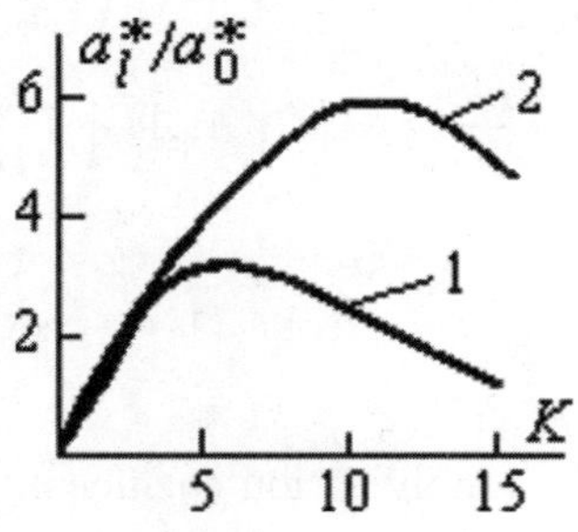

Fig. 4.17.

The concentrator's vibration amplitude when under load can be found by using general formula (4.45) and the values of K_0 and ξ that were found above (see Fig. 4.15). Fig. 4.16, b shows how the exponential concentrator's vibration amplitude, which is matched to the transducer when under load, is dependant on the compressive force. The dependency is shown for several values of the attached tool's mass. The figures are plotted using the same parameter values as those used in Fig.4.16, a. In addition, $w_n = 4.084 \cdot 10^4 N \cdot s/m$ and $\psi_n = 0.05$.

The concentrator's vibration amplitude, when under load, can be found by using a more straightforward method of solving the equation system (4.44) and (4.45). This can be done in a similar way, and of course, gives the same final results.

4. Calculations for different concentrator types can be performed in a similar way. Now, as an example, a concentrator consisting of two cylindrical bars with lengths l_1 and l_2, and cross sectional areas S_1 and S_2, is considered. Its dynamic compliance was found in subsection 2.2.11. The theory behind stepped concentrators, with an absence of losses, was also given in the same subsection. By using the methods developed there, the concentrator's characteristics are found. Internal friction is considered to be present in its material, and the concentrator is considered to be subjected to inertial and nonlinear elastic-dissipative loads.

Expression (2.155) is written for the concentrators dynamic compliance,

$$L_{lx}(j\omega) = \\ = \begin{cases} -\frac{\lambda}{\rho \cdot \omega^2} \cdot \frac{\cosh \lambda x}{S_1 \sinh \lambda l_1 \cosh \lambda l_2 + S_2 \sinh \lambda l_2 \cosh \lambda l_1} & 0 \le x \le l_1 \\ -\frac{\lambda}{S_2 \rho \cdot \omega^2} \cdot \frac{S_1 \sinh \lambda (x-l_1) \sinh \lambda l_1 + S_2 \cosh \lambda (x-l_1) \cosh \lambda l_1}{S_1 \sinh \lambda l_1 \cosh \lambda l_2 + S_2 \sinh \lambda l_2 \cosh \lambda l_1} & l_1 \le x \le l \end{cases} \quad (4.54)$$

where, $l = l_1 + l_2$ and the quantity λ has the same meaning as in formula (4.46), in which $c_\phi = c$ $(\gamma = 0)$.

It is assumed, as before, that absorption coefficient ψ is small. The following dynamic stiffness expressions are obtained from (4.54):

$$W_{l0}(j\omega) = -\frac{ES_2}{l_2} \xi_2 \sin \xi_1 \sin \xi_2 \left(A_1 + j\frac{\psi}{4\pi} \Phi_1 \right) \quad (4.55)$$

$$W_{ll}(j\omega) = -\frac{ES_2}{l_2} \xi_2 A_2^{-2} \left(A_1 A_2 + j\frac{\psi}{4\pi} \Phi_2 \right) \quad (4.56)$$

where, $\xi_i = \frac{\omega l_i}{c} (i = 1, 2)$, $A_1 = \cot \xi_1 + \frac{S_1}{S_2} \cot \xi_2$, $A_2 = \cot \xi_1 \cot \xi_2 - \frac{S_1}{S_2}$, $\Phi_1 = \xi_1 + \cot \xi_1 (1 - \xi_2 \cot \xi_2) + \frac{S_1}{S_2} [\xi_2 + \cot \xi_2 (1 - \xi_1 \cot \xi_1)]$ and $\Phi_2 = A_1 A_2 - \frac{S_1}{S_2} \frac{\xi_1}{\sin^2 \xi_1 \sin^2 \xi_2} - \frac{\xi_2}{\sin^2 \xi_2} \left(\cot^2 \xi_1 + \frac{S_1^2}{S_2^2} \right)$.

Using (4.55) and (4.56), amplification coefficient expression (4.38) can be written in the form,

$$K = \frac{a_l}{a_0} = |\cos \xi_1 \cos \xi_2 - (S_1/S_2) \sin \xi_1 \sin \xi_2| \quad (4.57)$$

which coincides with formula (2.159) obtained previously.

In subsection 2.2.11 it was shown that a free unloaded concentrator ($M = 0, P = 0$) attains its maximum amplification, $K_0 = S_1/S_2$, when $\xi_1 = \xi_2 = \pi(2n - 1)/2$ $(n = 1, 2, \ldots)$, i.e. when its steps have length $l_1 = l_2 = \pi(2n - 1)c/2\omega_n$.

In the presence of a load, the conditions by which the concentrator is matched to the actuator can be found by solving the equation system (4.57) and (4.44). The result takes the following form by virtue of (4.56):

$$a_l = \frac{2P}{\omega w_2}\left(\mu + \frac{\sin\xi_2\cos\xi_1 + K_0\sin\xi_1\cos\xi_2}{\cos\xi_1\cos\xi_2 - K_0\sin\xi_1\sin\xi_2}\right)^{-1} \tag{4.58}$$

The solution of equations (4.57) and (4.58) determines the unknown values ξ_i and a_l for a concentrator that is matched to a transducer when under load. The solution is obtained for a given amplitude a_0 at the concentrator's driven end $x = 0$. Subsequently, the concentrator's step lengths can be found from the formula,

$$l_i = \frac{c\xi_i}{\omega} \tag{4.59}$$

Fig. 4.18, a shows the dependence of the concentrator's length l on the compressive force P. It is shown for various values of the attached tool's mass M. In this instance, the concentrator is matched to the transducer when under load, and the calculations were performed for a concentrator with equal step lengths ($l_1 = l_2, \xi_1 = \xi_2$); l_0 is the length of the free unloaded concentrator and is given by formula (4.59) when $\xi = \pi n$; $\mu = M\omega/w_L$, in which $w_L = S_L\sqrt{E\rho}$ is the wave resistance at the concentrator's working end. The calculations were performed for the following parameters: the vibration frequency is equal to $f = \omega/2\pi = 20kHz, K_0 = 4, w_L = 3.14\cdot 10^3 N\cdot s/m, \psi = 0.01$ and $D = 5\cdot 10^4 N$. The curves in Fig. 4.18, a are plotted for the following values of μ: $1 - \mu = 0, 2 - \mu = 0.2, 3 - \mu = 0.4, 4 - \mu = 0.6, 5 - \mu = 0.8$ and $6 - \mu = 1$.

The vibration amplitude of an unloaded concentrator which is connected to a resonant actuator can be determined from formulae (4.42), (4.52) and (4.56),

$$a_l^* = a_0^* K_0\left[1 + \frac{\psi w_2 K_0(1+K_0)}{\psi_n w_n}\right]^{-1} \tag{4.60}$$

As follows from (4.60), the vibration amplitude of a stepped concentrator with internal losses is reduced from that of an ideal concentrator. The difference increases with an increase in losses, and the change in amplitude at the concentrator's free end is dependant on its amplification coefficient K_0. These results are similar to those obtained previously for an exponential concentrator. By using (4.60) in which $\psi_n w_n/\psi w_2 = 65$, the dependence of the stepped concentrator's free end's vibration amplitude on the amplification coefficient K_0 is determined. This is shown by curve 2 in Fig. 4.17. It can be seen that a specific value of the amplification coefficient K_0 exists which corresponds to a maximum vibration amplitude at the concentrator's working end.

The stepped concentrator's vibration amplitude, when under load, can be found by using the general formula (4.45) and the values of K_0 and ξ found previously (Fig. 4.15). Fig. 4.18, b shows how the stepped concentrator's vibration amplitude, which is matched to the transducer when under load, is dependant on the compressive force. The dependency is shown for several values of the attached tool's mass. The figures are plotted for the same parameters as used in Fig. 4.18, a. In addition, $w_n = 4.084\cdot 10^4 N\cdot s/m$ and $\psi_n = 0.05$.

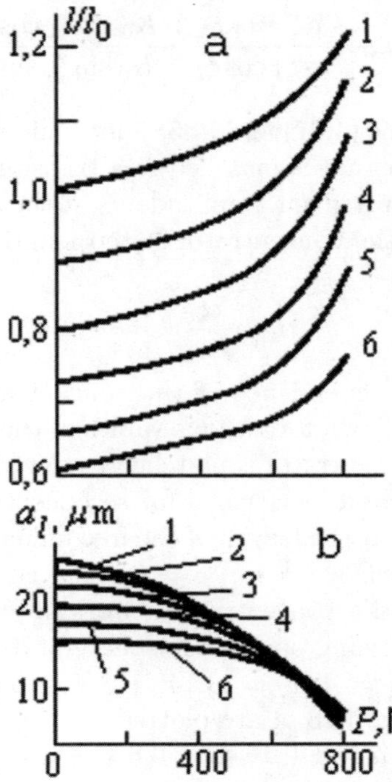

Fig. 4.18.

5. To conclude, the efficiency of the concentrator's matching with the actuator and load will be estimated. This will be done using formula (4.7) which relates the ultrasonic cutting speed v with the main process parameters,

$$v = 2a_l f \sin^2 \frac{\pi P}{D} \tag{4.61}$$

where, $f = \omega/2\pi$ is the tool's vibration frequency.

Fig. 4.19 shows the dependence of the ultrasonic treatment speed on the compressive force. It is plotted using data from Fig. 4.16, b and 4.18, b. Curves 1 and 2 in Fig. 4.19 correspond to a stepped concentrator in which $\mu = 0$ and $\mu = 1$ respectively. Curve 3 corresponds to an exponential concentrator when $\mu = 0.2$.

For comparison, the machining speeds are found for conventional ideal concentrators that are matched to the transducer, with no losses, and are operating in their idling regime. Previously, in section 4.1, it was shown that, in this instance, the maximal machining speed achievable is,

$$v_0 = 0.3a_l^{*2}\omega \left(\frac{V_0}{D}\right)^2 \tag{4.62}$$

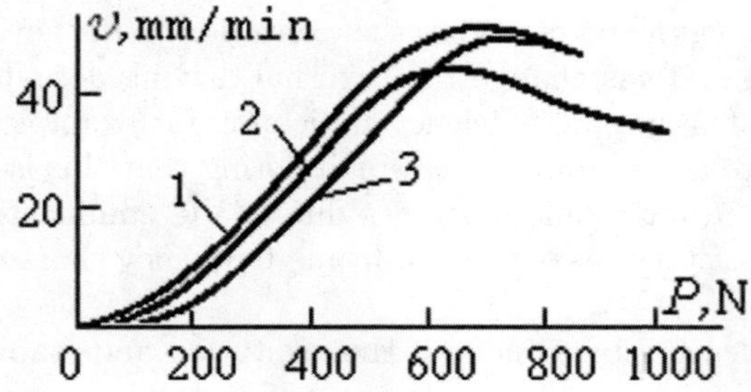

Fig. 4.19.

for a compressive force of,

$$P_0 = 0.4a_l^* V_0 \tag{4.63}$$

where, $V_0 = V_{l l} + V_n / K_0^2$.

For a system with the same parameters as considered previously, it is found that $V_0 = 46300 N/cm$ for an exponential concentrator, and $V_0 = 52900 N/cm$ for a stepped concentrator. Calculations using formulae (4.62) and (4.63) reveal that $P_0 = 45.8N, v_0 = 2.4mm/min$ and $P_0 = 46.6N, v_0 = 2.8mm/min$ respectively. These values, as well as the initial data, correspond to model 4770, the commercially available ultrasonic machine.

The analysis of this data shows that by ensuring that the concentrator is correctly matched to the resonant vibration actuator when under the influence of the systems nonlinear load, an increase in the ultrasonic machine's output capacity of up to 15 to 20 times can be revealed. These output capacity values have the same order of magnitude as those obtained when the system is tuned to resonance by adjusting the whole vibrating system's vibration frequency (see section 4.1). This is not surprising since the method under consideration is an alternative way in which the whole vibrating system may be tuned to resonance when under load. It is achieved by choosing the geometrical parameters of its parts, and therefore, their natural frequencies.

4.3 Autoresonant excitation of ultrasonic machines

1. As shown previously in section 4.1 by studying ultrasonic machine vibration systems, a pronounced increase in output capacity along with a considerable decrease in energy consumption can be achieved by increasing the vibrating system's Q-factor and tuning it to resonance whilst under load. However, the dynamic characteristic's typical nonlinear distortion, as caused by the tool – workpiece interaction, make the realization of this type of resonant tuning difficult.

The machine's dynamic properties can only be used to their full extent when an automatic frequency tuning system is present. This would support the system's resonant regime by changing the excitation frequency. Conventional automatic control systems change the external driving actuator's frequency in such a way that a specific efficiency indicator, for example the vibration amplitude, tends to a maximum. It is worth noting that this is inefficient and difficult to realize in ultrasonic machines due to the nonlinearity of the load and the nonlinear distortions of the amplitude-frequency characteristic caused by it.

These difficulties can be avoided if the excitation and stabilization of the system's resonant regimes are supported by self-sustaining actuation (Teodorchik (1952)). This can be realized by the introduction of a regenerative loop which regulates the driving actuation by a nonlinear signal transformation which is proportional to the motion of the machine's working tool. This is applicable to all types of ultrasonic processes and machines as discussed in sections 1.1 and 2.1 (Astashev et al. (1973), Babitsky, Astashev & Kalashnikov (2004)).

The subjects considered in this section include the synthesis of regenerative loops, the study of self-sustaining oscillation system dynamics for ultrasonic machine excitation, and the analysis of vibration self-excitation conditions. The existence and stability of periodic regimes, the behaviour of systems during a change in vibration parameters, regenerative loops, and the wide range of nonlinear load's acting on the systems are also studied. In addition, experimental data is discussed.

The arrangement of an ultrasonic machine's self-sustaining excitation system is shown in Fig. 4.20. The motion $u_x(t)$ of the vibration system arbitrary element x may be written as,

$$u_x(t) = m_x + u_x^0(t) = m_x + a_x \exp j(\omega t - \varphi_x) \tag{4.64}$$

where, m_x is a constant component, and a_x, ω and φ_x are the amplitude, angular frequency and phase of the harmonic motion component $u_x^0(t)$.

The machine's working tool l interacts with the load, and is described by the nonlinear dynamic characteristic,

$$f_l(u_l, j\omega u_l) \approx f_0(m_l, a_l) + [k(m_l, a_l) + j\omega b(m_l, a_l)]u_l^0 \tag{4.65}$$

where, $f_0(m_l, a_l)$ is a constant component, and $k(m_l, a_l)$ and $b(m_l, a_l)$ are the load's elastic and dissipative components. Their values can be determined using the methods described in section 3.1.

The signal from feedback sensor 2 is proportional to the tool's displacement u_l. The signal is fed through phase-shifting element 3 and is subjected to a phase shift φ. It is then fed to nonlinear transducer 4 (incorporating a power amplifier) which creates the excitation force $f_s = f_s(u_l)$. This acts in the vibrating system at point $x = s$. By taking (4.64) into account, the feedback circuit's full nonlinear characteristic can be written,

$$f_s = f_s(u_l^0)\exp(j\varphi) \approx Z(a_l)u_l^0 \exp(j\varphi) \tag{4.66}$$

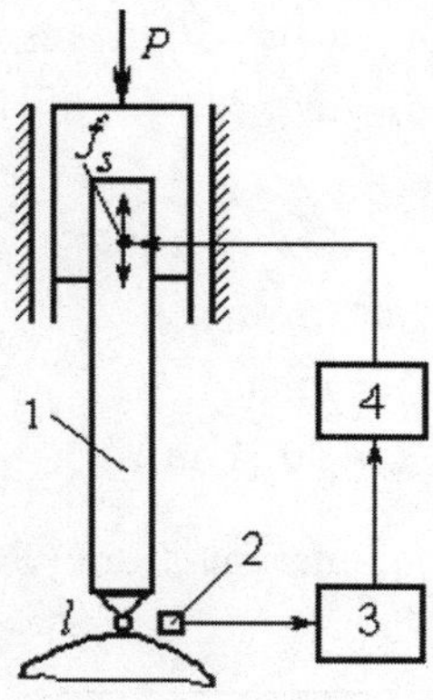

Fig. 4.20.

where, $Z(a_l)$ is the nonlinear transducer's harmonic amplification coefficient.

It is assumed that the feedback circuit only optimises the periodic component of element l's motion. The maximum actuation force which is generated by the actuator is limited,

$$f_s(u_l^0) \leq G \tag{4.67}$$

The periodic vibration actuation force F_s, which is limited in magnitude to G, can not have an amplitude greater than $4G/\pi$ for its first harmonic. As such, the following is obtained from (4.67),

$$a_l Z(a_l) \leq \frac{4G}{\pi} \tag{4.68}$$

The bar's cross sectional displacements can be determined from the dynamic compliance operator $L_{sx}(j\omega)$ which relates the motion u_x of element x to the forces acting on element s:

$$u_x(t) = L_{sx}(j\omega) f_s(u_l^0) - L_{lx}(j\omega) f_l(u_l, j\omega l) \tag{4.69}$$

By substituting (4.65) and (4.66) into (4.69) and separating the constant and periodic components when $x = l$, it is found that,

$$m_l = -L_{ll}(0) f_0(m_l, a_l) \tag{4.70}$$

$$1 + L_{ll}(j\omega)[k(m_l, a_l) + j\omega b(m_l, a_l)] -$$
$$-L_{sl}(j\omega) Z(a_l) \exp(j\varphi) = Q(m_l, a_l, j\omega) = 0 \tag{4.71}$$

By equating the real and imaginary components of characteristic equation (4.71) to zero, the conditions in which self-sustaining oscillations exist are found:

$$Z(a_l)\cos\varphi = U(\omega, m_l, a_l) \tag{4.72}$$
$$Z(a_l)\sin\varphi = V(\omega, m_l, a_l) \tag{4.73}$$

where,

$$U(\omega, m_l, a_l) = \mathrm{Re}\, W(j\omega, m_l, a_l),$$
$$V(\omega, m_l, a_l) = \mathrm{Im}\, W(j\omega, m_l, a_l),$$
$$W(j\omega, m_l, a_l) = \{1 + L_{ll}(j\omega)[k(m_l, a_l) + j\omega b(m_l, a_l)]\}/L_{sl}(j\omega)$$

By excluding trigonometric functions from (4.72) and (4.73), the following is obtained,

$$Z(a_l) = \sqrt{U^2(\omega, m_l, a_l) + V^2(\omega, m_l, a_l)} = |W(j\omega, m_l, a_l)| \tag{4.74}$$

This relation can be rewritten in the form,

$$a_l = \frac{a_l Z(a_l)}{(\sqrt{U^2(\omega, m_l, a_l) + V^2(\omega, m_l, a_l)})} \tag{4.75}$$

By comparing (4.75) with (4.68) it is seen that, at all frequencies, the maximum vibration amplitude is attained if $a_l Z(a_l) = \frac{4G}{\pi}$. This yields,

$$Z(a_l) = \frac{4G}{\pi a_l} \tag{4.76}$$

Condition (4.76) can be realized by using an excitation transducer 4 (Fig.4.20) that exhibits a relay characteristic $F_s = F_s(u_0)$. In order that the feedback circuit exhibits such a characteristic, the vibration amplitude is determined by the expression,

$$a_l = \frac{4G/\pi}{\sqrt{U^2(\omega, m_l, a_l) + V^2(\omega, m_l, a_l)}} \tag{4.77}$$

Previously, in section 3.2, expression (3.88) was obtained which describes the amplitude-frequency characteristic of the same system during forced vibration and when under the action of a periodic force $F_s = F_0 \exp j(\omega t - \varphi)$. This expression has the form,

$$a_l = \frac{F_s}{\sqrt{U^2(\omega, m_l, a_l) + V^2(\omega, m_l, a_l)}} = \frac{F_s}{|W(j\omega, m_l, a_l)|} \tag{4.78}$$

By comparing expressions (4.77) and (4.78) when $G = \pi F_0/4$, it is seen that the amplitudes and frequencies obtained for which self-sustaining vibration occurs, calculated for various values of phase, correspond to points on the machine's working tool resonant curves. Moreover, the oscillation frequency is determined by the system's phase-frequency characteristic which is described by equations (4.72) and (4.73).

Quantity $V(\omega, m_l, a_l)$ characterizes the vibrating system's reduced dissipative properties. Supposing that the dissipation is fairly small, it is natural to assume that the maximum vibration amplitude a_l^* is attained at frequencies ω^* which satisfy the condition,

$$U(\omega^*, m_l^*, a_l^*) = 0 \tag{4.79}$$

In accordance with (4.72) and (4.73), the value of phase φ which excites resonant vibration at its maximal amplitude is found to be,

$$\varphi = \varphi^* = \frac{\pi}{2} \tag{4.80}$$

Taking account of (4.79), and from (4.78), the maximum amplitude of self-sustaining oscillation is found,

$$a_l^* = \frac{4G}{\pi V(\omega^*, m_l^*, a_l^*)} \tag{4.81}$$

The equation system (4.79) and (4.81), together with relation (4.70), determines the regime's frequency ω^* and amplitude a_l^*. It should be noted that these equations coincide with the equations obtained in subsection 3.2, which determine the resonant state of the system when subjected to forced vibration. Equation (4.79) determines the backbone curve, and equation (4.81) determines the limiting amplitude line.

Thus, if the phase is equal to the value given in (4.80), any variations in the vibrating system's parameters, or those of the load, are compensated for, and the feedback circuit promotes an self-sustaining oscillation regime in which the maximum achievable amplitude is realized. The maximum achievable amplitude is determined, as found previously, by the limiting amplitude line. This means that the system's most efficient resonant state is automatically promoted. The type of excitation under consideration is designated as autoresonant (Andronov et al. (1966), Astashev et al. (1973), Babitsky (1995)).

In order to study the stability of the autoresonant regimes found above, the method described in section 3.2 can be used. It is assumed that in the vicinity of steady state periodic solution (4.64), the machine's working tool's vibration is of the form,

$$\breve{u}_l(t) = \breve{m}_l(t) + \breve{a}_l(t) \exp j[\omega t - \breve{\varphi}_l(t)] \tag{4.82}$$

where, $\breve{m}_l(t), \breve{a}_l(t)$ and $\breve{\varphi}_l(t)$ are slowly varying functions of time. Considering equality (4.73) as the reduced energy balance condition for the dissipative and exciting forces of the steady state motion under investigation, the motion's stability criterion, with respect to small perturbations in amplitude, can be formulated. It takes the form,

$$\left(\frac{d\mathrm{Im}\ Q}{d\breve{a}_l}\right)_{\breve{a}_l = a_l} > 0 \tag{4.83}$$

Inequality (4.83) means that if the amplitude $\breve{a}_l$ deviates from a stationary value $a_l = const$, the energy is unbalanced in such a way that it compensates for the deviation. Using the notation introduced earlier, this condition can be written in the form:

$$\frac{d}{d\breve{a}_l}[\omega b(\breve{a}_l) - K(\omega)Z(\breve{a}_l)\sin\varphi]_{\breve{a}_l=a_l} > 0 \tag{4.84}$$

where, $K(\omega) = U_{sl}(\omega)/U_{ll}(\omega)$ is a coefficient that is dependant on the point at which the excitation force is applied, and,

$$\text{sgn } K(\omega)\sin\varphi \geq 0 \tag{4.85}$$

Condition (4.84) is used in order to study the system's stability in some specific cases. Firstly, it is noted that according to (4.76),

$$\left[\frac{\partial Z}{\partial \breve{a}_l}\right]_{\breve{a}_l=a_l} < 0 \tag{4.86}$$

If the vibrating system works in an idling motion regime ($b = 0$), or under load with a linear ($b = b_0 = const$) dissipative component, stability criterion (4.84) is valid for all of the vibrational regimes possible and for the whole domain for which an amplitude-frequency characteristic exists. Specifically, this means that by changing the feedback's phase in an autoresonant regime, all of the resonant curve's branches can be realized. This includes those that are unstable in forced vibration situations (see sections 3.2 and 4.1).

Some instances are now considered in which nonlinear loading is present.

It is supposed that the working tool is subjected to a dry friction force $f_l(\dot{u}_l) = Q\text{sgn } \dot{u}_l$. The equivalent resistance coefficient (3.62) for such a load was found in subsection 3.1.6,

$$b(a_l) = 4Q/\pi\omega a_l$$

and the stability condition (4.84), when $s = l$ ($K(\omega) = 1$), takes the form,

$$G\sin\varphi > Q$$

It is easy to show that this condition coincides with the condition by which forced vibration exists in systems with dry friction (Timoshenko (1955)). In particular, autoresonant vibration can exist when $G > Q$.

Suppose that the working tool is subjected to a load that is described by a rigid-plastic material's characteristic. For relatively small driving forces, the load's dissipative component is described by expression (3.73),

$$b(a_l) = \frac{2\pi P^2}{\omega a_l D}$$

and the stability condition takes the form,

$$G \sin\varphi > \frac{(\pi P)^2}{2D}$$

When $\varphi = \pi/2$, this condition coincides with the energy condition by which resonant vibro-impact regimes exist. This is the case for a given system excitation level. If this condition is satisfied when the system is operating in an autoresonant vibration regime, both of the amplitude-frequency characteristic's branches can be realized within the range of frequencies given by the inequality. This is achieved by changing the phase of the system's feedback.

The simplest way by which self-sustaining oscillation is realized is to satisfy the condition by which vibration within the system is self-excited. Self-excitation occurs if the system's static equilibrium state loses its stability. The stability criterion for static equilibrium is now found by using the method of D-partitions (Neimark (1978)). In order to do this, functions (4.65) and (4.66) are linearized in the vicinity of equilibrium,

$$f(u_l, pu_l) \approx f_0 + (k_0 + p\beta_0)u_l^0, F_s(u_l^0) \approx Z_0 u_l^0 \qquad (4.87)$$

where, Z_0, k_0 and β_0 are constant coefficients that exist in front of the first terms in the Taylor series expansions of functions (4.65) and (4.66).

Similarly to equation (4.71),

$$[1 + L_{ll}(p)(k_0 + p\beta_0) - L_{sl}(p)Z_0 \exp(p\varphi/\omega)]u_l^0 = Q(p)u_l^0 = 0 \qquad (4.88)$$

The appearance of a zero root, or a pair of purely imaginary roots, in the characteristic equation $Q(p) = 0$, corresponds to the linear system's loss of stability (4.88). By substituting $p = 0$ and $p = j\omega$ into the characteristic equation, the following parametric equations that describe the D-partition boundaries of the parameter plane Z_0, φ are obtained:

$$Z_0 = [1 + k_0 L_{ll}(0)]/L_{sl}(0), \quad Z_0 = U_0(\omega)/\cos\varphi,$$
$$Z_0 = V_0(\omega)/\sin\varphi \qquad (4.89)$$

where,

$$U_0(\omega) = \mathrm{Re}\, W_0(j\omega), \quad V_0(\omega) = \mathrm{Im}\, W_0(j\omega),$$
$$W_0(j\omega) = [1 + L_{ll}(j\omega)(k_0 + j\omega\beta_0)]/L_{sl}(j\omega).$$

The last two relations in (4.89) can be obtained directly from (4.72) and (4.73), by substituting harmonic linearization coefficients values (4.87). For a given value of frequency ω, equations (4.89) allow Z_0 and φ, which belong to the boundary of self-excitation, to be found. By virtue of (4.80), the right-hand sides of equations (4.89) are finite. Therefore, the synthesized relay characteristic in which $Z_0 = \infty$, causes a system to self-excite, and a periodic regime to be achieved. Further types of feedback circuit characteristics are considered below.

2. Previously in chapter 2, when studying a number of ultrasonic machine vibration systems that were subjected to a load, the analysis led to the consideration of a viscoelastic bar's longitudinal vibration which acted against a limiter. By using this model, the main properties and capabilities of autoresonant vibration excitation within such machines are considered here.

The dynamic model of the system under consideration is shown in Fig.4.21, a.

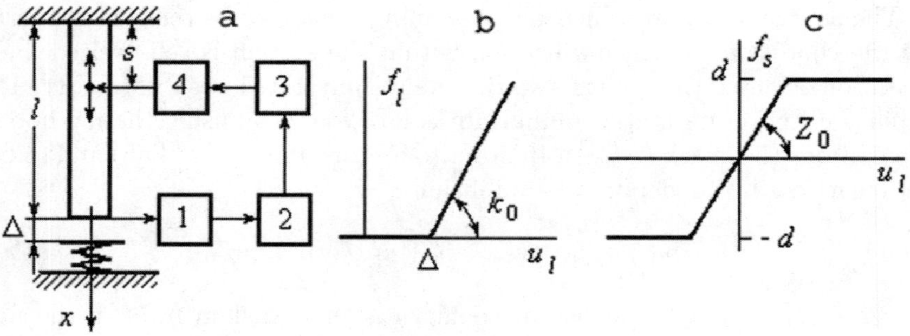

Fig. 4.21.

A viscoelastic bar with length l performs longitudinal vibration and is subjected to force F_s which is applied at a specific cross section $x = s$. The bar's upper end $x = 0$ is fixed and its lower end $x = l$ interacts with an elastic limiter which is positioned away from the bar with an initial clearance Δ. Δ is measured when the bar is in its static equilibrium state and a negative value of Δ corresponds to an initial interference between the bar and limiter. The excitation force is generated by the feedback circuit which creates a closed loop system. The circuit consists of sensor 1 which registers cross section $x = l$'s displacement, an amplifier 2, a phase changer 3 and transducer 4 which creates the actuation force.

The bar's self-excitation conditions, and periodic motion parameters, are sought for when the system interacts with an elastic limiter. The interaction occurs via an initial clearance (interference) and the elastic limiter has a stiffness k_0 (Fig. 4.21, b). It is supposed that the excitation force is applied to the bar's lower cross section ($s = l$). The feedback circuit is assumed to exhibit the characteristic of a linear amplifier with saturation (Fig. 4.21, c).

Using expressions (3.95) and (3.96) from section 3.2, the following is obtained for the case under consideration,

$$W(j\omega) = \frac{ES}{l}\left[\xi \cot \xi + q + j\frac{\psi\xi}{4\pi}\frac{(\xi + 0.5 \sin 2\xi)}{\sin^2 \xi}\right] \tag{4.90}$$

where, $\xi = \omega l/c$, $c = \sqrt{E/\rho}$ is the speed of sound in the bar's material, E and ρ are the elastic modulus and density of the bar's material, S is its cross sectional area, ψ is the absorption coefficient, and $q = kl/ES$.

By substituting the expressions which follow from (4.90) into relations (4.89), the following parametric equations which determine the D-partition's boundaries are obtained:

$$z_0 = 1 + Q_0 \tag{4.91}$$

$$z_0 \cos\varphi = \xi \cot\xi + Q_0, \quad z_0 \sin\varphi = \frac{\psi\xi(\xi + 0.5\sin 2\xi)}{(4\pi \sin^2\xi)} \tag{4.92}$$

where, $z_0 = Z_0 l/ES, Q_0 = 0$ when $\Delta > 0$ and $Q_0 = q_0$ when $\Delta < 0$, and $q_0 = k_0 l/ES$.

Fig. 4.22, a shows the D-partitions in the z_0, φ parameter plane for a system with an initial clearance ($\Delta > 0$) and when ξ changes from 0 to 2π. Hatching indicates the equilibrium state's stability region. In this situation, the characteristic equation has no roots with positive real parts.

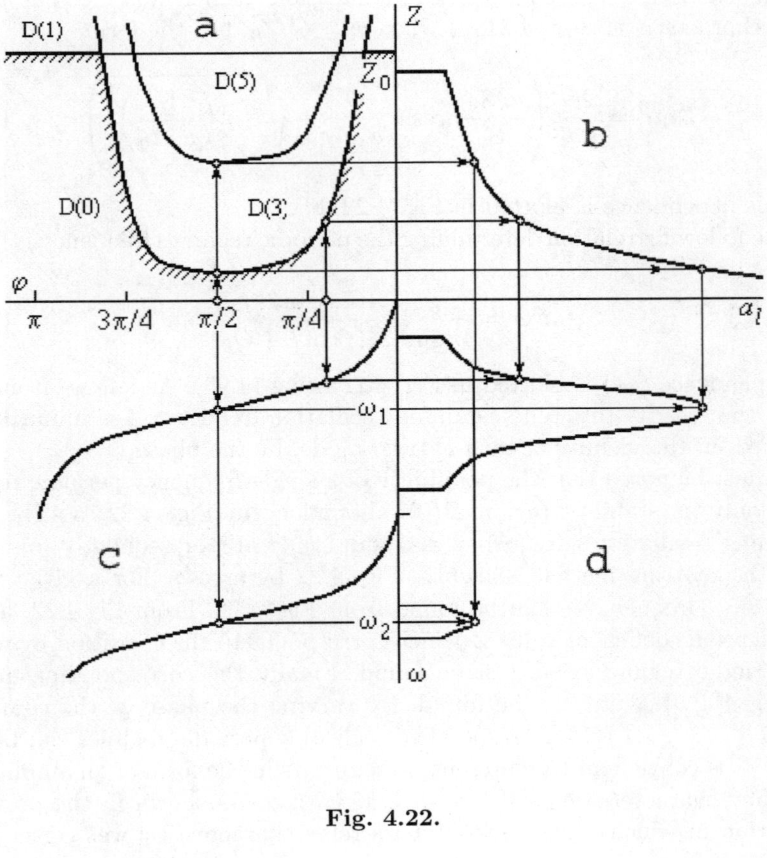

Fig. 4.22.

When the D-partition boundary (4.91) is crossed, in the direction in which z_0 increases, the characteristic equation's one positive root emerges. When boundary (4.92) is crossed however, a pair of complex conjugate roots with positive real parts appear. The parameter space domain in which the characteristic equation has i roots with positive real parts is denoted by $D(i)$. The self-excitation of vibration which results in an increase in amplitude takes place in the regions $i > 1$. In this process, the feedback circuit's nonlinearity ensures amplitude limitation and leads to steady-state periodic vibration.

Steady-state motion parameters are sought for which the amplitude does not exceed the limit of the limiter's static characteristic's linear zone (Fig. 14.2, b), i.e. $a_l < \Delta$ when $\Delta > 0$ and $a_l < -\Delta/(1+q_0)$ when $\Delta < 0$.

Using relations (4.72) and (4.73) and taking account of (4.90), the following is obtained,

$$\begin{aligned} z(a_l)\cos\varphi &= \xi\cot\xi + Q_0, \\ z(a_l)\sin\varphi &= \frac{\psi\xi(\xi + 0.5\sin 2\xi)}{(4\pi\sin^2\xi)} \end{aligned} \tag{4.93}$$

The feedback circuit's harmonic amplification coefficient, which has a nonlinear characteristic (Fig. 4.21, c) when $a_l \geq G/Z_0$, has the form,

$$Z(a_l) = \frac{2G}{\pi a_l}\left(\frac{a_a Z_0}{G}\arcsin\frac{G}{a_a Z_0} + \sqrt{1 - \left(\frac{G}{a_a Z_0}\right)^2}\right) \tag{4.94}$$

This dependence is plotted in Fig. 4.22, b.

The following relation determining the periodic regime's frequency is found from (4.93):

$$\tan\varphi = \frac{\psi\xi(\xi + 0.5\sin 2\xi)}{4\pi\sin^2 2\xi(\xi\cot\xi + Q_0)} \tag{4.95}$$

Dependence $\xi(\tau)$ is plotted in Fig. 4.22 using (4.95). As follows from this figure, the steady-state self-sustaining oscillation frequency ξ is uniquely determined, in the vicinity of each natural mode, by the phase.

It must be noted that the possibility of a single-frequency periodic regime exists only at stability region $D(0)$'s boundary in Fig. 4.22, a. The self-sustaining oscillation's frequency and amplitude are subsequently found by using the plotting method shown in Fig. 4.22 by arrows. For a given value of phase φ, frequency ξ can be found from Fig. 4.22. From Fig.4.22, a, the amplification coefficient value Z which corresponds to the condition by which the periodic regime exists can be found. Finally the corresponding amplitude a_l (Fig. 4.22, b) can be found. By varying the phase φ, the complete parameter set ξ, a_l that correspond to realizable periodic regimes can be obtained. It is convenient to represent these results in the form of an amplitude-frequency characteristic as shown in Fig. 4.22, d. As shown in the previous subsection in which a feedback circuit's relay characteristic was considered, the amplitude-frequency characteristic coincides with the system's resonant curve when working in a forced vibration regime and subject to the driving force $F_l = 4G/\pi$. Since the harmonic amplification coefficients for both the relay characteristics and the characteristics under consideration are similar in the regions where the self-excitation conditions are satisfied, their corresponding amplitude-frequency characteristics are also similar. Therefore, in order to find and analyse the resulting periodic motion, it would make sense to consider the simplest and clearest instance of an amplifier's relay characteristic.

Fig. 4.22, d allows a clear understanding to be gained of how the system's vibration develops during a smooth change in phase. Each value of φ only corresponds to one point on the amplitude-frequency characteristic (Fig. 4.22, d). By changing the delay the resonant curve's branch, which corresponds to the system's first vibration mode, can be realized (the relevant branches for the system's two lowest modes are shown in Fig. 4.22, d). When the amplification coefficient Z_0 is reduced, the periodic motion's frequency region reduces further and further in the vicinity of the system's resonant frequency. The use of an amplifier which has a large amplification coefficient leads to a broadening of the frequency range. This happens mainly at higher frequencies.

All of the above is equally true, for a system in which an initial interference ($\Delta < 0$) is present, when the vibration is limited to the limiter's initial deformation $a_l < -\Delta/(1+q_0)$.

It's worth noting that an increase in the initial amplification coefficient Z_0 yields the possibility of multi-frequency regimes appearing. In order to exclude them during oscillation excitation in the vicinity of the system's first natural frequency ω_l, coefficient Z_0 can be reduced, which in turn leads to a decrease in the excited vibration's frequency range. Conversely, when designing ultrasonic systems, it is often necessary to excite higher vibration modes. In order to realize these higher modes, an increase in the feedback circuit's amplification coefficient Z_0 is required. Therefore, in order to extract the necessary vibration mode into the feedback circuit, a band-pass filter must be used. A band-pass filter can strongly distort the system's phase-frequency characteristic and degrade the properties by which the system stabilizes its resonant regimes. Although this problem can be solved successfully by the use of modern electronics, such a method leads to specific complications in the system's electrical makeup, and complicates its tuning.

3. In some cases, the necessary vibration mode's extraction can be achieved by alternative means. A delay element may be employed in order to change the feedback circuit's phase. This element can be naturally formed by using a microphone as the system's feedback sensor. This can be used to register the sound's radiation at one of the vibrating system's elements. The feedback circuit's delay t_0 is determined by the acoustic wave's propagation speed between the radiating element and the microphone which receives its acoustic signal.

In this instance, the problem under consideration is reduced to the study of a distributed nonlinear system's oscillation which incorporates a delay in its feedback circuit. A linear oscillator's self-oscillation when subjected to the action of a delayed force has been considered in Andronov & Maier (1946), Gonorovsky (1958), Perepeliatnik (1961), Rubanik (1969). All the relations obtained above, as the result of the problems considered in this subsection, transpire to be true when $\varphi = -\omega t_0$.

Fig. 4.23, a shows, for a system with an initial clearance ($\Delta > 0$), the D-partitions in the parameter plane Z_0, τ ($\tau = t_0 c/l$) as the frequency ξ is changed from 0 to 2π. Hatching denotes the equilibrium state's stable regions for which the characteristic equation has no roots with positive real parts. Crossing this boundary, in the direction of an increasing Z_0 coefficient, yields the appearance of two complex conjugate roots within the characteristic equation. These roots have positive real parts. In regions $D(i)$, where i is the number of roots which have positive real parts within a given region, vibration self-excitation occurs and gives rise to an increase in amplitude. The amplitude increase is limited by the amplifier's nonlinearity .

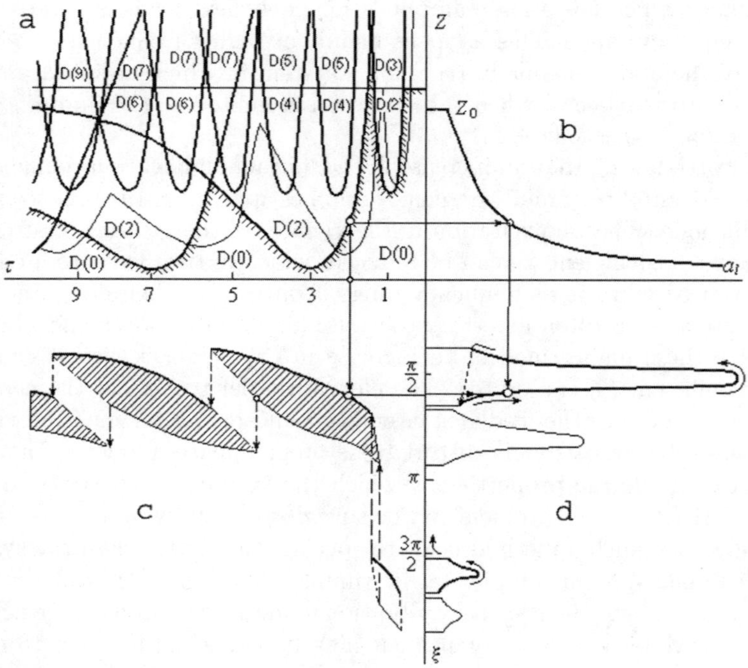

Fig. 4.23.

Fig. 4.23, b shows the dependence of the feedback circuit's harmonic amplification coefficient on the working tool's vibration amplitude a_l.

The existence of a single-frequency periodic regime is only possible on a stability region's boundary. In Fig. 4.23, c, the parts of dependence $\xi(\tau)$ which are plotted in accordance with relations (4.92) and satisfy the condition of existence mentioned above can be seen. As such, frequency ξ is uniquely determined by the delay τ for all values except those that are bifurcated for which the frequency jumps. The vibration amplitude is determined by using the plotting method which is shown in Fig. 4.23 by arrows: for a given value of τ_0 the values Z_0 (Fig. 4.23, a) and ξ (Fig. 4.23, c) are found, then the vibration amplitude a (Fig. 4.23, b) is obtained along with the corresponding point on the amplitude-frequency characteristic (Fig. 4.23, d).

As follows from the plot shown, different resonant curve branches, which correspond to different vibration modes, can be realized by changing the delay (Fig. 4.23, d shows resonant curve branches for the system's two lowest modes). An increase in τ causes vibration to be excited in an increasingly narrow band in the vicinity of the system's first natural frequency. A sufficiently small amplification coefficient value Z_0 can cause zones of periodic motion to alternate with areas of static stability. Conversely, sufficiently large values of Z_0, at the moment when bifurcated values of τ are passed, cause vibration jumps either from one mode to another, or inside the same mode as shown in Fig. 4.23, c, d by the dashed arrows. The use of an amplifier with a large amplification coefficient leads to a widening of the frequency range. This happens mainly at high frequencies. The largest frequency range realizable occurs when an amplifier exhibiting a relay characteristic is employed.

All of the above remains true for systems which have an initial interference ($\Delta < 0$) so long as the vibration is limited to the limiter's initial deformation ($a_l \leq -\Delta/(1+q_0)$). In this instance, Fig. 4.23 shows the self-excitation boundary (the thin lines in Fig. 4.23, a), the periodic vibration frequency's dependence on the delay (Fig. 4.23, c), and the amplitude-frequency characteristic (Fig. 4.23, d). The hatching in Fig. 4.23, c denotes the domains in which periodic regimes exist as the limiter's stiffness value varies from 0 to q. The dashed lines enclosing these areas are the boundaries due to the bifurcation present.

4. Periodic regime parameters are now sought for which the bar's lower cross section performs vibration which is accompanied by a separation from the limiter. Consideration is restricted to situations in which the feedback circuit's amplifier exhibits a relay characteristic. By using harmonic amplification coefficient expression (4.76), and reduced dynamic stiffness expression (4.90), equations (4.72) and (4.73) take the form,

$$a_l = \frac{\delta \cos \xi\tau}{\xi \cot \xi + q} \tag{4.96}$$

$$a_l = -\frac{4\pi\delta}{\psi\xi} \frac{\sin^2 \xi \sin \xi\tau}{\xi + 0.5 \sin 2\xi} \tag{4.97}$$

where, $\delta = 4Gl/\pi ES$.

From (4.96) and (4.97), the self-oscillation amplitude is found,

$$a_l = \delta \left\{ (\xi \cot \xi + q)^2 + \left[\frac{\psi \xi}{4\pi} \cdot \frac{(\xi + 0.5 \sin 2\xi)}{\sin^2 \xi} \right]^2 \right\}^{-1/2} \tag{4.98}$$

Expression (4.98) coincides with equation (3.98) which determines a bar's amplitude-frequency characteristic as it interacts with a limiter whilst subject to a harmonic force $F_l = 4G/\pi$ which acts in the bar's lower cross section. Fig. 4.24, a, shows amplitude-frequency characteristics for various limiter positions. It was created in accordance with the methodology described in section 3.2.

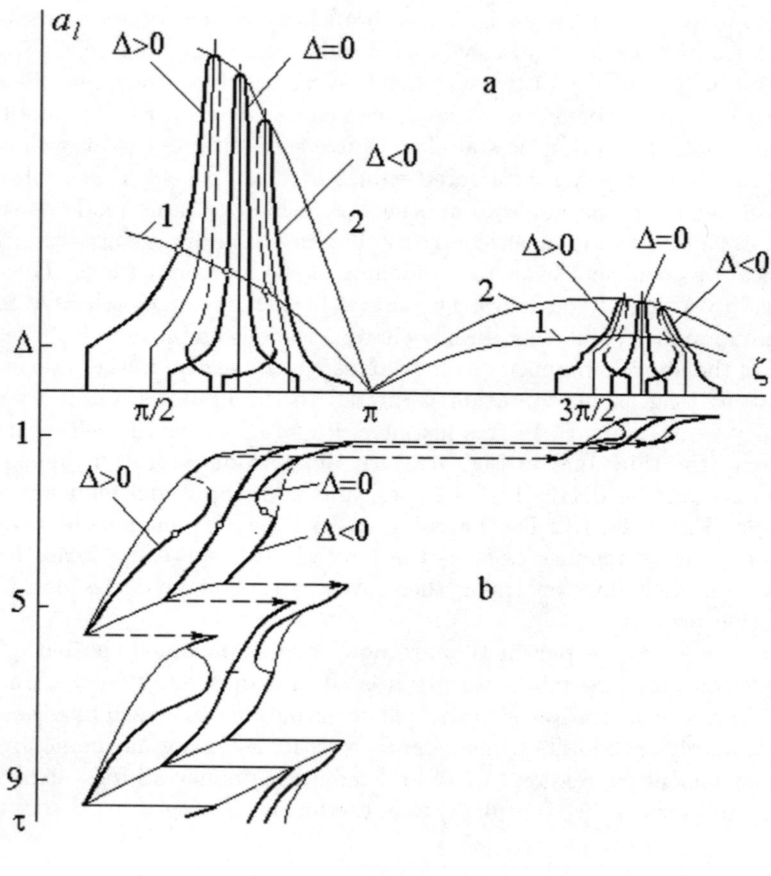

Fig. 4.24.

The way in which the oscillation frequency ξ is dependant on the delay τ can be found by solving the equation system (4.97) and (4.98). It must be noted, that according to (4.98), the equality $\xi \cot \xi + q = 0$ determines the backbone curve configuration which is shown in Fig. 4.24, a by dashed-dotted lines. By using (4.96) and (4.97), it can be concluded that values $\pi \leq \xi\tau \leq 3\pi/2$ correspond to resonant curve branches which are situated to the right of the backbone curve, and values $3\pi/2 \leq \xi\tau \leq 2\pi$ correspond to branches which are situated to the left of the backbone curve. Therefore, the points at which the resonant curve intersects (Fig. 4.24, a) with curve 1, plotted in accordance with dependence (4.97) in which $\xi\tau = const$, determine the self-sustaining oscillation frequency ξ which corresponds to the given phase. The resulting dependence of the oscillation frequency ξ on the feedback circuit's delay τ is shown in Fig. 4.24, b.

When $\xi\tau = 3\pi/2$, and from (4.99), the equation which determines the limiting amplitude curve 2 is obtained,

$$a_l = \frac{4\pi\delta}{\psi\xi} \frac{\sin^2 \xi}{\xi + 0.5 \sin 2\xi} \tag{4.99}$$

Its intersection with the backbone curve determines, according to (4.98), the highest points on the amplitude-frequency characteristic. It is worth noting that the harmonic linearization coefficient q has a value within the range $0 \leq q \leq q_0$. As such, the system can only realize motion which satisfies the conditions by which the motion can exist. These conditions were found at the end of the previous subsection. The domains of existence, which are denoted in Fig. 4.23 by hatching, are also shown in Fig. 4.23, b.

In this instance, as follows from Fig. 4.23, it is again possible to realize parts of the resonant curve which correspond to alternative vibration modes. This includes the possible vibration jumps from one mode to another, and jumps within a single mode. These jumps can occur on the bifurcated boundaries that were found above.

5. Autoresonantly excited systems that are designed in accordance with the arrangement shown in Fig. 4.20 have been realized on experimental machines which treat brittle materials ultrasonically and perform ultrasonic turning (see section 5.1).

Fig. 4.25, a shows the arrangement of an ultrasonic machine's vibration system. The magnetostrictive transducer 1 transmits vibration through the stepped concentrator 2 to tool 3. The concentrator is connected to carriage 4 at a vibration node, and is forced against the treated workpiece 5 by a constant force P. The power supply is fed from the delayed feedback circuit. In this case, the voltage in the magnetostrictive winding is generated by amplifier 6 which has a saturated characteristic and transforms the signal from the feedback sensor 7. The sensor, which in this instance is a microphone, is positioned a specific distance from magnetostrictive transducer's free end. The time taken for the acoustic wave to propagate from the transducer's radiating end to sensor 7 determines the delay in the feedback circuit. By changing the amplifier's saturation level and the time delay, the vibration's amplitude and frequency can be regulated. Adhering to the specific relations between amplitude and frequency, the system's resonant regime can be stabilized in the presence of a varying load which is determined by the tool's mass and driving force.

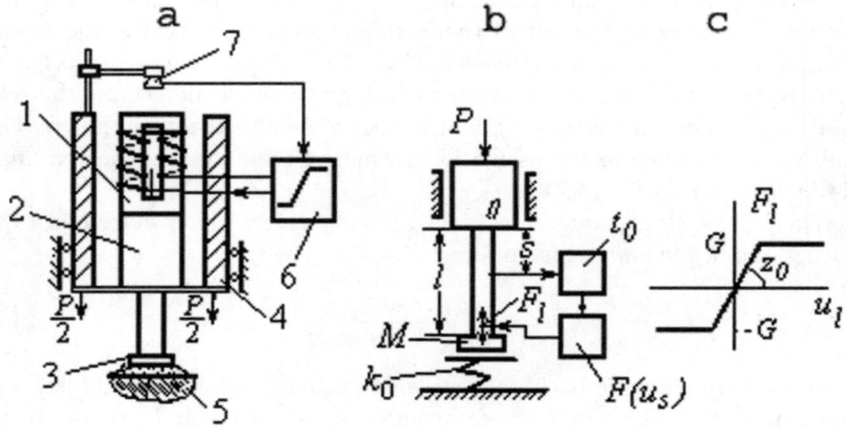

Fig. 4.25.

The system's dynamics can be described using a dynamical model (Fig. 4.25, b). The model consists of a homogeneous bar and has a length l which is equal to the length of the concentrator's final step. A mass M is attached at its end. The vibration in the bar is excited by a force F_l that is applied to its lower cross section. The force is generated by the feedback circuit as a result of the signal's nonlinear transformation $F_l = F_l(u_r)$ which is proportional to cross section $x = r$'s displacement. It is also subjected to the delay t_0. All of the general relations obtained in this section are valid for this model.

The system's amplitude-frequency characteristics for its two lowest vibration modes and varying values of compressive force P are plotted in Fig. 4.26, a using formulae (4.78) and (4.90). In this example it is assumed that the limiter's stiffness $k_0 \to \infty$, i.e. the situation in which it has the most pronounced influence on the system's vibrational character, is considered. The harmonic linearization coefficient is $k(a) = 2P/a_l$. The dashed-dotted lines show the backbone curves that relate the free vibration's frequency and amplitude and separate the resonant curve branches. Curve 1 shows the limiting amplitude line which shows the resonant curve's upper limits.

As follows from Fig. 4.26, a, when $\delta_p = Pl/ES = 0$ the system is linear and the limiter does not influence its vibration. The application of constant force P shifts the resonant curve's maximum into a higher frequency region and when $\delta_p > \delta = Fl/ES$ the curve's character is drastically changed: it

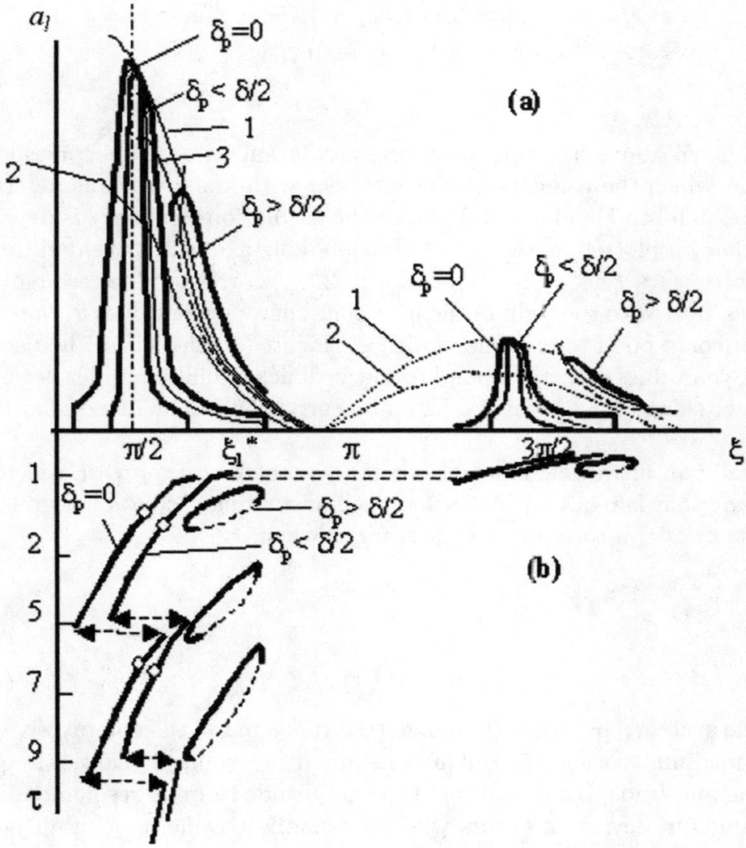

Fig. 4.26.

becomes double-valued, and an increase in δ_p causes its frequency range to be reduced.

The relation between the system's resonant frequency and constant force is given by the expression:

$$\frac{\delta_p}{\delta} = \frac{2\pi}{\psi} \frac{(\sin 2\xi - \mu\xi \sin^2 \xi)}{(2\xi + \sin 2\xi)} \tag{4.100}$$

The maximum values of function (4.100) correspond to the limiting values at which the system's resonant curves degenerate into points. For example, when $\mu = 0$, the values for the first vibration mode are as follows:

$$\xi_1 = 0.71\pi, \delta_{p1} = 1.75\delta/\psi;$$
$$\xi_2 = 1.74\pi, \delta_{p1} = 0.76\delta/\psi.$$

As shown above, the self-sustaining oscillation parameters correspond to points at which the resonant curves intersect with line 2 which is determined by expression (4.97). The way in which the oscillation frequency is dependant on the delay is plotted in Fig. 4.26, b. It should be taken into consideration that values $\pi(n-1) \leq \xi\tau < \pi(2n-1)/2$ $(n = 1, 2, \ldots)$ correspond to resonant curve branches located to the right of the backbone curve, while values $\pi(2n-1)/2 < \xi\tau \leq \pi n$ correspond to branches which are located to the left of the backbone curve. Even values of n correspond to the feedback circuit amplifier performing direct switching $(\delta > 0)$ whilst odd values correspond to its inverted switching $(\delta < 0)$.

Usually, in ultrasonic machines, it is resonant vibration with a maximum amplitude that is required. A resonant self-sustaining oscillation regime, with a frequency ω^* (autoresonance), is realized when,

$$|\sin \omega^* t_0| = 1$$

or,

$$\tau = (2n-1)\pi/2\xi^* \qquad (4.101)$$

If the system's parameters change (the tool's mass, the compressive force, or the medium's properties) at a constant delay value, it causes frequency detuning and leads to a deviation of the amplitude from its resonant value a^*. The minimum deviation occurs when the smallest value, among all possible values gained from (4.101), is employed. By assigning $n = 1$, and assuming that the resonant frequency exists at its limiting value $\xi^* = \xi_l = 0.71\pi$, adjustment parameter $\tau = 0.71$ is found from (4.101). Fig. 4.26, a shows curve 3 which is plotted in accordance with equation (4.97) and determines the self-oscillation amplitude for the given state of tune. Under this state of tune, the amplitude's deviation from its resonant value a^* is no greater than 10% throughout its whole working range. As such, the system under consideration can be effectively employed to stabilize the system's resonant regimes when subjected to a significantly changing external load.

6. In experiments investigating the system's idling motion, a piezoelectric sensor was glued to the concentrator's end and was used to register its amplitude. As a result of studying the vibration system in its forced vibration regime, its amplitude-frequency characteristic was plotted, and was used to find the resonant frequency $l = 20.4kHz$, and absorption coefficient $\psi = 2\pi/Q$. $Q = 126$ is the system's Q-factor.

When considering the dynamical model (Fig. 4.25, b), the resonant frequency and idling motion amplitude are determined by the expressions $f = c/4l, a_l = 16Fl/\pi\psi ESandF = 4G/\pi$, in which the bar's effective length is $l = 63.5mm$. All further experiments were conducted at a constant excitation level for which the idling motion's amplitude, when operating in its resonant regime, was $a = 10\mu m$. The power consumed by the magnetostrictive transducer was $60W$. The excitation force, whcse effect was relocated to the bar's lower cross section, was considered to be $F = 100N$, and the feedback circuit's saturation level was $G = 79N$. (Calculations were made for the following parameter values: $c = 5.17 \cdot 10^5 cm/s, E = 2 \cdot 10^7 N/cm^2$ and $S = 3.14cm^2$).

The self-excitation conditions and steady-state vibration parameters were studied in the self-sustaining oscillation regime, during idling motion, for a wide range of delay values. The results of this study are shown in Fig. 4.27 in which the points represent experimental data that is superimposed on to the calculated curves. Fig. 4.27, a shows the self-excitation boundary for a saturated system's characteristic. The system is similar to that plotted in Fig. 4.25, c.

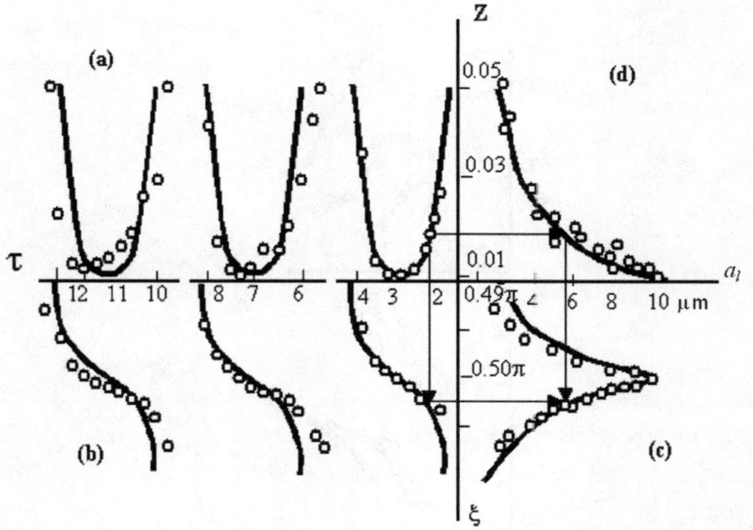

Fig. 4.27.

The boundary's parametric equations have the form (4.92) when $\varphi = -\xi\tau$. By excluding Z_0 from these equations, an expression (4.95) is obtained which determines the steady-state oscillation frequency. Fig. 4.27, b shows how the oscillation frequency ξ is dependant on the delay value τ. The realizable amplitudes are shown in Fig. 4.27, c. They correspond strictly to the amplitude-frequency characteristic that was obtained for a forced vibration regime. The theoretical resonance curve is plotted using equations (4.78) and (4.90) when $q = 0$. Fig. 4.27, c shows the graph of harmonic amplification coefficient (4.94) and the experimental values $Z = k\nu_M/\nu_\Omega$. ν_M and ν_Ω are the voltage across the magnetostrictive transducer's winding, and the voltage registered by the feedback sensor. $k = 2 \cdot 10^{-6}$ is a constant coefficient which is calculated from values which correspond to autoresonance. The curves shown in Fig. 4.27 are related to each other by the plotting method shown by arrows.

The interaction force between the tool and the workpiece was recorded during ultrasonic cutting when the vibration system was loaded by the constant force P. A piezoelectric element was used as a sensor. It was glued in between the massive base and the glass plate being treated, and placed within the abrasive slurry (see section 4.1). The experimental results, which were obtained for different values of force P, are shown in Fig. 4.28. They show the change in character of the impact interaction force's amplitude F_m (Fig. 4.28, a), and the self-oscillation frequency f (Fig. 4.28, b), during a smooth change in delay value.

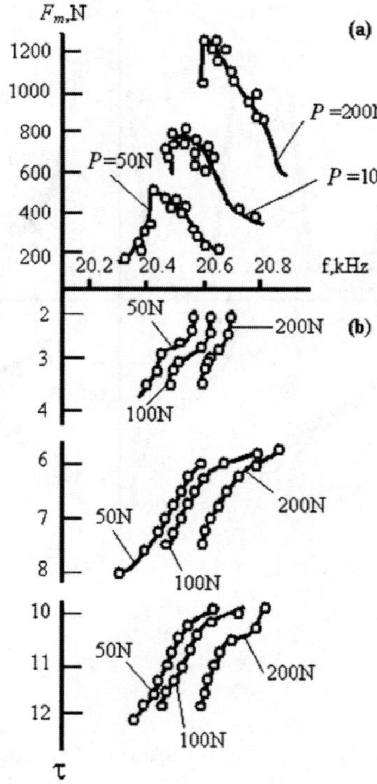

Fig. 4.28.

It is worth noting that experiments conducted during both idling motion, and when the system is under load, have confirmed the theoretical conclusions about how the self-sustaining oscillation's amplitudes and frequencies correspond to the resonant curves obtained when the system is operating in a forced vibration regime. The nonlinear effects as described above were also observed. The force value $P = 50N$, for the given excitation level $F = 100N$, is the limiting value for which the resonant curve (Fig. 4.26, a) remains single-valued in the whole domain of existence. Forces in excess of this value yield the appearance of a second branch, and vibration jumps at specific frequencies.

Resonant vibration which corresponds to the amplitude-frequency characteristic's maximum value was realized at the constant delay $\tau = 3$. The dependencies, as found from experimentation, of the oscillation frequency and impact force amplitude on the constant force are shown in Fig. 4.29. The calculated result, found by using formula (4.100) when $\mu = 0$, is also shown (line 1) for comparison.

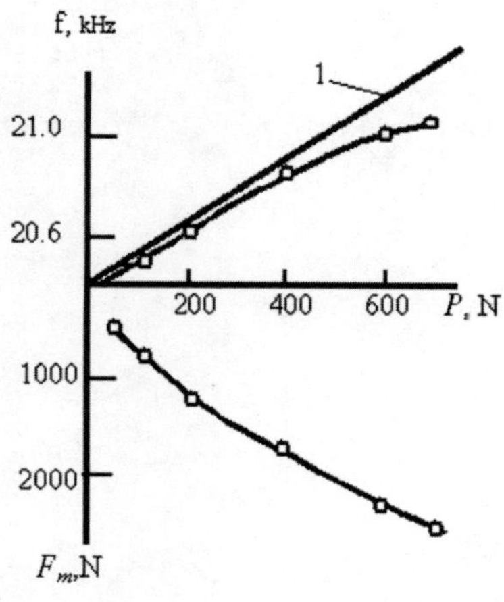

Fig. 4.29.

It is emphasized that the system self-excites and reaches its resonant regime, with no additional tuning, for all of the constant force values that are shown in the graph. The results achieved are shown in column AR of Table 4.1 (see section 4.1). Tests have demonstrated that the system suggested for exciting and automatically stabilizing the machine's resonant regimes also works effectively when other parameters are changed: the attached tool's mass, the workpiece's stiffness and size, the temperature, etc.

5

Ultrasonically assisted machining

It is like a barber's chair, that fits all buttocks.

William Shakespeare (1564-1616): All's Well That Ends Well

5.1 Ultrasonically assisted cutting as a nonlinear process

1. Many modern materials, such as high-strength, heat resistant and stainless steals and alloys, titanium, ceramics and other non-metallic materials, are hardly suitable for traditional machining methods. Treating parts which are made from these materials using traditional cutting machines lead to the chipping or destruction of the part's surface layer, and results in a poor surface quality. Conversely, it is also known that when cutting soft materials such as pure copper and aluminium, the cutting process is accompanied by a process by which the workpiece material sticks to the cutting tool's edge. This also has the effect of deteriorating the workpiece surface quality.

The treatment quality can be improved by transforming the cutting process. This is done by exciting high frequency (ultrasonic) vibration at the tool's cutting edge. This is precisely why such a process is designated as being ultrasonic (Markov (1962, 1966, 1980)) (or vibration (Kumabe (1979)) cutting. The arrangement of an ultrasonic cutting process and some of the effects that were found experimentally during its realization were described in section 1.4. One of these effects is the decrease in cutting force in the presence of vibration. This effect is most noticeable when vibration occurs in the direction of cut. As previously, a constant cutting force component is meant here. This can be estimated by measuring the torque M which is developed by the machine's drive.

In the present section these effects are explained using a proposed rheological model of the ultrasonic cutting process. It is established again, as in the processes considered previously, that the influence of high frequency vibration leads to a phenomenological transformation of the material's elastic-plastic characteristic into that of a visco-plastic material. The dry friction present also appears to transform into viscous friction. The dynamical characteristics of the transformed cutting process are constructed. It is shown that the vibro-impact process plays a special role and is the most efficient process when present in the cutting zone. Also, machine dynamics during ultrasonic cutting is investigated, and amplitude-frequency characteristics are constructed.

Fig.5.1 shows the arrangement of a device that allows ultrasonic vibration to be superimposed on to a cutting process. The workpiece 1 being treated rotates with a constant angular speed Ω and is driven by a lathe which develops a constant torque M. The treatment is undertaken by a cutter 2. The cutter vibrates $u(t) = a\cos\omega t$ at an angular frequency ω and amplitude a, which is generated by the ultrasonic vibration system. Mechanical vibration is excited due to the piezoelectric transducer 3 being fed an alternating current from generator 4. The vibration is transmitted to the cutter with an increase in amplitude via concentrator 5. The vibration system is fixed in frame 6, and attached to the lathe by means of holder 7.

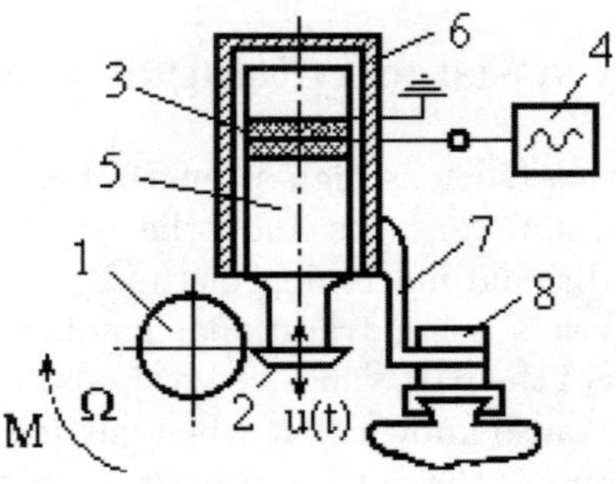

Fig. 5.1.

In order to achieve the maximum effect from the vibration that is superimposed on to the cutting process, the vibration system needs to be tuned to resonance. However, as was the case in processes considered previously, the system resonant frequencies are largely dependant on the process parameters, and vary under different cutting regimes. Tuning to resonance is further complicated by the fact that the load experienced by the vibration system during cutting is strongly nonlinear. In the following, it is shown that the system dynamical behaviour becomes increasingly saturated by nonlinear effects under vibration cutting conditions. Consequently, the amplitude-frequency characteristic nonlinear distortions, which are caused by the load, become increasingly non traditional.

2. During turning, cutting forces are mainly generated due to the plastic deformation of the layer being cut and overcoming the frictional forces which act on the cutters working surfaces (Granovsky & Granovsky (1985), Shaw (2005), Tlusty (2000)). As the cutter vibrates, an alternating cutting force $F(t)$ appears on its cutting edges, and has a direction which is opposite to the speed $\dot{u}(t)$ of the cutters relative motion,

$$u(t) = vt + u^0(t) = vt + a\cos\omega t \tag{5.1}$$

where, $v = \Omega r$ is the cutting speed, and r is the radius of the workpiece being turned. The alternating ultrasonic cutting force shall be described using the contact interaction's dynamic characteristic $f = f(u,\dot{u})$. This can be represented as a sum of two components,

$$f(u,\dot{u}) = f_1(u,\dot{u}) + f_2(u,\dot{u}) \tag{5.2}$$

The first component $f_1(u,\dot{u})$ is dependant on the elastic-plastic properties of the material being treated. During traditional turning, i.e. with no vibration, this cutting force component is determined by the plastic deformation force. When considering a vibration cutting process, the load cyclic nature has to be accounted for, and the material elastic deformation zone at the points of contact must be introduced into its dynamic characteristic. In order to construct a dynamic characteristic it is assumed, as previously, that for every cutting speed value v , there is a steady-state periodic regime $u_0(t)$ for the cutters vibration. Therefore, consideration can be restricted to the cutter's interaction with the workpiece during one vibration cycle.

In order to construct a dynamic characteristic, the following scenario is thought through: by slowly increasing the workpiece torque M against a static tool, the workpiece is seen to rotate. If the torque M is sufficiently small, the workpiece will turn by a specific angle and reach a position of dynamic equilibrium and the rotation speed will become equal to $v = 0$. This means that the deformation in the contact zone does not exceed the limits of the material elastic characteristic. In this situation, the dynamic characteristics component $f_1(u,\dot{u})$ can be represented in the form,

$$f_1(u,\dot{u}) = f_1(u) = k_0(u-\Delta)\eta(u-\Delta) \tag{5.3}$$

where, k_0 is the static stiffness in the contact zone between the cutter and the workpiece, Δ is the cutting edges relative coordinate at the moment of contact interaction, and $\eta(u)$ is a unit function.

This component does not depend on the deformation speed, and when within the elastic zone, it describes both the loading and unloading cycles.

If the torque M is increased, a situation will occur in which the deformation present in the contact zone will enter the materials plastic deformation zone. By using an elastic-plastic model for the material under consideration, the ultrasonic cutting force component $f_1(u,\dot{u})$ can be written in the following way:

$$f_1(u,\dot{u}) = k_0 \left\{ \begin{array}{l} \eta(\dot{u})[(u-\Delta)\eta(u-\Delta)- \\ -(u-\Delta-d)\eta(u-\Delta-d)]+ \\ +\eta(-\dot{u})(u+d-u_m)\eta(u+d-u_m) \end{array} \right\} \tag{5.4}$$

alternatively, this can be written as,

$$f_1(u,\dot{u}) = \begin{cases} 0 & u \le \Delta & \dot{u} \ge 0 \\ k_0(u-\Delta) & \Delta \le u \le \Delta + d & \dot{u} \ge 0 \\ D_0 & \Delta + d \le u \le u_m & \dot{u} \ge 0 \\ D_0 + k_0(u-u_m) & u_m - d \le u \le u_m & \dot{u} \le 0 \\ 0 & u \le u_m - d & \dot{u} \le 0 \end{cases} \tag{5.5}$$

where, D_0 is the force that is present in the contact zone that corresponds to the materials yield limit, $d = D_0/k_0$, and

$$u_0 = a[\sqrt{1-(v/a\omega)^2} + (v/a\omega)\arcsin(v/a\omega)$$

is the maximum value of function (5.1) over the period considered.

The ultrasonic cutting process dynamical characteristic component (5.4) is shown in Fig.5.2 by thin line 1. As can be seen, it is entirely analogous to the elastic-plastic deformation process characteristic, which was considered in detail in Chapter 3.

The second component $f_2(u,\dot{u})$ of dynamic characteristic (5.2) is determined by frictional forces which accompany the cutting process. It is assumed that this component is proportional to $f_1(u,\dot{u})$ and has a direction opposite to that of the cutters speed relative to the workpiece. As such, this component is described using the following expression:

$$f_2(u,\dot{u}) = \vartheta f_1(u,\dot{u})\mathrm{sgn}\dot{u} \tag{5.6}$$

where, ϑ is a coefficient of dry friction.

This component is shown in Fig.5.2 by thin line 2.

The total interaction characteristic (5.2) (the thick line in Fig.5.2) can be represented as follows:

$$f(u,\dot{u}) = f_1(u,\dot{u})(1 + \vartheta \cdot \mathrm{sgn}\dot{u}) \tag{5.7}$$

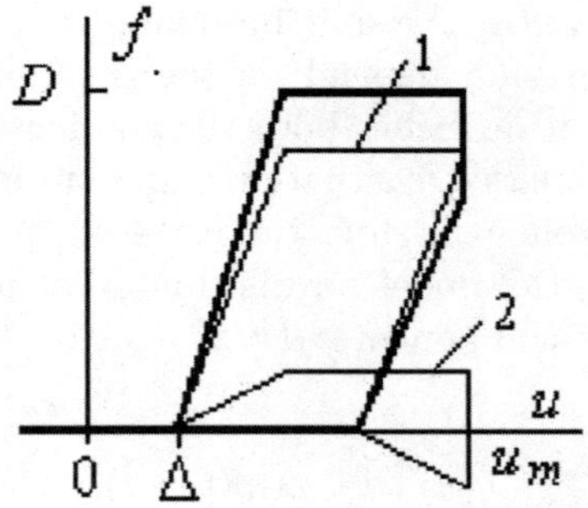

Fig. 5.2.

Characteristics of this form have been discussed in section 3.1.

Cyclic interactions between the cutter and the workpiece occur according to the following scenario. As the tool moves upwards (see Fig.5.1), no interaction force is present until the cutter reaches the workpiece being treated at which point $u = \Delta$. Subsequent to this, a linear loading occurs in the area of interaction and $\Delta \leq u \leq \Delta + d$. In this instance, the material in the contact zone obeys Hook law. After the materials yield limit is reached, the material plastically deforms, i.e. the cutting process, along with the appearance of cutting waste, takes place (cutting in its conventional meaning). This process continues until the cutting tip is about to change direction. At this point in time, the sign of $\dot{u}(t)$ changes and the displacement $u(t)$ reaches its maximum value u_m (see (5.5)). During this process the interaction force between the cutter and the workpiece remains constant and equal to the conventional turning cutting force $f = D_0(1 + \vartheta)$, which occurs in the absence of ultrasonic vibration. Subsequently, elastic unloading occurs until the cutter separates from the workpiece ($f = 0$).

3. The influence of ultrasonic vibration on the cutting process can now be estimated. It is taken into account that, as a consequence of the tools motion being periodic, the interaction force $F(t) = f[u(t), \dot{u}(t)]$ is a periodic function of time with a period $T = 2\pi/\omega$. As was done previously in sections 1.2 and 1.3, the momentum theorem can be used to obtain formulae that relate the constant force P with the cutters motion parameters and the process elastic-plastic characteristics,

$$P = \frac{1}{T} \int_0^T F(t)dt = \frac{1}{T} \int_0^T f_1[u(t), \dot{u}(t)]dt \tag{5.8}$$

From (5.4)-(5.7) it follows that, for conventional turning with no vibration ($a = 0$), and also for vibration cutting in which the speed v exceeds the vibration speed ($v \geq a\omega$), the interaction force is equal to $F(t) = D_0(1+\vartheta) = D$. According to (5.8), a constant force $P = D$ is needed to overcome this. If the cutter vibration is such that $a\omega > v$, then in the interval $0 \leq t \leq T$, the interaction force $F(t) \leq D$, and the constant force required becomes $P < D$, i.e. there is a decrease in cutting force. This was observed through experimentation.

The static force P dependence on the tool vibration parameters and cutting speed, during vibration cutting, is sought. By substituting (5.5) - (5.7) into (5.8), and integrating, the following is obtained,

$$P = \frac{1}{2\pi} \left\{ \begin{array}{l} k_0(1+\vartheta)[\frac{v}{2\omega}(\tau_2^2 - \tau_1^2) - \Delta(\tau_2 - \tau_1) + a(\sin\tau_2 - \tau_1)] + \\ +k_0(1-\vartheta)[\frac{v}{2\omega}(\tau_4^2 - \tau_3^2) - (\Delta + \delta)(\tau_4 - \tau_3) + a(\sin\tau_4 - \tau_3)] + \\ +D(\tau_3 - \tau_2) \end{array} \right\} \tag{5.9}$$

where, according to (5.2),

$$\delta = u_m - \Delta - d \tag{5.10}$$

is the arc length of the surface that was treated during one period of vibration, $\tau_i = \omega t_i$ (i=1,...,4), and t_i are the switching moments between the characteristics parts (5.5) which are determined by the following equations:

$$\begin{gathered}\cos\tau_1 = \tfrac{1}{a}\left(\Delta - \tfrac{v}{\omega}\tau_1\right), \quad \cos\tau_2 = \tfrac{1}{a}\left(\Delta + d - \tfrac{v}{\omega}\tau_2\right), \\ \tau_3 = \arcsin\tfrac{v}{a\omega}, \qquad \cos\tau_4 = \tfrac{1}{a}\left(\Delta + \delta - \tfrac{v}{\omega}\tau_4\right), \\ \tau_3 \in [0, \pi/2], \;\; \tau_{1,2} \in [-(\pi + \tau_3), \tau_3] \text{ and } \tau_4 \in [\tau_3, (\pi - \tau_3)].\end{gathered} \tag{5.11}$$

The average cutting speed can now be calculated as follows:

$$v = \frac{\delta}{T} = \frac{\omega}{2\pi}(u_m - \Delta - d) \tag{5.12}$$

From (5.12), the coordinate at which the interaction begins is found,

$$\Delta = u_m - d - 2\pi\frac{v}{\omega} \tag{5.13}$$

By substituting the solutions of equation (5.11) into equality (5.10), and by virtue of (5.13), the static force P in regimes when the cutter contact with the workpiece is periodically interrupted, is determined. In this instance, the interaction force consists of a sequence of impulses, and it is designated as being a vibro-impact process.

With an increase in cutting speed v, the contact time between the tool and the workpiece is also increased. At some value of cutting speed, the vibro-impact regimes is transformed into continuous vibration cutting regimes, in which the tool and workpiece remain in contact throughout the whole process. The cutting force in continuous cutting regimes is described by the same expression (5.9) in which the switching moments limiting values should be used. This corresponds to the following expressions, which determine the switching moments in equations (5.11):

$$\tau_1 = -\left(\pi + \arcsin\frac{v}{a\omega}\right), \qquad \tau_4 = \tau_1 + 2\pi = \pi - \arcsin\frac{v}{a\omega}$$

Fig.5.3 shows how the static force P is dependant on the cutting speed v. It is plotted for various values of a/d. The dashed line divides the regions of continuous cutting (upper region), and vibro-impact cutting (lower region). As can be seen from this figure, an increase in cutting speed is accompanied by an increase in cutting force until the value $P = D$ is reached, at which point the speed $v \geq a\omega$.

These results are in agreement with the experimental data (Ji et al. (1991), Kumabe (1979), Kremer (1995), Markov (1962, 1966, 1980)) obtained whilst studying the effect by which the cutting force is decreased when ultrasonic vibration is superimposed on to the cutting tools motion. This can be seen by

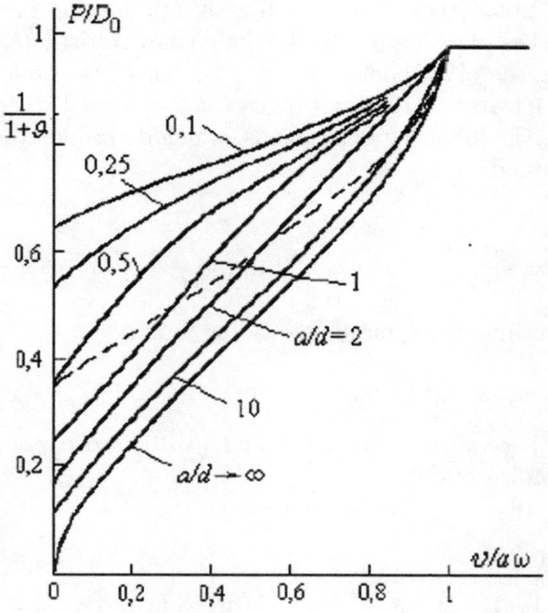

Fig. 5.3.

comparing Fig.5.3 with the experimental curve shown in Fig.1.5, which was taken from paper (Isaev & Anokhin (1961)).

The maximum decrease in static cutting force is achieved when rigid-plastic materials ($k_0 \to \infty$) are treated. This corresponds to the curve, which is denoted by the index $a/d \to \infty$ in Fig.5.3. In this case, the static force is described by the expression,

$$P = \frac{D}{2\pi}(\tau_3 - \tau_1) \tag{5.14}$$

where the switching moment $\tau_3 = \arcsin \frac{v}{a\omega}$, and τ_1 is given by the root of the equation,

$$\cos \tau_1 = \frac{1}{a}\left(u_m - \frac{v}{\omega}(2\pi + \tau_1)\right)$$

As follows from Fig.5.3, for a given speed v, force P decreases with an increase in the vibration amplitude a. For small vibration amplitudes, $a \leqslant v/\omega$, the cutting force $P = D$, and the cutting process is not influenced by the ultrasound present. This feature has been observed in all known experimental studies ((Isaev & Anokhin (1961), Komaraiah et al. (1988), Markov (1962, 1966, 1980)))

It should be noted that when $v = 0$ a situation exists in which the cutters vibration takes place inside the dynamic characteristic (5.5) elastic zone without entering the plastic deformation zone. This leads to a dynamic equilibrium regime. Clearly, in the limiting case $u_m = a$, and when $v = 0$, (5.11) and (5.13) allow the following expressions, which determine the vibro-impact regimes, to be found,

$$\tau_1 = -\arccos\frac{a-d}{a}, \quad \tau_2 = \tau_3 = 0, \quad \tau_4 = -\tau_1 = \arccos\frac{a-d}{a} \tag{5.15}$$

those for the continuous regimes may also be found,

$$\tau_1 = -\pi, \qquad \tau_2 = \tau_3 = 0, \qquad \tau_4 = -\tau_1 = \pi \tag{5.16}$$

By substitution values from (5.15) and (5.16) into expression (5.9), the following relation is obtained:

$$P = \begin{cases} D_0(1-\frac{a}{d}) & a \le d/2 \\ \frac{D_0 a}{\pi d}\left[\sqrt{1-(1-\frac{d}{a})^2} - (1-\frac{d}{a})\arccos(1-\frac{d}{a})\right] & a \ge d/2 \end{cases} \tag{5.17}$$

Expression (5.17) determines the static force value, that after which, causes the appearance of residual plastic deformation. This, consequently, signifies the beginning of the cutting process. The way in which limiting force (5.17) is dependant on the cutters vibration amplitude a is shown in Fig.5.4. It should be noted that, as $a \to 0$, the static force P jumps from $P = D_0$ to $P = D = (1+\vartheta)D_0$. This can be explained as follows. Any vibration, however small, alleviates the "hanging" as caused by dry friction, transforming it into viscous friction for small relative speeds of motion (see section 1.3).

4. As in all of the ultrasonic machines considered above, when performed ultrasonically, the interaction force between the tool and workpiece causes a load to act on the device vibration system. This is shown in Fig.5.1. Methods for finding the loads parameters were considered in detail in section 3.1. Accounting for the motion under considerations periodic nature, its dynamic characteristic can be written approximately in the form,

$$f(u, \dot{u}) \cong P(v, a) + k(v, a)u^0 + b(v, a)\dot{u}^0 \tag{5.18}$$

where, $P(v, a)$, $k(v, a)$ and $b(v, a)$ characterize the constant, elastic, and dissipative components of the load.

These quantities can be found in accordance with the method described in subsection 3.1.6 by using the relations,

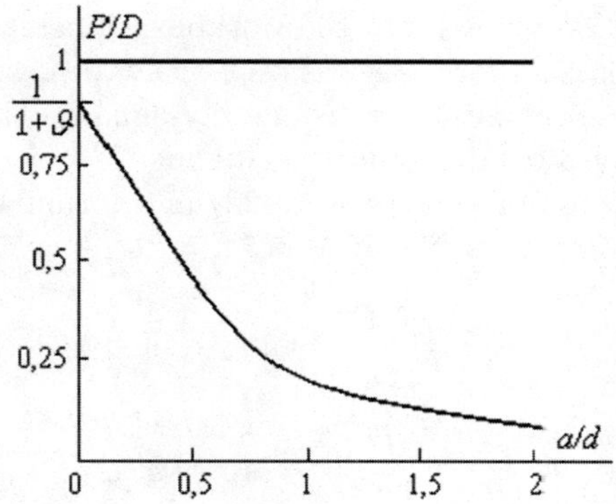

Fig. 5.4.

$$P(v,a) = \frac{1}{T}\int_0^T f[u(t),\dot{u}(t)]dt$$

$$k(v,a) = \frac{2}{Ta}\int_0^T f[u(t),\dot{u}(t)]\cos\omega t\,dt \tag{5.19}$$

$$b(v,a) = -\frac{2}{Ta}\int_0^T f[u(t),\dot{u}(t)]\sin\omega t\,dt$$

where, $T = 2\pi/\omega$ is the tool vibration period.

Note that the first formula in (5.19) coincides with (5.8) and describes the influence of vibration on the cutting force constant component. This was studied in detail in the previous section. The results obtained show that the vibration effect is most pronounced when treating rigid-plastic materials, $k \to \infty$. Conversely, the tool's interaction with the workpiece influences the vibration system dynamics the greatest in the same instance.

In this situation, the contact interaction's dynamic characteristic is given by,

$$f(u,\dot{u}) = \frac{D}{2}\eta(u - \Delta)(1 + \mathrm{sgn}\dot{u}) \tag{5.20}$$

The reader is reminded that characteristics of this type have been considered previously whilst studying ultrasonic machining systems for brittle materials (see sections 4.1 and 4.2). The ultrasonic cutting considered now has two pronounced peculiarities. Firstly, the cutting speed cannot be assumed to be small when compared to the tool's vibration speed. Secondly, in this instance, cutting is performed at a given cutting speed, not for a given driving force. In other words, the tool is now fed into the workpiece kinematically as opposed to it being fed by a constant force, as was previously the case. In the following, it is shown that these peculiarities considerably enhance the general picture, which describes the systems dynamical behaviour.

After the substitution of expression (5.20) into formulae (5.19), the integration coefficients k and b are found to be,

$$k(v,a) = \frac{D}{\pi a}(\sin\tau_3 - \sin\tau_1) \tag{5.21}$$

$$b(v,a) = \frac{D}{\pi a\omega}(\cos\tau_3 - \cos\tau_1) \tag{5.22}$$

where, τ_1 and τ_3 are switching moments determined by equations (5.14).

Equations (5.21) and (5.22), and relations that determine the switching moments, should be complemented by the condition that reflects the equality of the layer being cut, and the workpiece displacement during one vibration period, i.e.

$$u_m - \Delta = 2\pi v/\omega \tag{5.23}$$

where, $u_m = u(\tau_3) = (v\tau_3/\omega) + acos\tau_3$. From (5.23), and after some transformations, it is found that,

$$\frac{\Delta}{a} = \sqrt{1-\left(\frac{v}{a\omega}\right)^2} - \frac{v}{a\omega}\left(\pi - \arcsin\frac{v}{a\omega}\right) \tag{5.24}$$

The substitution of values τ_1 and τ_3 into (5.21) and (5.22), by virtue of (5.24), give values for coefficients k and b and can be represented in the form,

$$k(v,a) = \frac{D\omega}{\pi v}K\left(\frac{v}{a\omega}\right), \quad b(v,a) = \frac{D}{\pi v}B\left(\frac{v}{a\omega}\right) \tag{5.25}$$

where,

$$K(v/a\omega) = (v/a\omega)(\sin\tau_3 - \sin\tau_1), \quad B(v/a\omega) = (v/a\omega)(\cos\tau_3 - \cos\tau_1)$$

Fig.5.5 shows graphs of the functions $K(v/a\omega)$ and $B(v/a\omega)$ which determine the elastic and dissipative load components. As is seen, both load components are non-monotone functions and are equal to zero when $v/a\omega = 0$, $v/a\omega \geq 1$ and as $a\omega \to \infty$. These functions have maximum values when $v/a\omega = 0.5$ and $v/a\omega = 0.8$ respectively. These load peculiarities determine the specific behaviour of the ultrasonic cutting device.

5. As shown in section 2.1, and in the devices analysed above, ultrasonic cutting device dynamics can be considered using generalized dynamic models. In these models, all forces and dynamic stiffnesses are reduced to one element. In this situation, the natural element choice is the cutter. Its vibration is described by the equation,

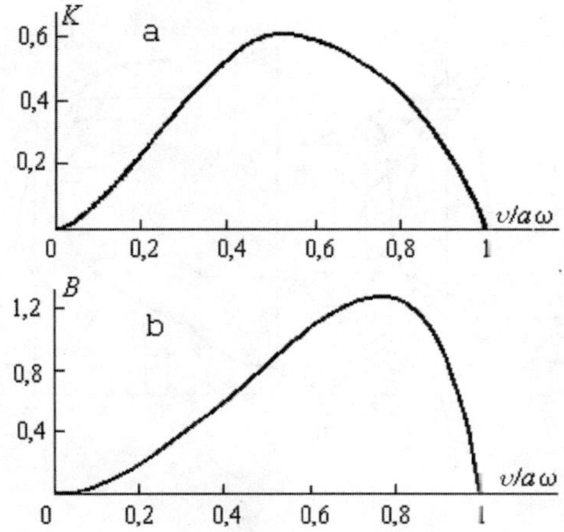

Fig. 5.5.

$$u(t) = W^{-1}(j\omega)[N(j\omega)e^{j\omega t} - f(u, \dot{u})] \tag{5.26}$$

where, $N(j\omega)e^{j\omega t}$ is the exciting force's complex amplitude whose effect has been relocated to the tool, and $W(j\omega)$ is the vibration system relocated dynamic stiffness.

After substituting the linearized characteristic (5.18) into (5.26), the complex amplitude of the cutter motions periodic component is found,

$$\tilde{a} = \frac{N(j\omega)}{W(j\omega) + k(v, a) + j\omega b(v, a)} \tag{5.27}$$

Taking account of the fact that the load coefficients (5.25) are dependant on the vibration speed, from (5.27) the following equation is obtained:

$$a\omega = \left| \frac{\omega N(j\omega)}{U(\omega) + k(v, a\omega) + j\omega[V(\omega) + b(v, a\omega)]} \right| \tag{5.28}$$

where, $U(\omega) = ReW(j\omega)$ and $V(\omega) = \omega^{-1}ImW(j\omega)$. Transcendental equation (5.28) can be solved using the following graphical method (Fig.5.6).

Assuming that k is the independent variable, and using equation (5.28), graphs can be plotted at the fixed frequency values ω, that show $a\omega$ dependence on k. In Fig.5.6,a these curves, denoted by the numbers 1, 2 and 3, are plotted for three frequency values where $\omega_1 < \omega_2 < \omega_3$. Curves of type 1 correspond to frequencies below the resonance of the systems linear part for which $U(\omega) \geq 0$. Curves of type 2 and 3 are obtained for frequencies situated in the region above resonance where $U(\omega) < 0$. When plotting these curves, the correspondence between each value of k and b, in accordance with the dependencies shown in Fig.5.5, should be found.

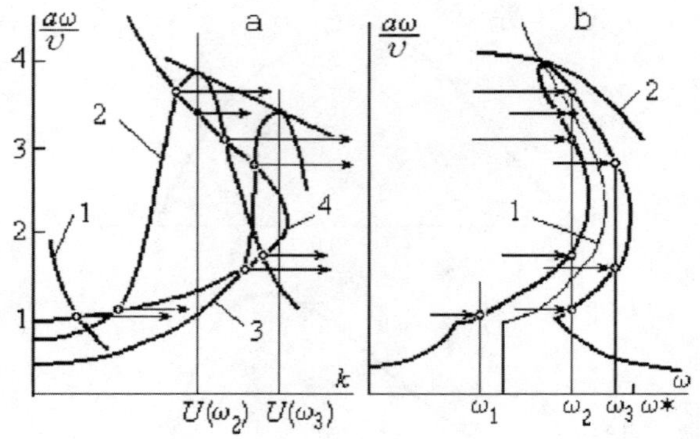

Fig. 5.6.

The points at which these curves intersect with graph 4, which shows the inverse function $\omega a(k)$ and is plotted using Fig.5.5, give the required solutions.

The procedure described allows the tool vibration amplitude-frequency characteristic in the presence of a load to be constructed; it is plotted in Fig.5.6,b using a thick line. Fig.5.6,b shows that, depending on the given frequency and specific parameter combinations, the system under consideration can have one, three or five solutions, which belong to different resonant curve branches. If the cutter vibration speed is sufficiently small, such that $a\omega \leq v$, the realizable parts of the linear system resonant curve correspond to the points at which curves (5.28) intersect with the vertical line $k = 0$ in Fig.5.6.

Accounting for the fact that the ultrasonic vibration system has a high Q-factor, meaning that the dissipative losses which are determined by quantities $V(\omega)$ and $b(v, a\omega)$ are small, it can be supposed that the maximum values of curves 2 and 3 in Fig.5.6,a, and consequently, the resonant curve in Fig.5.6,b, are achieved under the condition,

$$U(\omega) + k(v, a\omega) = 0 \tag{5.29}$$

and therefore, are determined by the expression (see equation (5.28)):

$$a\omega = \frac{|N(j\omega)|}{V(\omega) + b(v, a\omega)} \tag{5.30}$$

Equation (5.29) determines the amplitude-frequency characteristics backbone curve 1 (Fig.5.6, b). The backbone curves plotting is shown in Fig.5.6 by closed circles. As can be seen, the resonant curve for the situation when $a\omega > v$ envelopes the backbone curve. Equation (5.30) determines limiting amplitude line 2, which intersects the backbone curve at the same points as the amplitude-frequency characteristic.

The backbone curve represents the variation of the ultrasonic system's natural frequencies in the presence of a nonlinear load. Following from Fig.5.6,b, the system natural frequency equals a linear system's natural frequency $\omega_0 = const$ if the ultrasonic vibration intensity is small: $a\omega \leq v$. As the vibration intensity is increased, the system's natural frequency rises initially to the value ω^* when $v < a\omega \leq a^*\omega^*$. Subsequently, it begins to fall, and $\omega \to \omega_0$ as $a\omega \to \infty$. As such, the system is isochronous when $a\omega \leq v$, stiff nonlinear when $v < a\omega \leq a^*\omega^*$, and soft nonlinear when $a\omega > a^*\omega^*$.

The complexity of the vibration system's natural properties when a nonlinear load is present determines the diversity of the resonant curve configurations possible. Their specific form is dependant on the cutting speed, the relation between the elastic and dissipative parameters of the vibrating system and load, and the level of external excitation. Fig.5.7 shows the resonant curves transformation when the cutting speed v is increased from $v = 0$ to a value that exceeds the cutter vibration speed in idling resonant regime ($v > a_0\omega_0$).

Fig.5.7,a shows the system's amplitude-frequency characteristic during idling motion. The resonant curve has a form that is typical for a linear system. Maximum amplitude is obtained at the idling motions resonant frequency $\omega = \omega_0$.

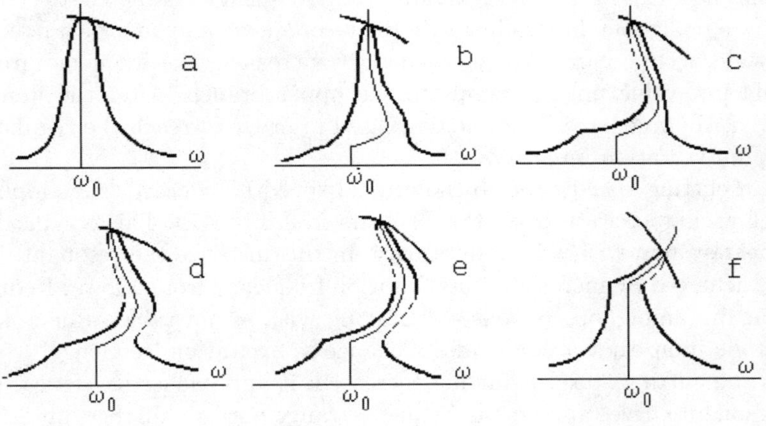

Fig. 5.7.

For small cutting speeds $v \ll a_0\omega_0$, the resonant curves form (Fig.5.7,b) changes subtly, but the system resonant frequency is shifted to a higher frequency. As the cutting speed is increased, the branch present below resonance becomes much steeper while the branch present above resonance becomes increasingly flat, enveloping the backbone curve.

A further increase in speed changes the resonant curves character (Fig.5.7,c). This then takes the form of a typical nonlinear system with soft nonlinearity. A range of frequencies appears, where three resonant curve branches are present. The middle branch, shown by a dashed line, corresponds to unstable solutions and is physically unrealizable. The most effective resonant regime, at which maximum amplitude occurs, can be realized by reducing the system's frequency from a higher frequency. A jump from one curve branch to another is accompanied by a jump in vibration amplitude. When the vibration speed $a\omega \leq v$, the branch corresponding to the linear system's resonant curve can also be realized.

The next qualitative change in the amplitude-frequency characteristic occurs at the point $a\omega = v$. At this point, the linear resonance branch described above hits the frequency range at which the backbone curve exists (Fig.5.7,d). In this instance, the amplitude-frequency characteristic has two domains of existence with three branches. The branches shown by dashed lines are unstable. For a smooth change in excitation frequency, a jump from one branch to another is accompanied by a jump in vibration amplitude. The resonant regime can be reached by reducing the frequency from a higher value, or increasing it from a lower value until a jump occurs and then reducing it back again to the system's resonance regime.

The situation shown in Fig.5.7,e differs from the previous situation. Five branches now exist in the system amplitude-frequency characteristic, three of which are realizable. In this instance the resonant regime, at which maximum amplitude occurs, can only be reached by increasing the frequency from an initially low value until it jumps to the upper branch. After the jump has occurred, the frequency must be decreased in order to reach the regime with maximum vibration intensity.

At a cutting speed close to the critical speed ($v = a_0\omega_0$), the amplitude-frequency characteristic takes the form shown in Fig.5.7,f. This is typical for a nonlinear system with stiff nonlinearity. In this instance the resonant regime can be achieved by increasing the vibration frequency from a lower frequency. Here again, ambiguous regions exist, along with an unstable branch and an amplitude jump that accompanies a change in excitation frequency.

As the cutting speed is further increased, it approaches its critical value. The resonant curves ambiguous frequency range narrows further, until finally at speeds $v \geq a_0\omega$, the amplitude-frequency characteristic regains the form shown in Fig.5.7,a which is typical for a linear system.

Therefore it is seen, that a vibration cutting device exhibits different behaviour when subjected to different cutting conditions: it can behave in a similar manner to a vibro-impact system with an initial clearance, an initial interference or to a system which is subjected to a compressive force, or a combination of these.

5.2 Autoresonant control of transduser nonlinear vibration

1. As was shown in section 5.1, in order to make the ultrasonic cutting process effective, cutter vibration is required to be at its maximum possible amplitude. Realizing a regime of this type requires resonant vibration to be excited and maintained when the system is subjected to a load. A picture of how the amplitude-frequency characteristic changes, as described above, shows clearly the difficulties that arise when exciting resonant vibration regimes by external actuation. Briefly speaking, the difficulties are as follows: the complex and ambiguous tuning required in order to achieve a resonant regime when the excitation frequency is changed, the system's high sensitivity to changes in load, and the danger of vibration breakdown due to unpredictable load changes. It appears that these very reasons restrain the industrial use of ultrasonic cutting processes.

The passage devoted to autoresonant excitation, which is based on self-sustaining oscillation excitation by means of positive feedback, allows the problems related to the excitation and stabilization of resonant vibration during ultrasonic cutting to be solved in the most efficient way. The main principles and possibilities of exciting vibration within ultrasonic systems have been considered in the previous chapter. Fig.5.8 shows the arrangement of an ultrasonic cutting device that employs an autoresonant excitation system (Astashev et al. (1977)).

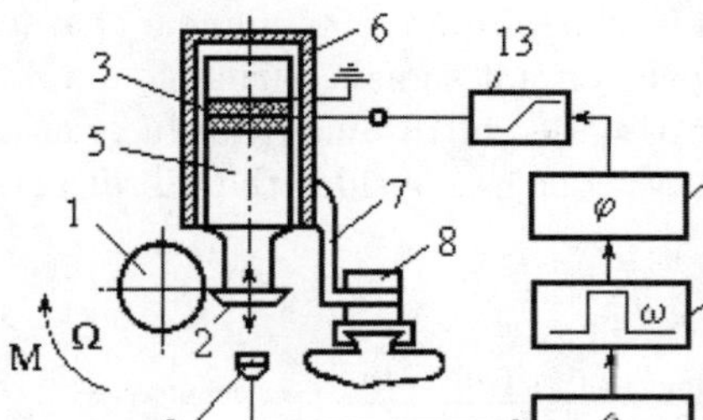

Fig. 5.8.

This system includes several additional elements to that shown in Fig.5.1: 9 is a feedback sensor, 10 is a preliminary amplifier, 11 is a band-pass filter, 12 is a phase shifting element, and 13 is a nonlinear amplifier. The amplifier has a saturated characteristic with a linear initial trait.

The autoresonant system solves the following problems:

- it provides self-excitation of vibration under any system loading condition,
- it automatically promotes resonant regimes for a widely varying load,

- it allows the cutter vibration amplitude to be regulated to within a specified range.

The first problem is solved by selecting the feedback circuit initial amplification coefficient. It was shown above that self-excitation conditions are always satisfied if an amplifier with a relay characteristic, or a characteristic similar to it, is used. It must be considered however, that when an self-sustaining oscillation regime is used to excite vibration within a complicated vibration system, which has a set of natural frequencies, it is possible that an unwanted frequency may be excited. As a rule, the regime occurs which corresponds to the maximum amplitude of a feedback sensor signal. This signal is largely dependant on the sensor positioning (Babitsky, Kalashnikov & Molodtsov (2004), Hertz (1979)). In order to extract the working frequency range of interest within the autoresonant system's feedback circuit, a band-pass filter 11 is employed. The feedback circuit amplification coefficient should be chosen in accordance with the recommendations given in the previous section. The second problem of resonant regime realization, when the system is influenced by a varying load, can be solved by adjusting the phase characteristics of the drive, vibration system and feedback circuit. This function is performed by the phase shifting element 12. If the phase characteristics are chosen correctly, the autoresonant system promotes regimes that correspond to the limiting amplitude line (see section 4.4) which envelopes all possible resonant curves. As such, the most effective resonant vibration is automatically realized for a changing speed, feed, cut depth, vibration system temperature, cutter mass (due to its wear), etc. Finally, vibration amplitude regulation is achieved by changing the saturation level in amplifier 5 nonlinear characteristic. This leads to a change in the feedback circuit signals harmonic amplification coefficient.

2. A transmissibility of loaded ultrasonic transducer as an object of control can be obtained as follows. Let us consider the vibrating system detailed in Fig.5.9.

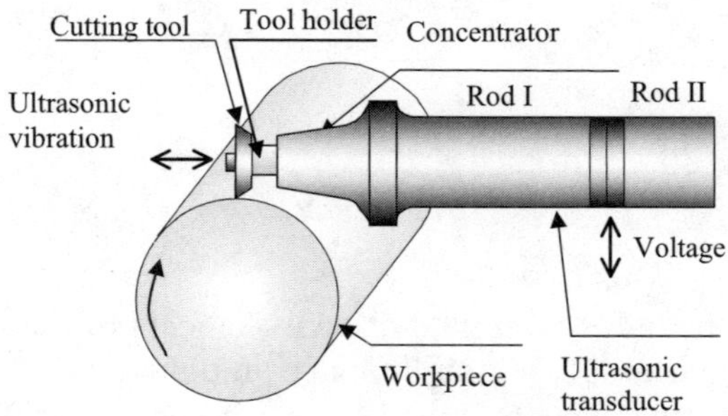

Fig. 5.9.

It consists of two circular plates of piezoceramic material with thickness $l_0/2$ bolted between a cylindrical concentrator (rod I), with length l, and a cylindrical rod II with length l_2. The piezoceramic plates form a piezo element of the transducer. They are connected electrically in parallel and supplied with an alternating voltage $\nu(t) = \nu_0 e^{j\omega t}$. Taking (5.1) into consideration we can rewrite the equation (5.26) for transducer vibration as follows:

$$u(t) = W^{-1}(j\omega)\{N(j\omega)\nu(t) - f[u(t), \dot{u}(t)] + M/r\} \tag{5.31}$$

In the equation (5.31): $F_1(t) = N(j\omega)\nu(t)$ is the force of the ultrasonic excitation of the cutter, $N(j\omega)$ is a linear operator, which reflects the transfer of the ultrasonic excitation from the electronic excitation signal $\nu(t)$ to the cutter and includes the transfer functions of the amplifier, the match box and the concentrator; M is a torque of a lathe main drive; r is the radius of the workpiece; $W(j\omega)$ is the reduced dynamic stiffness of the tool fixed in tool holder of the concentrator, ω is the frequency of the ultrasonic vibration.

The transformation of the variables in equation (5.31) is represented by the block diagram in Fig.5.10. The use of the resonant band of the linear operator $W^{-1}(j\omega)$ for intensification of ultrasonic vibration amplifies essentially the main harmonic component in spectrum of $u(t)$. This permits the searching for an approximate solution in one-mode form: $u(t) = vt + u^0(t) = vt + a\cos\omega t$, where a, ω are amplitude and frequency of the main harmonic component of ultrasonic vibration.

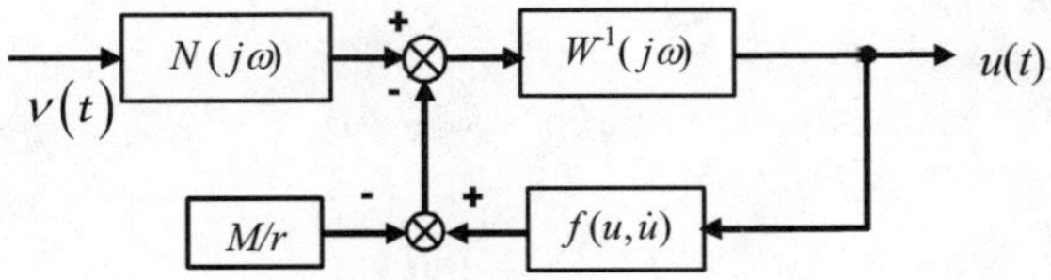

Fig. 5.10.

Using the above-mentioned approximations, the dynamics of the system can be investigated with the use of the describing function method (harmonic linearisation) (Babitsky (1998)).

A transmissibility of ultrasonic cutting unit $\Im[u(t)]/\Im[\nu(t)]$ ($\Im[\ ]$ is an operator of Fourier transform) under the load can be obtained from solution of the equation (5.31) (Babitsky, Astashev & Kalashnikov (2004)). It is proposed that piezo element has a thickness $l_0 \ll \Lambda_0$ (where Λ_0 is the wavelength of the vibration within the ceramic) and the deformation of the piezo element can be treated as homogeneous with a relative strain, $\varepsilon = \Delta l_0/l_0$, where Δl_0 is the longitudinal deformation of the element. We will consider the steady state vibration of the system.

Let us define the forces acting on the two rods (at cross-sectional planes $x_1 = x_2 = 0$, which coincide with position of the piezo element, whose length can be neglected in comparison with rod's lengths) due to the actions of the piezoceramic element as f_0^I, f_0^{II}. Because the deformation of the piezoceramic element is homogeneous it can be written that, $f_0^I = -f_0^{II} = f_0$. The reactions of the rods will therefore be equal in value and opposite in direction, $f_0^I = -f_0^{II} = -f_0$.

Taking into consideration a dominating effect of the main harmonic component of vibration process, the periodic displacement of the cross-sectional area with coordinate x of rod i (i = I, II) can be written as $u_x^i(t) = \tilde{a}_x^i e^{j\omega t}$, where $\tilde{a}_x^i$ is the complex amplitude of vibration. The left end of rod I having co-ordinate $x = l$ is loaded with a non-linear cutting process that has the dynamic characteristic $f_l = f\left(u_l^I, \dot{u}_l^I\right)$.

Using harmonic linearisation (see Section 3.1), we have

$$
\begin{aligned}
f_l = f(u_l^I, \dot{u}_l^I) &\approx [k(v, a_l^I) + j\omega b(v, a_l^I)]u_l^I + P(v, a_l^I) = \\
&= \tilde{k}(v, a_l^I)u_l^I + P(v, a_l^I)
\end{aligned}
\tag{5.32}
$$

where, P s the non-variable component of the load generated due to the ultrasonic cutting process (vibro-induced force). The force P acts as an average permanent load on the main drive, which has to be equalised by machine tool motor torque $M = Pr$.

Let us now write the equations for the complex vibration amplitudes of the transducer rods I and II clumping the piezoceramic rings. For the cross-sectional areas of rod I with coordinates $x = 0$ and $x = l$ we have

$$
\tilde{a}_0^I = L_{00}^I \tilde{F}_0 - L_{l0}^I \tilde{k} \tilde{a}_{1l} \tag{5.33}
$$

$$
\tilde{a}_l^I = L_{0l}^I \tilde{F}_0 - L_{ll}^I \tilde{k} \tilde{a}_{1l} \tag{5.34}
$$

Here $\tilde{F}_0$ is used to denote the complex amplitude of the force $f_0^I = f_0$, and $L_{sr}^I = L_{sr}^I(j\omega)$ is the dynamic compliance (receptance) connecting the force acting in the cross section with a coordinate s with the displacement of the cross section with a coordinate r. The upper index denotes the number of the rod being considered.

At the cross-sectional plane $x = 0$ of rod II we have

$$
\tilde{a}_0^{II} = L_{00}^{II} \tilde{F}_0 \tag{5.35}
$$

The state of the piezoceramic elements is described by the equations that couple the forces applied with the deformation of the elements, voltage $\tilde{\nu}$ and charge $\tilde{q}$ on their plates (see Chapter 2),

$$
-\tilde{F}_0 = K_1 \tilde{a} + \Phi \tilde{q} \tag{5.36}
$$

$$\tilde{\nu} = \Phi \tilde{a} + \frac{\tilde{q}}{C} \tag{5.37}$$

where $K_1 = E_0 S / l_0$ is the stiffness of a piezoceramic element without charge (short-circuited contacts), E_0 is the elastic modulus of the piezoceramic, S is the area of a piezoceramic plate,

$$\tilde{a}_0 = \left(\tilde{a}_0^I + \tilde{a}_0^{II}\right)/2 \tag{5.38}$$

is the complex amplitude of deformation for a single piezoceramic plate, C is the capacitance of the piezoceramic element without deformation ($\tilde{a}_0 = 0$), Φ is the coefficient of electro-mechanic coupling, $\tilde{\nu}$ and $\tilde{q}$ are the complex amplitudes of voltage and charge.

Equations (5.33)-(5.38) permit finding a transmissibility expression. From (5.34) we find,

$$\tilde{a}_l^I = \tilde{F}_0 \frac{L_{0l}^I}{1 + \tilde{k} L_{ll}^I} \tag{5.39}$$

From (5.33) we have the expression for the amplitude $\tilde{a}_0^I$, using (5.39),

$$\tilde{a}_0^I = \tilde{F}_0 \left[L_{00}^I - \tilde{k} \frac{\left(L_{0l}^I\right)^2}{1 + \tilde{k} L_{ll}^I} \right] \tag{5.40}$$

It must be noted that the equality $L_{0l}^I = L_{l0}^I$ was taken into consideration. Now, for the amplitude, $\tilde{a}_0$, of deformation in the piezoceramic plate, from (5.38), with account of (5.35) and (5.40),

$$\tilde{a}_0 = \frac{\tilde{F}_0}{2} \left[L_0 - \tilde{k} \frac{\left(L_{0l}^I\right)^2}{1 + \tilde{k} L_{ll}^I} \right] \tag{5.41}$$

where $L_0 = L_{00}^I + L_{00}^{II}$.

Excluding the variable $\tilde{q}$ from equations (5.36) and (5.37), we can write,

$$\tilde{F}_0 = -\Phi C \tilde{\nu} - K_0 \tilde{a}_0 \tag{5.42}$$

where $K_0 = K_1 - \Phi^2 C$ is the stiffness of a piezoceramic element with an open circuit.

After substitution of (5.41) into (5.42) and transformations, we can write the complex amplitude of the force, $\tilde{F}_0$, of interaction between the piezoceramic transducer and the wave guide,

$$\tilde{F}_0 = \frac{-2\Phi C \tilde{\nu} \left(1 + \tilde{k} L_{ll}^I\right)}{2 + K_0 L_0 + \tilde{k} \left[2 L_{ll}^I + K_0 L_0 L_{ll}^I - K_0 \left(L_{0l}^I\right)^2 \right]} \tag{5.43}$$

Substituting (5.43) into (5.39), we write an equation linking the amplitude of vibration $\tilde{a}_l^I$, for section $x = l$ of rod 1, with the voltage $\tilde{\nu}$ applied to the piezoceramic plate,

$$\tilde{a}_l^I \triangleq \tilde{a}_l = \frac{-2\Phi C L_{0l}^I \tilde{\nu}}{2 + K_0 L_0 + \tilde{k}\left[2L_{ll}^I + K_0 L_0 L_{ll}^I - K_0 \left(L_{0l}^I\right)^2\right]} \tag{5.44}$$

The transmissibility $\frac{\tilde{a}_l}{\tilde{\nu}}$ of a loaded ultrasonic transducer, as every nonlinear system, depends generally on level of vibration a_l due to coefficient $\tilde{k}(v, a_l)$. Amplitude-frequency characteristics of loaded transducer based on equation (5.44) follow the configurations in Fig.5.7. For their analysis the equation (5.44) has to be complemented with expressions for dynamic compliances as will be obtained below in subsection 3.

Let us now consider some special cases. Idle regime: $\tilde{k} = 0$.

$$\tilde{a}_l = \frac{-2\Phi C L_{0l}^I \tilde{\nu}}{2 + \left(L_{00}^I + L_{00}^{II}\right) K_0}$$

In the case when rod I and rod II are similar, i.e. a symmetrical system, the dynamic compliances for both rods are equal. In this situation,

$$\tilde{a}_l = \frac{-\Phi C L_{0l} \tilde{\nu}}{1 + K_0 L_{00}}, \quad \tilde{F}_0 = \frac{-\Phi C \tilde{\nu}}{1 + K_0 L_{00}}, \quad \tilde{a}_0 = \frac{-\Phi C L_{00} \tilde{\nu}}{1 + K_0 L_{00}}$$

Neglecting the energy losses in the system, it can be written that, $L \equiv \mathrm{Re}L = U^{-1}$, where U is the dynamic stiffness. It is possible to distinguish here two possible tunings.

I. Tuning to a rod's resonant frequency: $U_{00} = -U_{0l} = 0$. As a result, $F_0 = 0$ and $\tilde{a}_0 = -\tilde{a}_l = -\Phi C \tilde{\nu}/K_0$. This means that there is no loading on the transducer. Its deformation follows to the variation of the applied voltage only. This shows that joining a resonant rod, which exhibits free vibration with the same amplitude as the piezoceramic element, to a transducer will not load the piezoceramic element. This type of tuning does not amplify the amplitude of the transducer. In this case, rod I and rod II have to be with lengths $l = n\Lambda/2$, where Λ is the wavelength of the vibrations in rod, n is an arbitrary number.

II. Tuning to a rod's antiresonant frequency: $\mathrm{Re}L_{00} = 0$. Neglecting energy losses, we have $L_{00} = \mathrm{Re}L_{00} = 0$. Therefore, $a = a_0 = 0$, $F_0 = -\Phi C \nu$ and $a_l = -\Phi C L_{0l} \nu$. For a rod with uniform area in cross section and length $l = \Lambda/4$,

$$L_{0l} = -\frac{2l}{\pi E S} \quad \text{and} \quad L_{0l} = -\frac{2l}{\pi E S}$$

In this case, the rod action is similar to a dynamic absorber 'neutralising' vibration of the piezo element.

Let us compare the amplitudes achieved by the above tuning states. Utilising the formula $K_0 \approx \frac{\pi E_0 S}{2 l_0}$ we have

$$\frac{a_l^{II}}{a_l^I} = \frac{2 K_0 l}{\pi E S} \approx \frac{l E_0}{l_0 E}.$$

The moduli of rod, E, and the piezoceramic element, E_0, have values of a similar order and $l \gg l_0$. The second method of tuning produces amplitudes of vibration that are essentially bigger than that of the first method. The second method is therefore utilised within the design of piezoceramic transducers.

3. The dynamic stiffness $W(j\omega)$ can be found from a solution of the boundary problem for steady-state waves in one-dimensional wave-guides with variable cross section as was described in section 2.2. The equation of longitudinal vibration of the wave-guide is

$$\rho \frac{\partial^2 u\,(x,t)}{\partial t^2} = \tilde{E} \left[\frac{\partial^2 u\,(x,t)}{\partial x^2} + \frac{S'_x}{S_x} \frac{\partial u\,(x,t)}{\partial x} \right] \tag{5.45}$$

where $u(x,t)$ is a motion of the cross section x, $\tilde{E} = E\left(1 + j\frac{\psi}{2\pi}\right)$, E is the elastic modulus, ψ is a coefficient of inner damping in material, ρ is the density of the material, $S(x)$ is an area of the cross section x, $S'_x\,(x) = dS\,(x)\,/dx$, $j = \sqrt{-1}$.

Let $x = s$ $(s = (0,l))$ be a co-ordinate of the application of the exciting force $f_s\,(t) = \tilde{F}_s e^{j\omega t}$. For a wave guide of length l with free ends the boundary conditions are

$$\left. \frac{\partial u\,(x,t)}{\partial x} \right|_{x=l-s} = 0, \quad \tilde{E} S_x \left. \frac{\partial u\,(x,t)}{\partial x} \right|_{x=s} = (-1)^{(1-s/l)}\,\tilde{F}_s e^{j\omega t} \tag{5.46}$$

Considering a harmonic vibration of the wave guide $u\,(x,t) = \tilde{a}_x e^{j\omega t} = a_x e^{j(\omega t - \varphi_x)}$, where a_x, φ_x are the amplitude and phase of the vibration of a cross section x, we have instead of (5.45) and (5.46):

$$\begin{aligned} &\tilde{E}\left[\tilde{a}''_x + \left(S'_x/S\right)\tilde{a}'_x\right] + \rho\omega^2 \tilde{a}_x = 0 \\ &\tilde{a}'_x\big|_{x=l-s} = 0, \quad \tilde{E} S_x \tilde{a}'_x\Big|_{x=s} = (-1)^{(1-s/l)}\,\tilde{F}_s \end{aligned} \tag{5.47}$$

Further the solution depends on a form of the wave-guide defined by $S(x)$. As an example, for a uniform waveguide $S(x) = S = const$ we obtain from (5.47):

$$L_{sx}\,(j\omega) = W^{-1}\,(j\omega) = \frac{\tilde{a}_x}{\tilde{F}_s} = (-1)^{s/l} \frac{\lambda}{\omega^2 \rho S} \frac{e^{\lambda x} + e^{-\lambda x} e^{2\lambda(l-s)}}{e^{\lambda s} - e^{-\lambda s} e^{2\lambda(l-s)}} \tag{5.48}$$

where $\lambda = j\omega\sqrt{\frac{\rho}{\tilde{E}}} = \frac{j\omega}{c\sqrt{1+j\psi/2\pi}}$, $c = \sqrt{\frac{E}{\rho}}$.

Usually waveguides are composed of parts with different properties. The parts may be made of different materials. The cross sectional areas of these different parts may vary according to different forms. Some parts can be represented as elements with lumped parameters. Different compound concentrators with a tool attached used for ultrasonically assisted cutting belong to such systems. To describe the compound vibration system, it is practical to express its dynamic compliance of concentrator (rod I, Fig.5.9) via the dynamic compliances of its parts as was shown in Chapter 2.

Consider, as an example, it waveguide I (see Fig.5.9), which consists of two steps of length l_1 and l_2 , where $l_1 + l_2 = l$, as is widely used in ultrasonic equipment. Dividing the waveguide at the step, the interaction of its parts may be represented by a reaction force with a complex amplitude $\tilde{F}_{12}$. New coordinates $x_1 = x$ ($x_1 \in [0, l_1]$), $x_2 = x - l_1$ ($x_2 \in [0, l_2]$) may now be introduced. A complex displacement amplitude for the rod cross sections x_q ($q = 1, 2$) may be written for the system, which is under the action of force $\tilde{F}_0$ applied at the end $x = x_1 = 0$,

$$\begin{aligned} \tilde{a}_{x_1}^{(1)} &= L_{0x_1}^{(1)}(j\omega)\,\tilde{F}_0 - L_{l_1x_1}^{(1)}(j\omega)\,\tilde{F}_{12} \\ \tilde{a}_{x_2}^{(2)} &= L_{0x_2}^{(2)}(j\omega)\,\tilde{F}_{12} \end{aligned} \tag{5.49}$$

where the upper index (in brackets) denotes the number of the step to which the expression corresponds. Equality $\tilde{a}_{l_1}^{(1)} = \tilde{a}_0^{(2)}$ reflects the continuity of the system at cross section $x = l_1$ ($x_1 = l_1$; $x_2 = 0$), and allows the reaction amplitude to be found from (5.49)

$$\tilde{F}_{12} = L_{0l_1}^{(1)}(j\omega)\,\tilde{F}_0 \Big/ \left[L_{l_1l_1}^{(1)}(j\omega) + L_{00}^{(2)}(j\omega)\right] \tag{5.50}$$

Substituting (5.50) into (5.49), the dynamic compliance of the compound system may be found. Its expressions for the end cross sections and of the compound system are:

$$\begin{aligned} L_{00}^{I}(j\omega) &= L_{00}^{(1)}(j\omega) - \frac{L_{0l_1}^{(1)}(j\omega)L_{l_10}^{(1)}(j\omega)}{L_{l_1l_1}^{(1)}(j\omega)+L_{00}^{(2)}(j\omega)} \\ L_{0l}^{I}(j\omega) &= \frac{L_{0l_1}^{(1)}(j\omega)L_{0l_2}^{(2)}(j\omega)}{L_{l_1l_1}^{(1)}(j\omega)+L_{00}^{(2)}(j\omega)} \end{aligned} \tag{5.51}$$

Let, as usual, the steps be homogeneous rods with cross sections S_1 and S_2. We are assuming also that the steps are made from the same material. The dynamic compliances for the homogeneous rod under excitations acting at the end cross sections $s_i = l_i$ and $s_i = 0$ can be found from (5.48):

$$L_{l_ix_i}^{(i)}(j\omega) = -\frac{\lambda}{\rho S_i\omega^2}\frac{ch\lambda x_i}{sh\lambda l_i} \tag{5.52}$$

$$L_{0x_i}^{(i)}(j\omega) = -\frac{\lambda}{\rho S_i\omega^2}\frac{ch\lambda\left(l_i - x_i\right)}{sh\lambda l_i} \tag{5.53}$$

Substituting into (5.51) the proper expressions from (5.52) and (5.53), we have

$$L_{00}^{I}(j\omega) = -\frac{\lambda}{S_1\rho\omega^2}\frac{S_1ch\lambda l_1ch\lambda l_2 + S_2sh\lambda l_1sh\lambda l_2}{S_1sh\lambda l_1ch\lambda l_2 + S_2sh\lambda l_2ch\lambda l_1} \tag{5.54}$$

$$L_{0l}^{I}(j\omega) = -\frac{\lambda}{\rho\omega^2}\frac{1}{S_1 sh\lambda l_1 ch\lambda l_2 + S_2 sh\lambda l_2 ch\lambda l_1} \tag{5.55}$$

Neglecting the dissipation in the material ($\psi = 0$), we can simplify expressions (5.54), (5.55) as follows

$$L_{00}^{I}(j\omega) = -\frac{l}{S_1 E\xi}\frac{S_1 \cos\xi_1 \cos\xi_2 - S_2 \sin\xi_1 \sin\xi_2}{S_1 \sin\xi_1 \cos\xi_2 + S_2 \sin\xi_2 \cos\xi_1} \tag{5.56}$$

$$L_{0l}^{I}(j\omega) = -\frac{l}{E\xi}\frac{1}{S_1 \sin\xi_1 \cos\xi_2 + S_2 \sin\xi_2 \cos\xi_1} \tag{5.57}$$

where $\xi = \omega l/c$ and $\xi_i = \omega l_i/c$.

With the use of the piezoceramic transducer it is necessary to obtain a maximum dynamic loading of the ceramic rings. This corresponds to the fixed ends of the piezo transducer. When combined with a concentrator, this can be obtained by organisation of a node in a cross section $x = 0$. Condition of such tuning is

$$L_{00}^{I}(j\omega) = 0 \tag{5.58}$$

or according to (5.56)

$$S_1 \cos\xi_1 \cos\xi_2 - S_2 \sin\xi_1 \sin\xi_2 = 0 \tag{5.59}$$

It is possible to show that this expression defines the natural frequencies of a rod with a fixed end $x = 0$. In this case the concentrator works like a dynamic absorber of the ceramic ring vibration and the amplitude of the cross section $x = l$ can be found with the help of (5.57)

$$\tilde{a}_l^I \triangleq \tilde{a}_l = L_{0l}^{I}(j\omega)\,\tilde{F}_0 \tag{5.60}$$

whereas ξ_1 and ξ_2 follows from (5.59). Let us consider two special cases for condition (5.59):

I.

$$\cos\xi_1 = 0, \quad \sin\xi_2 = 0 \tag{5.61}$$

or $\xi_1 = (2m-1)\,\pi/2$, $\xi_2 = n\pi$ $(m, n = 1, 2, \ldots)$. In this case the first step has a length equal to an odd number of quarter wave lengths: $l_1 = (2m+1)\,\pi c/2\omega$, but the second step's length is equal to an even number of quarter wave lengths: $l_2 = n\pi c/\omega$. From (5.60) with account of (5.57) and (5.61), we have an amplitude of a free end of the concentrator:

$$a_l = \frac{F_0 l}{E S_1 \xi} \tag{5.62}$$

II.

$$\cos\xi_2 = 0, \quad \sin\xi_1 = 0 \tag{5.63}$$

or $\xi_2 = m\pi$, $\xi_2 = (2n-1)\pi/2$. In this case the first step has a length equal to an even number of quarter wave lengths: $l_1 = m\pi c/\omega$, but a length of the second step is equal to an even number of quarter wave lengths: $l_2 = (2n-1)\pi c/2\omega$. From (5.60), with account of (5.57) and (5.63), we have an amplitude of a free end of the concentrator for this case:

$$a_l = \frac{F_0 l}{E S_2 \xi} \tag{5.64}$$

For equal lengths of the concentrators, the second case gives S_1/S_2 increase of amplitude, which is used in practice.

4. An effective influence of ultrasonic vibrations on cutting demands, as shown above, an excitation of ultrasonic vibrations with maximum amplitude. The complexity of the amplitude response makes the excitation and stabilisation of the resonant oscillation in conditions of variable technological loads difficult. The improvement of the excitation efficiency can be achieved with the transition of the excitation into an autoresonant one (Fig.5.8).

Introducing the proper transfer functions for electro-acousto-mechanical and control system, we can write the equation of cutter motion of autoresonant system by analogy with (5.31) in the following operator form:

$$u_l^I(t) = W^{-1}(j\omega)\left\{\begin{array}{l} N(j\omega)\Psi\left[T(j\omega)u_l^I(t)\right] - \\ -f\left[u_l^I(t), \dot{u}_l^I(t)\right] + M/r \end{array}\right\} \tag{5.65}$$

In equation (5.65) $F_1(t) = N(j\omega)\Psi\left[T(j\omega)u_l^I(t)\right]$ is the force of the ultrasonic excitation of the cutter, which is produced by an ultrasonic transducer with the positive feedback. The feedback includes (see Fig.5.8) a sensor, a band-pass filter and a phase shifter with a common linear operator $T(j\omega)$ and an electronic limiter with a nonlinear characteristic $\Psi\left[T(j\omega)u_l^I(t)\right]$; $N(j\omega)$ is a linear operator, which reflects the transfer of the ultrasonic excitation from the limiter to the cutter and includes the transfer functions of the amplifier, the piezo transducer and the concentrator; $W(j\omega)$ is the reduced dynamic stiffness of the tool, which includes the proper operators for the concentrator and the tool; ω is the frequency of the ultrasonic vibration, M is a torque of a lathe main drive; r is the radius of the workpiece. The transformation of the variables in (5.65) is represented by the block diagram in Fig.5.11 (compare with Fig.5.10).

Autoresonant excitation of stable ultrasonic vibration corresponds to the existence and stability of the periodic solution of (5.65). Due to the strong filtering effect of the mechanical structure and the filter, the harmonic approximation of the solution can be found by using technique of harmonic linearisation. We suppose as in section 4.3 that the periodic displacement of the cross-sectional area with coordinate x of rod i $(i = I, II)$ (see Fig.5.9) can be presented as $u_x^i(t) = \tilde{a}_x^i e^{j\omega t}$, where $\tilde{a}_x^i$ is the complex amplitude of

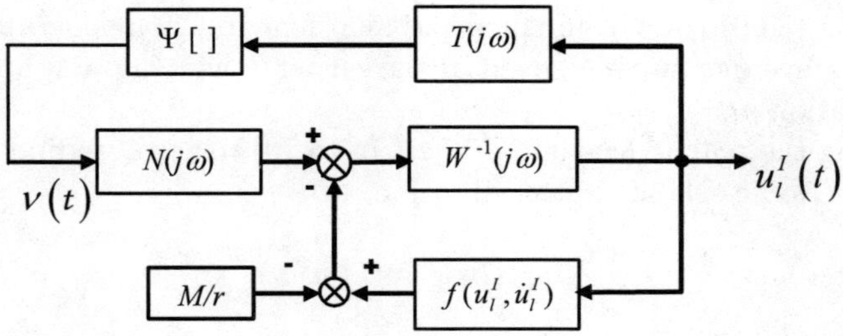

Fig. 5.11.

vibration. The left end of the rod I having co-ordinate $x = l$ is loaded with a non-linear cutting force $f_l(t) = f\left[u_l^I(t), \dot{u}_l^I(t)\right]$.

Filter action permits consideration of the main harmonic component only of the process in the feedback circuit. As a result, the voltage $\nu(t)$ applied to the piezo transducer can be described as

$$\nu(t) = \Psi\left[T(j\omega)\, u_l^I(t)\right]$$

After harmonic linearisation of this expression we have a complex amplitude of vibration:

$$\tilde{\nu} = Z(a_l)\, \tilde{a}_l e^{j\varphi} \tag{5.66}$$

where $Z(a_l)$ is a coefficient of harmonic linearisation for the electronic limiter, φ is a phase shift due to the action of phase shifter. We suppose that a maximum response of the limiter is G, i.e.

$$\left|\Psi\left[T(j\omega)\, u_l^I(t)\right]\right| \le G \tag{5.67}$$

The periodic function $\nu(t)$ at this case cannot have an amplitude of fundamental exceeding $4G/\pi$. As a result, it follows from (5.66)

$$Z(a_l)\, a_l \le \frac{4G}{\pi} \tag{5.68}$$

After harmonic linearisation of (5.65) with account of (5.32) and (5.66) we have the following equations for the constant and periodic component of cutter vibration

$$P(v, a_l) = M/r \tag{5.69}$$

$$1 + W^{-1}(j\omega)\left[\tilde{k}(v, a_l) - N(j\omega)\, Z(a_l)\, e^{j\varphi}\right] = 0 \tag{5.70}$$

Equation (5.69) couples in the expression for vibration generated average force P, the average cutting speed of the cutter v with an amplitude of the cutter vibration a_l.

Equating the real and imaginary part of (5.70) to zero, we find the conditions of a periodic self-sustained vibration

$$Z(a_l)\cos\varphi = U(\omega, a_l) \tag{5.71}$$

$$Z(a_l)\sin\varphi = V(\omega, a_l) \tag{5.72}$$

where
$U(\omega, a_l) = \mathrm{Re}Q(j\omega, a_l)\,,\ V(\omega, a_l) = \mathrm{Im}Q(j\omega, a_l)$
$Q(j\omega, a_l) = [1 + W(j\omega)]\,\tilde{k}(v, a_l)/N(j\omega)$

Excluding the trigonometric functions from (5.71) and (5.72), we have finally

$$a_l = a_l Z(a_l)\left[U^2(j\omega, a_l) + V^2(j\omega, a_l)\right]^{-1/2} \tag{5.73}$$

Comparing (5.73) and (5.68) we can conclude that the maximum amplitude of the vibration is achieved when $a_l Z(a_l) = 4G/\pi$. This means that

$$Z(a_l) = 4G/\pi a_l \tag{5.74}$$

Relationship (5.74) can be fulfilled by the use of a relay characteristic $\Psi(u_l)$ in feedback: $\Psi(u_l) = G\mathrm{sgn}u_l$. Due to the variation of the phase φ with the help of a phase shifter the amplitude of the cutter vibration a_l and its frequency ω, as follows from (5.71)-(5.73), belong to the amplitude-frequency characteristic $a_l = \nu_0\left[U^2(j\omega, a_l) + V^2(j\omega, a_l)\right]^{-1/2}$ of forced vibration for loaded transducer described by (5.31) when $\nu_0 = 4G/\pi$.

In order to study the stability of the autoresonant regimes the method described in (Babitsky (1998)) can be used as in section 4.3. It is assumed that in the vicinity of steady-state periodic solution the cutter vibration is of the form

$$\breve{u}_l(t) = \breve{v}(t)\,t - \breve{a}_l(t)\exp j\left[\omega t + \breve{\varphi}(t)\right] \tag{5.75}$$

where $\breve{v}(t)$, $\breve{a}_l(t)$ and $\breve{\varphi}(t)$ are slowly varying functions of time.

Considering equality (5.72) as the reduced energy balance condition for the dissipative and exciting forces of the steady state motion, the motion stability criterion with respect to small perturbation in amplitude, can be formulated as

$$\left(\frac{d\mathrm{Im}Q}{d\breve{a}_l}\right)_{\breve{a}_l = a_l} > 0 \tag{5.76}$$

Inequality (5.76) means that if the amplitude $\breve{a}_l$ deviates from a stationary value $a_l = const$ due to perturbation, the energy is unbalanced in such a way that it compensates for the deviation. In (5.75) three unknown slow functions introduced instead of one, and we can connect them in turn to two arbitrary conditions. Let us assume that in the disturbed motion $\breve{v}(t)$ and $\breve{a}(t)$ as well as $\breve{\varphi}(t)$ and $\breve{a}(t)$ are related in the same way as in steady-state case.

Using the notation introduced above, the condition (5.76) can be rewritten in the form

$$\frac{d}{d\breve{a}_l}\left[\omega b\left(v,\breve{a}_l\right) - K(\omega)\, Z\left(\breve{a}_l\right)\sin\varphi\right]_{\breve{a}_l=a_l} > 0$$

where $K(\omega) = U_{sl}(\omega)/U_{ll}(\omega)$ is a coefficient that is dependant on the point at which the excitation force is applied, and, $\operatorname{sgn}K(\omega)\sin\varphi \geq 0$. According to (5.74), $\left[dZ/d\breve{a}_l\right]_{\breve{a}_l=a_l} < 0$, for real processing regimes $\omega b\left(v,\breve{a}_l\right) > 0$. From this it follows that under autoresonant excitation all points of the amplitude-frequency characteristic of the loaded transducer correspond to stable vibration as opposed to the forced excitation, when some branches of the characteristic are unstable due to the jump phenomena (see Fig.5.7).

5. The experiments with various configurations of the autoresonant control system showed that optimal performance, in terms of energy consumption and reliability of self-excitation, was achieved for the system with a schematic diagram presented in Fig.5.12 (Babitsky, Kalashnikov & Molodtsov (2004)).

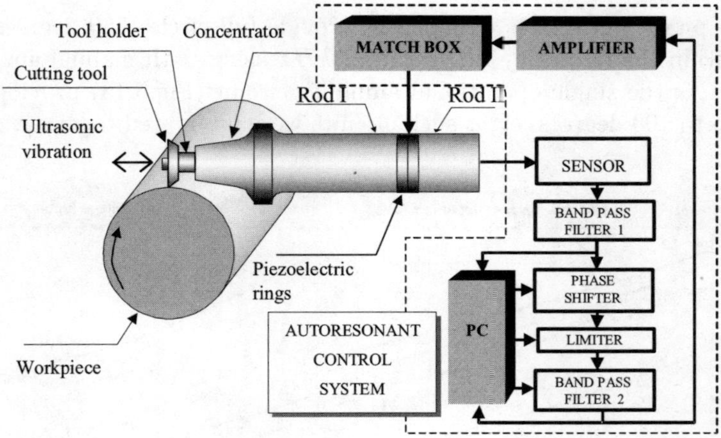

Fig. 5.12.

The comparative analysis of the system performance using different vibration sensors concluded that the current sensor and accelerometer are the most suitable sources of reference signal for the autoresonant excitation within industrial environments. The accelerometer should be installed at the end face of the concentrator along the direction of vibration and near one of an antinode point of the transducer vibration mode. This allows a safety disposition of the sensor outside the cutting area.

A first band-pass filter provides elimination of high-frequency components of the closed loop signal, which could cause energy inefficient high-frequency modes to be self-excited. These modes could suppress the principal resonant mode, in spite of the latter being more energy efficient. A second band-pass filter was used for purifying the electronic limiter output signal, which reduced higher frequency losses in the matchbox. Both filters were identically designed as multiple feedback band-pass filters (Floyd (1992)).

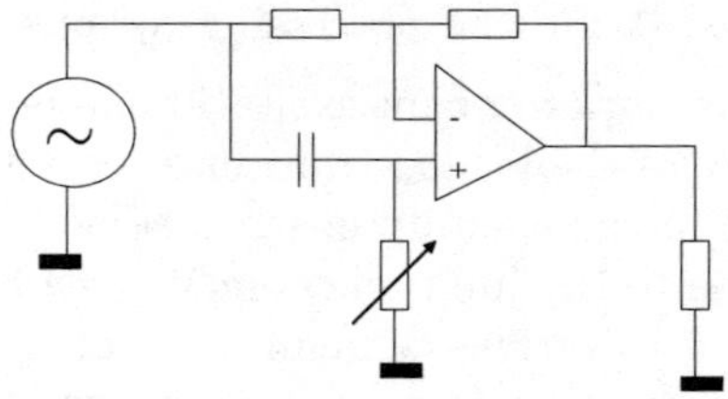

Fig. 5.13.

The phase shifter was designed to provide full circle (360 degrees) phase shift within the frequency range $15 - 25kHz$ using both manual and remote control. As the single operational amplifier circuit (Fig.5.13) provides phase shift up to 100 degrees, eight such circuits were integrated together.

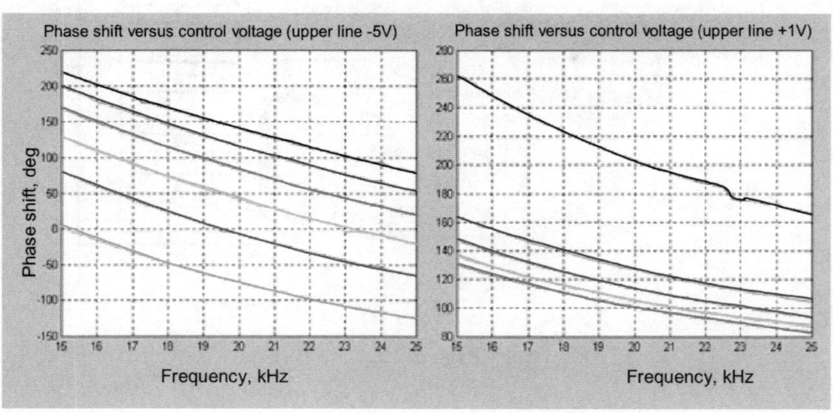

Fig. 5.14.

The responses of the phase shifter are presented in Fig.5.14. Each line represents a different control voltage with $0.1V$ increments. The picture shows that the control curves are nonlinear, but they cover whole the phase shift range (0 - 360 deg). The use of the device allows adjustment of the self-excitation frequency. Additionally the phase shifter gain (not shown) is frequency dependent. That is why its output is connected to the input of the limiter in order to keep the amplitude of the signal under control.

The limiter in Fig.5.15 was designed to provide a controlled maximum level of output signal up to $5V$ by truncating the input signal manually and remotely. The diodes for both the half waves of the input signal are connected with the operational amplifier feedback, which improves the characteristics of the diodes for low input voltages.

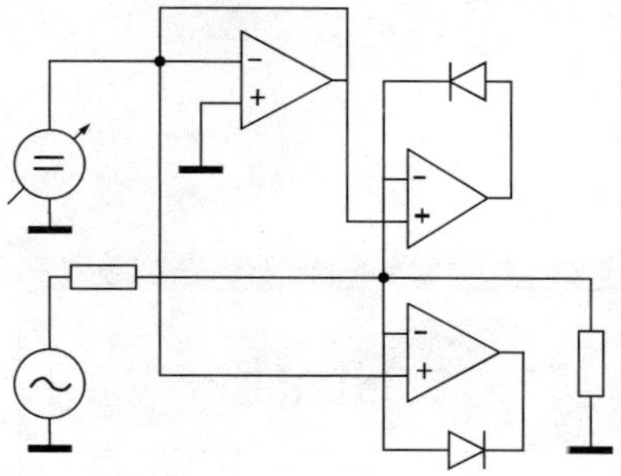

Fig. 5.15.

The response of the limiter is shown in Fig.5.16. Each line represents a different control voltage with $0.1V$ increments. The device allows the control of vibration levels during operation.

Additionally an amplifier (for amplification of the control signal) and a matchbox (for interfacing between the amplifier and transducer) were used to deliver the control signal to the transducer (see Fig.5.12). A digital-to-analog board (for provision of the analog control and test signals from the computer), analog-to-digital (ADC) board (for computer acquisition of the analog signals) and PC were used for monitoring and supervisory control purposes.

The software for control and monitoring was created within the Hewlett-Packard Visual Engineering Environment and allows monitoring, storage, display and processing data from different sensors, and provides manual and automatic supervisory control. The front panel of the principal control program is presented in Fig.5.17 and allows the operator to choose the supervisory control algorithm, adjust the phase shift and threshold in manual mode and monitor the state of the system. After finishing the test, the information is saved and is available for further analysis.

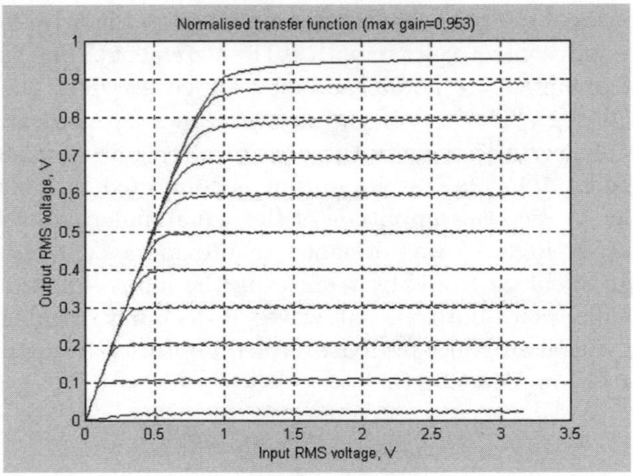

Fig. 5.16.

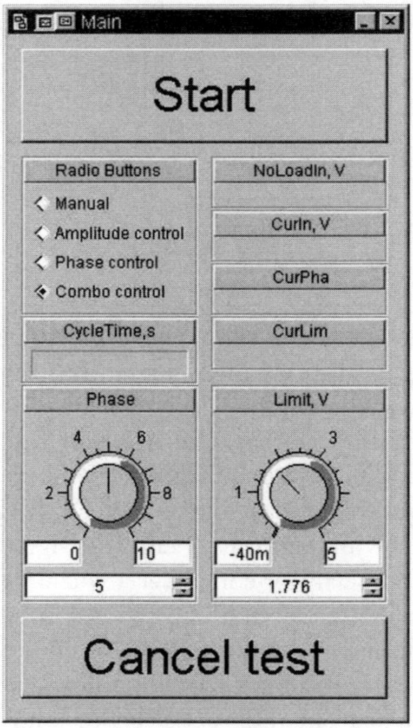

Fig. 5.17.

The hardware was fabricated and matched to provide the possibility for real time computer control. The phase shifter and limiter can be controlled from a computer on the basis of the reference signal properties. The principal aims of supervisory control are to provide the most efficient state of the electro-acousto-mechanical system and identify the control signal interval, if there have been any changes within the mechanical system (e.g. fitting a new cutting tip or a different tool holder). The control system keeps a stable level of vibration under variable cutting load. Changing the phase shift within the autoresonant loop allows efficient control of the vibrating state. Changing the limiter threshold could control the vibration level.

The underlying principle of supervisory computer control is to acquire samples of the reference signal after band pass filtering, evaluate the efficiency of oscillations (the RMS value of the reference signal was used as a measure), elaborate new control values according to the specific algorithm, and set them up.

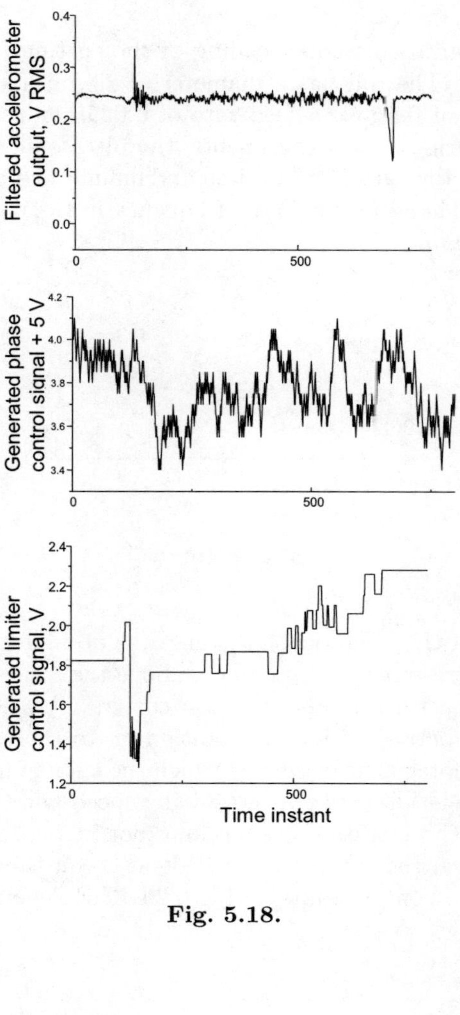

Fig. 5.18.

Fig. 5.18 demonstrates the operation of the autoresonant system in real working conditions. Here the horizontal axis corresponds to the time (each unit is about $50ms$ which is the control cycle time). The vertical axis reflects (from top to bottom) the RMS value of the reference signal in V (the accelerometer was used), the phase shifter control signal in the control units, and the limiter control signal in V respectively. The autoresonant system keeps the stable vibration of the cutter under variable processing load.

6. Now, we describe experiments on validation of the nonlinear theory of ultrasonically assisted cutting described in this chapter. An ultrasonic cutting device which was manufactured in accordance with the arrangements shown in Fig.5.1 was attached to a lathe. The cutters vibration was either excited in a forced vibration regime by an external generator, or in an self-oscillation regime using the autoresonant control. During experimentation, the cutting forces constant component (vibro-induced force) was registered by strain gauges attached to tool holder 7 (Fig.5.1). The signal, which is proportional to the cutters vibration speed, was taken from the microphone situated under the cutter.

Fig.5.19 shows an oscilloscope reading of the cutting force present when turning a steal rod. The rod has a diameter of $29mm$ and cutting was performed at a depth of $0.5mm$, a feed rate of $0.025mm/rev$, and various rotational speeds. During these experiments, the ultrasonic vibration that was superimposed on to the cutter was excited and maintained by an autoresonant excitation system. The vibration had a frequency in the region of $20kHz$ and an amplitude of $10\mu m$.

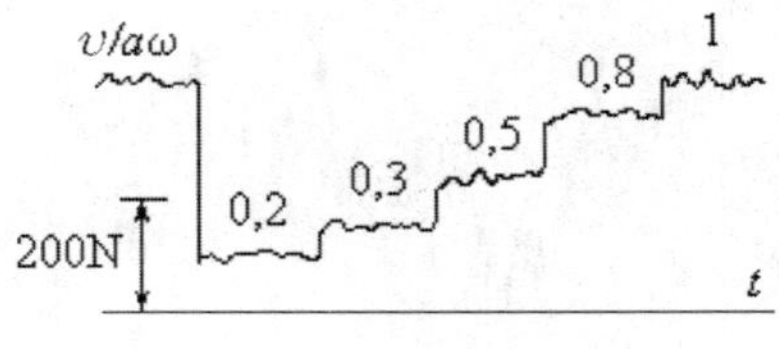

Fig. 5.19.

The first part of the oscilloscope reading was obtained during traditional turning with no ultrasound. The numbers above the other sections show the ratio between the cutting speed v and the cutters vibrational speed $a\omega$. As can be seen, the superimposition of vibration on to the turning process decreases the cutting force. The degree by which the cutting force is decreased is dependant on the relation between the cutting speed and the vibration speed $a\omega$. When $v > a\omega$, the vibration present does not influence the cutting force.

The way in which the cutting force P depends on the cutting speed v is plotted, using experimental results, in Fig.5.20. This dependence is similar to the theoretical curve plotted in Fig.5.3 denoted by the index ($ak_0/D_0 = 5$). Therefore, the assumptions made when modelling the process during the theoretical analysis of the device dynamics when under load are justified.

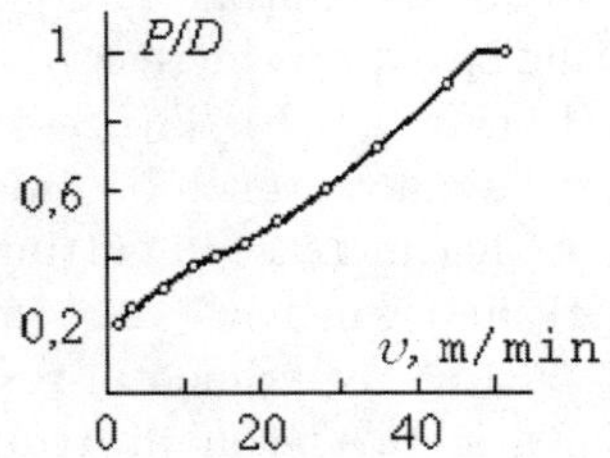

Fig. 5.20.

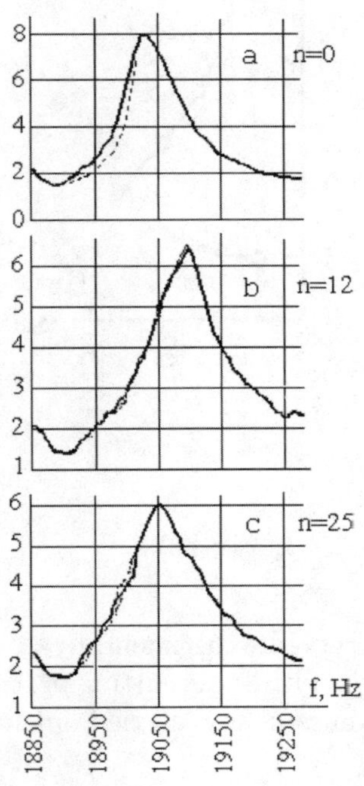

Fig. 5.21.

The vibrating system amplitude-frequency characteristics when under load were recorded when performing experiments in forced vibration regimes. They were recorded when turning steal specimens that had a diameter of $30mm$. A cutting depth of $0.25mm$ and a feed rate of $0.06mm/rev$ was used for various spindle rotation speeds. Examples of the resonance curves obtained are shown in Fig.5.21.

The system resonant frequency during idling motion (Fig.5.21,a) was $f_0 = 19030Hz$. As the cutting speed was increased, the resonant frequency at first increases up to $19125Hz$ (Fig.5.21,b). The resonant curves form changes in such a way that its upper resonant branch becomes steeper, while its lower branch becomes flatter. A further increase in cutting speed decreases the resonant frequency (Fig.5.21,c), and when $v \geq a\omega$, the resonant curve returns to the form shown in Fig. 5.21,a. Experimental results are used in Fig.5.22 to plot the way in which the system's resonant frequencies, when under load, are dependant on the ratio of vibration speed $a\omega$ to cutting speed v. This dependence represents the backbone curve, which is similar to the theoretical backbone curve plotted above (Fig.5.6).

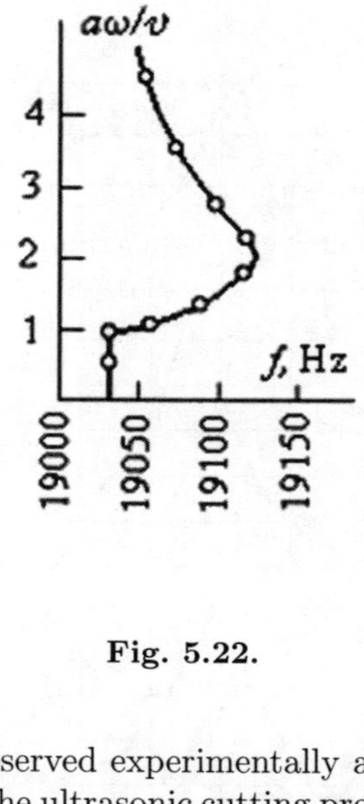

Fig. 5.22.

As such, the effects observed experimentally are in good agreement with the results obtained from the ultrasonic cutting process dynamical model analysis and the device's dynamics as was carried out in section 5.1.

5.3 Ultrasonically assisted turning

1. There are many reported applications of ultrasonic turning that have been developed for brittle and ductile materials (glass, ceramic, fibres, plastic (Kim & Choi (1997, 1998), Kim & Lee (1996), Moriwaki et al. (1992), Takeyama & Lijima (1998), Weber et al. (1984), Whitehouse (1997)), brass

(Hansen (1997)), red copper (Ji et al. (1991)), and mild steel (Astashev (1992), Kremer (1995), Skelton (1969))) which produce qualitative improvements in the cutting process (e.g. reducing the probability of the workpiece damage). Quantitative improvements of surface quality have also been observed for some conventional materials (carbon steel (Wang & Zhao (1987), Weber et al. (1984)), stainless steel (Astashev (1992), Kremer (1995), Moriwaki & Shamoto (1991))). Most of the experiments employed tangential vibration. When the cutting tip is vibrated ultrasonically in cutting velocity direction, the following limitation for the cutting velocity v is imposed (see Fig.5.3): $v = \pi nd < v_t = 2\pi af$. This permits separation of a tool from a workpiece in every cycle of vibration and transform the cutting process into an effective vibro-impact regime. The calculations show that for the contemporary commercially available bolted Langevin type transducers ($a \approx 20\mu m$, $f \approx 20kHz$) the vibration tip velocity must not exceed about $150m/min$. Moreover, reduction of tip velocity also occurs during the cutting process due to the cutting tip interaction with the workpiece so the upper limit on surface speed is further reduced. Thus the efficient ultrasonic cutting with vibration applied in the direction of cutting velocity ($v_t \gg v$) can be achieved only for low diameter workpiece or low rotational speeds. For example, n is limited not to exceed approximately $160rev/min$ when $d = 60mm$ and condition $0.2v_t = v$ is required.

Due to it the rotational speeds were necessarily low relative to the recommended cutting speeds used routinely for manufacturing. Modern Ni-based alloys, though, (e.g. Inconel) are usually machined at lower cutting speeds than conventional materials, which makes it possible to realise the advantages of applying the ultrasonic vibration both in the direction of the cutting velocity (some results on ultrasonic turning of high alloy steel materials were presented in (Wang & Zhao (1987))). Therefore, the ultrasonic cutting of these materials may be viable for ultrasonic cutting in a modern manufacturing environment.

The results of ultrasonic machining are highly affected by the vibration inherent in the machine tool and its accuracy during metal cutting operations. Therefore, special ultra precision machines (Kim & Choi (1998), Moriwaki & Shamoto (1991), Moriwaki et al. (1992)) or special arrangement (Han et al. (1998)) were used for ultrasonic applications. However, even when ordinary machine tools were used, in most cases an improvement of surface finish was reported (Kim & Lee (1996), Kremer (1995), Skelton (1969), Wang & Zhao (1987), Weber et al. (1984)). Achieving surface finish improvement is relevant to industrial applications, as it might simplify or even eliminate some additional manufacturing operations.

The other important issue concerned with the ultrasonic machining is the choice of the transducer control system. In the most cases the generator drives the transducer directly. However, this simple implementation is not adequate for most industrial applications because of the following reasons. Fitting different tool holders or cutting tips would require re-adjustment of the oscillator frequency to match the changes in mechanical properties of the vibrating system. Besides, the cutting tip dynamic load is changed through the cutting process which affects the dynamics of the cutting process (e.g. it leads to the decrease of the vibration level (Skelton (1969), Zharkov (1986))). There is a possibility of generator frequency control by using the phase-lock system (Astashev et al. (2000)), but this possibility does not provide stable results when a controlled resonant system featured strong non-linearity (Babitsky (1998)). Therefore, an autoresonant control system was developed for driving the transducer to keep the cutting process under automatic control.

turning of some modern materials was conducted with a developed system by using universal lathe (Babitsky et al. (2003), Mitrofanov et al. (2004)). The transducer was fixed in the cross slide of the lathe (Fig.5.23) by a specially designed tool post attachment, so that it was possible to apply ultrasonic vibration in the direction tangential to the surface of the workpiece. The vibration amplitude of the piezo ceramic rings was intensified by the concentrator and transmitted to the tool holder mounted at the thin edge of the concentrator. The autoresonant excitation was employed. Resultant vibration of the cutting tip fixed in the tool holder usually reached $10\mu m$ (i.e. $20\mu m$ peak-to-peak) at the frequency of about $20kHz$. Cutting conditions were close to used for manufacture.

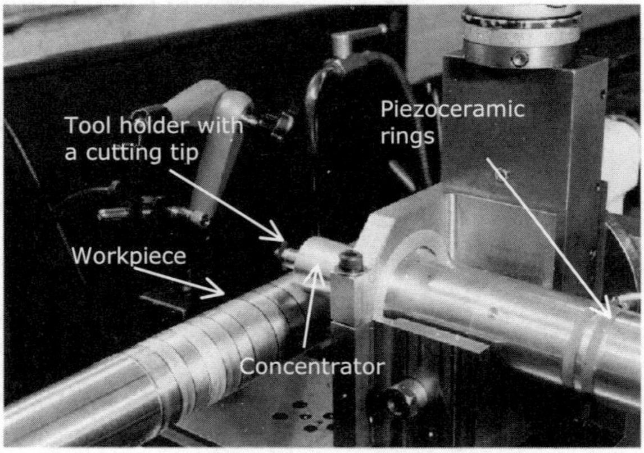

Fig. 5.23.

The clamping method used for securing the insert to the transducer is very important in order to obtain maximum amplitude of vibration at the tool tip. However, as any additional mass added to the vibrating system will change its properties, lightweight and high strength titanium is used to make the tool holder. It acts as an intermediate device to fix the cutting tool insert to the concentrator. The titanium tool holder is screwed into the tapped hole in the concentrator and seated on a brass washer. A cutting tip is also fixed to the tool holder by using a standard high tensile screw.

A decrease in cutting force with a superposition of the tangential ultrasonic vibration at cutting speeds $v < a\omega$ has been observed during the turning of all metals: aluminium, copper, bronze, brass, carbon steel, stainless steel, nickel-based alloys, titanium etc. Ultrasonic vibration changes drastically the cutting process character. As an example, when turning aluminium and copper using conventional methods the workpiece material sticks to the cutters surface (build-up edge) worsening the cutting conditions. When ultrasonic vibration was superimposed on to the cutting process, this unpleasant phenomenon is entirely alleviated.

During the ultrasonic turning, the tendency for self-excitation of a low-frequency machine-tool-workpiece system vibration (chatter) is eliminated (Astashev (1992)). This was clearly observed during the cutting of a cylindrical rod with a diameter of $30mm$ and a length of $660mm$. When turning conventionally the rod at a cut depth of $1mm$, a feed rate of $0.1mm/rev$ and at rotational ratio that exceeds $200rev/min$ the chatter was observed on approaching a middle of the rod. This reduced the quality of the surface. When ultrasonic vibration was activated, which generated vibration in the autoresonant mode, the chatter was entirely eliminated and a uniform surface along the whole length of the rod was obtained. It is worth noting at this point that due to the decrease in cutting force, ultrasonic cutting allows non-rigid workpieces to be machined with no intermediate supports.

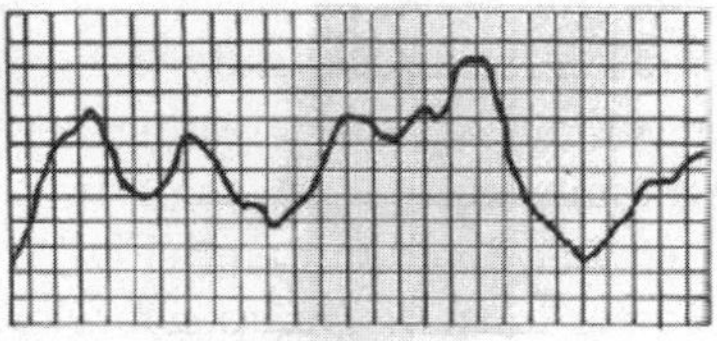

a

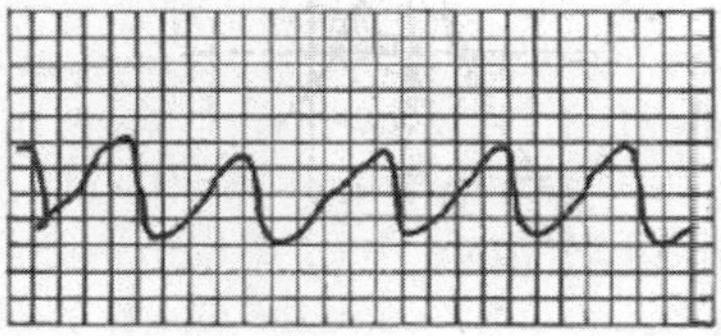

b

Fig. 5.24.

Ultrasonic cutting changes radically the quality of the surface treated. Under the microscope, after traditional turning, the surface looks like a roughly ploughed field, which indicated instability of cutting process. After ultrasonic turning, the surface is similar to a cultivated field which has a strictly regular fine-meshed surface. This is reflected in the surface profiles that were obtained when conventionally (Fig.5.24,a) and ultrasonically (Fig. 5.24,b) turning a steal blank with a diameter of $56mm$. In both experiments, cutting was performed at a feed rate of $0.1mm/rev$ and a cut depth of $1mm$. In subsection 2 we will analyse surface profiles with more details.

Ultrasonic vibration changes the character of the chip formation. Even when machining materials that produce brittle chips under conventional turning conditions, ultrasonic cutting is found to modify the process and chips with a plastic flow characteristic and no burrs or cautery are encountered. Resulting from this effect, when ultrasound is superimposed on to the cutting process, the minimum depth of cut may be reduced essentially. As an example, when turning carbon and stainless steel on a universal lathe, it was possible to use a feed rate of $0.05mm/rev$ and a cut depth of $0.015mm$.

The chip formation was further investigated in (Babitsky, Mitrofanov & Silberschmidt (2004), Mitrofanov, Babitsky & Silberschmidt (2005)) with the high-speed photography equipment. Fig.5.25 shows an experimental set up for ultrasonic turning filming. A high-speed digital camera was used in experiments. The camera was fixed on the lathe by a specially designed holder, enabling the distance to the film region to be adjusted. The cutting area was illuminated by an impulse laser. The laser pulses were synchronised with camera operation so that to provide maximum illumination during each subsequent frame exposure time. The laser beam was directed into the cutting area by means of the fibre-optic cable.

Figures 5.26 show frames from high-speed video recording during the cutting of a mild steel. Continuous video recording allows also estimating the visible size of the process zone, unclear in single frames. The distinction between chip formation in conventional cutting (Fig.5.26,a) and ultrasonically assisted one (Fig.5.26,b) was most noticeable for low rotational speed and becomes less evident for higher cutting speeds. This fact can be explained by approaching the critical cutting speed level (see Fig.5.3).

For studied cutting conditions, the chip formed in conventional cutting of the mild steel was segmented. The segments were turn from the workpiece and slipped along the shier plane with visible levels of deformation between segments being considerably higher than within each segment. Switching on the ultrasonic vibration drastically changes the tool-chip interaction and the process of chip formation. There were no more segments or steps in chip formation, instead continuous material separation in front of the cutter and slipping of the material along the shear plane with each impact of the cutting tool are observed. The process of ultrasonic turning looks completely different in comparison with conventional material deformation and separation. The material separated from a workpiece was considerably more plasticized, i.e. experiencing higher level of plastic deformation happening throughout the chip volumes. The chip formed in ultrasonic turning rises nearly in the radial direction very

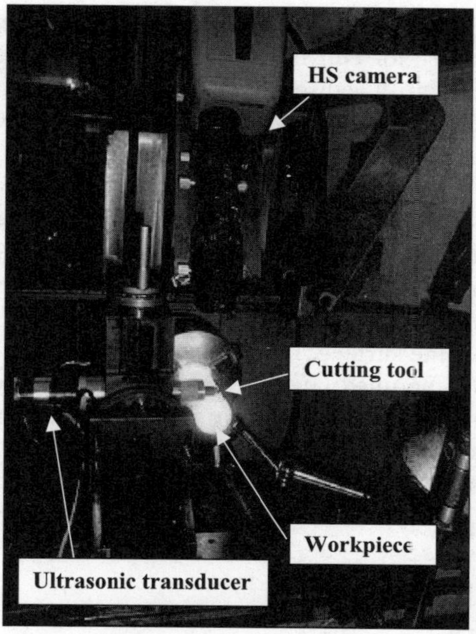

Fig. 5.25.

close to the cutting tool surface in the recorded frame (Fig.5.26,b), whereas the chip in conventional turning shifts much further from the cutting tool, leading to a more curled chip. This is in a good agreement with a subsequent macroscopic chip analysis, indicating the chip radius of curvature being several times greater with superimposing of ultrasonics.

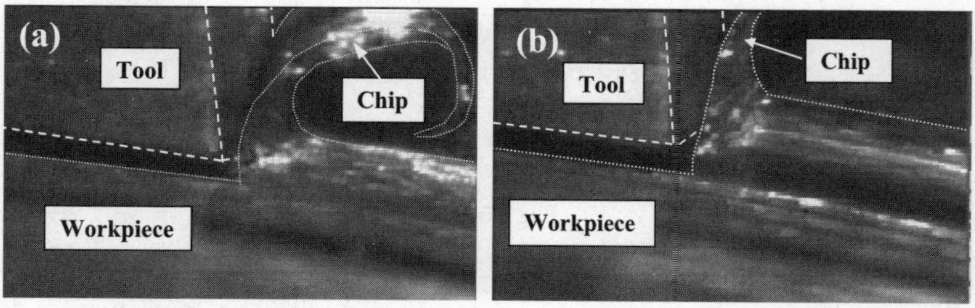

Fig. 5.26.

Differences between two cutting techniques were also observed in shapes as well as in sizes and areas of the process zones. The area of the visible process zone for ultrasonic turning and its width in the radial direction are considerably smaller than those for conventional turning. Deformation processes for the ultrasonic process are localised in the direct vicinity of the cutting edge along the surface of the workpiece and are not observed underneath the clearance face of the tool at the newly formed layer in contrast to the conventional process.

Nanoindentation tests were made to quantify differences between microstructures of surface layers of the workpieces machined conventionally and ultrasonically (Mitrofanov et al. (2004)). Fig.5.27 presents a comparable hardness of the surface layers for nickel based alloy Inconel 718 samples on the depth from the machine surface. According to the results of this analysis the average width of the modified layer is 70% higher for the conventionally machined specimen, than for the ultrasonically treated one. The average hardness of the hardened surface layer was reduced by ultrasonic treatment up to 40%. Ultrasonically assisted cutting better preserves the original microstructure of the materials.

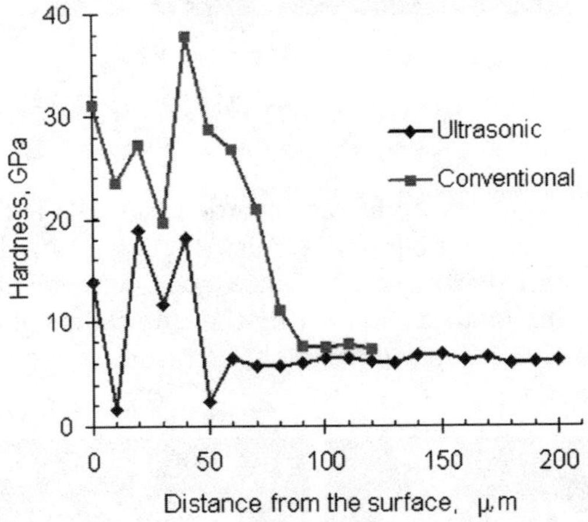

Fig. 5.27.

The results described are in a good correlation with the experimental data described in many works that are devoted to the study of ultrasonic cutting (Kremer (1995), Kumabe (1979), Markov (1962, 1966, 1980), etc.).

2. Application of ultrasonic vibration along the feed direction enables the cutting parameters used in manufacturing industry for most materials to be reached independently of the workpiece diameter. The vibro-impact process is preserved in this case under the condition $sn < v_t$, where s is a feed rate. For

example, n is limited not to exceed about $20,000 rev/min$ when $s = 1mm/rev$ and condition $0.2v_t = sn$ is required. Therefore tool vibration in the feed direction seems to be more suitable for industrial ultrasonic turning requiring high levels of productivity. In (Babitsky et al. (2003)) it was named as *sweep cutting*.

Fig. 5.28 shows an experimental set up of the ultrasonic vibration turning system used in these experiments. The workpiece is clamped by the three-jaw chuck of a universal lathe. A commercial ultrasonic piezoelectric transducer (unloaded resonant frequency $f \approx 20kHz$) is used to provide the vibration. Parallel placement of the transducer to the workpiece in the horizontal plane allows the insert to make the ultrasonic vibration movement in the feed direction. The vibration is measured by a laser vibrometer, and the signal is processed by a phase shifter, limiter and band-pass filter to form an autoresonant control (Babitsky, Astashev & Kalashnikov (2004), Babitsky, Kalashnikov & Molodtsov (2004)). The control signal is amplified using an amplifier and fed into the transducer via a passive matching unit (matchbox). This causes the transducer to vibrate at its resonant frequency. The autoresonant system maintains the resonant mode of vibration during the dynamic changes of the load. The maximal level of vibration of the cutting tip was about $20\mu m$ peak-to-peak at $17kHz$, and decreased during the cutting up to 30-40% depended on the cutting conditions. This level of vibration was found satisfactory and comparable with the levels used by other researchers (Astashev (1992), Han et al. (1998), Ji et al. (1991), Kim & Lee (1996), Kim & Choi (1997), Kim & Choi (1998)).

Fig. 5.28.

As the standard top slide of the universal lathe does not facilitate fixing the transducer in the manner necessary to conduct the turning experiments, it was required to design a special attachment to serve the purpose. The transducer, along with the tool holder attached to it, was fixed in the cross slide of the lathe by a specially designed tool post attachment. The transducer was clamped at its nodal point. This arrangement allows rotation of the transducer along its axis to keep the position of the insert in the horizontal plane passing through the axis of the workpiece required to conduct the cutting experiments. The position was checked by placing a level on the specially prepared flat surface on the tool holder, which is parallel to the insert base. The special attachment along with the transducer is placed on the cross slide of the lathe assuring the correct centre height using an 'L' shape mild steel part. The depth of cut was adjusted by rotating the hand wheel of the cross slide during the turning trials.

All surfaces of the workpieces, cylindrical and faces, were machined prior to the experiments. In order to facilitate smooth tip - workpiece contact at the beginning of the each test, a chamfer ($3mm \times 27.5°$) was cut at the free end of the workpiece. The length of a workpiece is limited to around $75mm$ but this was adequate to have four cuts, each one of $15mm$, in one workpiece. After fixing a workpiece in the three-jaw chuck, a finish cut with a very small depth of cut was performed using the same insert to be used in the test in order to eliminate any remaining eccentricity. This also allows the insert to reach the stable tool wear region before starting each test.

The first cut was made with the application of ultrasonic vibration, and as soon as the tool had traversed $15mm$ the vibration was switched off thus allowing the second cut to proceed under the same cutting conditions but without ultrasonic vibration. After changing the cutting speed by way of changing the spindle rotational speed the next two cuts were performed with and without application of ultrasonic vibration. With the new workpiece the insert was changed and the previous procedure was repeated.

Surface quality was assessed by measuring surface roughness along the axial direction of the workpieces as well as their roundness value. Both surfaces produced by application of ultrasonic vibration, and under conventional conditions, were evaluated using the 'Taylor Hobson - Talysurf 4' surface measuring instrument. The instrument was set to 'Gaussian filter' mode while the tracing length was set at $5.6mm$ and the filter cut-off value was $0.8mm$. The geometrical accuracy of components was tested by measuring the roundness value using the 'Taylor Hobson Talyrond - Model 51' measuring instrument.

Centre line average (R_a) value was used to analyse the surface roughness of machined workpieces. However, other major roughness parameters are also available. The perimeter of the workpiece was divided into five equal parts and five surface roughness measurements were performed on the cylindrical surface. For analysing roundness values three measurements were performed in each cut, and peak-to-valley values were used as an overall measure of roundness.

Two modern high grade nickel based alloys, C263 and Inconel 718, used at present in the aviation industry were tested. Free cutting mild steel workpieces were used as a control material. Recommended by Sandvik Coromant cutting inserts for machining nickel-based alloys were used in the cutting tests. These rhombic 55 degrees tips successfully withstand ultrasonic vibration, and therefore were used in machining free cutting mild steel workpieces too. Specifications of the inserts used in the tests and recommended cutting speeds for medium application of the materials used are presented in the Table 5.1.

Table 5.1.

Insert grades and recommended cutting speeds for workpiece materials

Workpiece material	Recommended cutting speed, m/min	Insert grade	Nose radius $r_\epsilon (mm)$	Clearance angle α_n
C263	20	H13A	0.4	7°
		KRH13A	1.2	7°
Inconel 718	20	KMH13A	1.2	7°
		H13AKRH13A	0.4	7°
		H13A	1.2	7°
Mild steel (230MO7)	325	KRH13A	1.2	7°

Cutting conditions used in machining workpieces are given in the Table 5.2. Feed rate s and depth of cut a_p, which affected the surface quality highly, were keep the same during all the experiments reported to provide compatibility of results. The presented values were chosen to investigate the effect of superimposed ultrasonic vibration on the medium depth of cut combined with low feed rate used for ensuring good surface quality. Synthetic cutting fluid 4% mixed with water was used as cutting fluid in every test.

Table 5.2.

Cutting conditions used in the tests

Material	$n, rev/min$	d, mm	$v, m/min$	$s, mm/rev$	a_p, mm
C263	125 - 260	34 - 38	12 - 25	0.05	0.08
Inconel 718	85 - 180	43 - 45	14 - 29	0.05	0.08
Mild steel	85 - 800	34 - 38	10 - 90	0.05	0.08

Fig.5.29 shows representative profiles of the machined surfaces of C263, Inconel-718 and mild steel workpieces measured in the axial direction at cutting velocities 14, 17 and 20m/min respectively. The upper graph belongs to the ultrasonic application and the lower graph represents the conventional

cutting in each figure. Other cutting conditions such as feed rate ($0.05mm/rev$) and depth of cut ($0.8mm$) were kept constant during the tests. Surface profiles are plotted in the same scale for easy comparison.

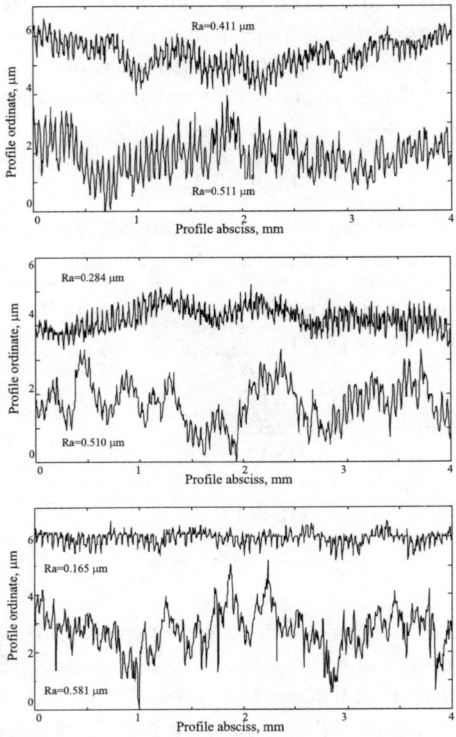

Fig. 5.29.

It is seen from the graphs that the magnitudes of roughness profiles are reduced nearly by 50% in the case of Inconel-718 and C263, whereas the reduction was even more in mild steel with the application of ultrasonic vibration. This result confirmed those obtained by previous researches (Kim & Choi (1997), Skelton (1969)). In all three cases, with application of ultrasonic vibration the regularity of roughness profile is improved considerably and the surface becomes smoother along the axial direction. Abolition of built up edge with the application of ultrasonic vibration in low cutting speeds especially in the case of cutting of the ductile materials leads to further stabilisation of cutting process.

Although a single parameter such as R_a could be used to distinguish differences of manufacturing processes, it may not be sufficient to pinpoint where the changes in the processes have occurred. From the numerous methods

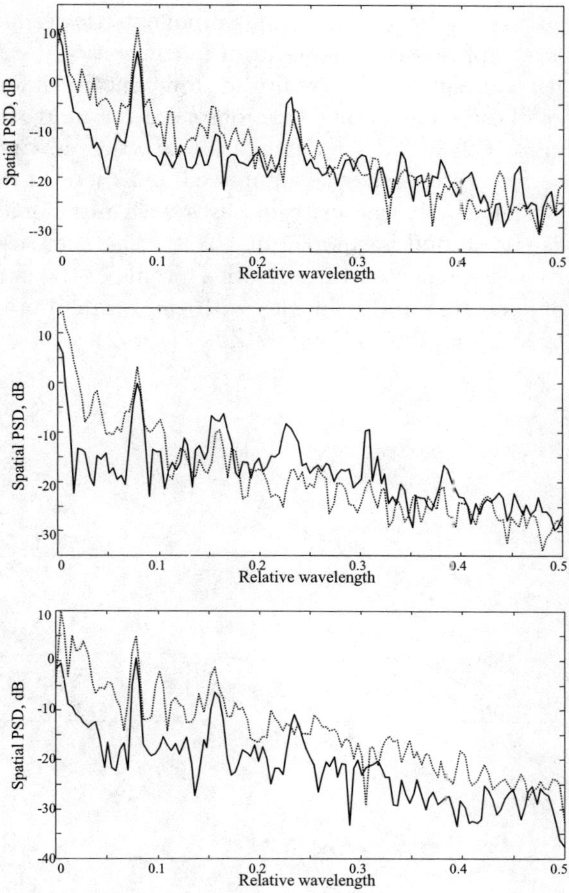

Fig. 5.30.

available for further analysis of surface profiles, spectral analysis can be used to test processes, which are fundamentally periodic or repetitive, as in turning (Whitehouse (1997)). Using this method macro and micro morphological difference between ultrasonically machined surfaces and surfaces produced under conventional cutting can be compared by observing the distribution of energy spectrum density at different wavelengths. Several researches have previously used the same method for analysis of ultrasonically machined surfaces (Kim & Choi (1997), Wang & Zhao (1987)).

Figures 5.30 give the corresponding wavelength spectra of surface profiles shown in Fig.5.29. The dotted lines represents the wavelength spectrum of surfaces machined under conventional condition whereas the continuous line represents ultrasonically machined surfaces. The analysis shows that during the ultrasonic cutting the low wavelength components (lower and around the feed rate peak) are suppressed compared to the conventional cutting. For the higher wavelength components the results are different for different materials (the reduction of these components were observed for mild steel, but there was no reduction for C263, and some increase was observed for Inconel). The magnitude of the feed rate peak was suppressed and the feed rate harmonics became clearly visible in the majority of cases with ultrasonic cutting (the only exception is the second harmonic for C263). The decrease in the lower wavelength components reflected the smoother profile of the surface achieved with the ultrasonic cutting and coincides with the results obtained by other researchers (Kim & Choi (1997), Wang & Zhao (1987)).

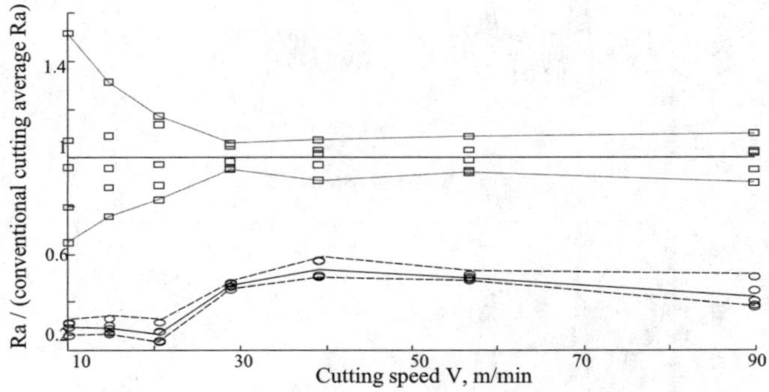

Fig. 5.31.

Fig.5.31 shows the relative surface roughness of machined surfaces of the mild steel both in conventional and cutting under application of ultrasonic vibration for different cutting speeds. The best data collected from a number of tests under the same cutting conditions in both cases are plotted in the graphs. Dotted lines and short dashed lines show the maximum and minimum levels of data fluctuations, while the continuous line shows the mean relative surface roughness in conventional and ultrasonic cutting.

In the range of the cutting speeds investigated, the achieved surface roughness of Inconel-718, C263 and mild steel workpieces machined under the application of ultrasonic vibration is superior to the surface roughness of workpieces machined by conventional cutting (as is illustrated by the mean lines drawn to represent these cases). Improvement of surface roughness of mild steel is more than 50% in the whole range of cutting speeds under investigation.

It appears that the fluctuations in the surface roughness obtained in conventional cutting are greater than those obtained for ultrasonic cutting especially over the range of speeds. The reason for being able to reduce the fluctuations is that the machining process becomes more stable with the application of ultrasonic vibration. The conventional turning process is an unstable cutting process and therefore the roughness values may vary in a large range (Chen (2000)).

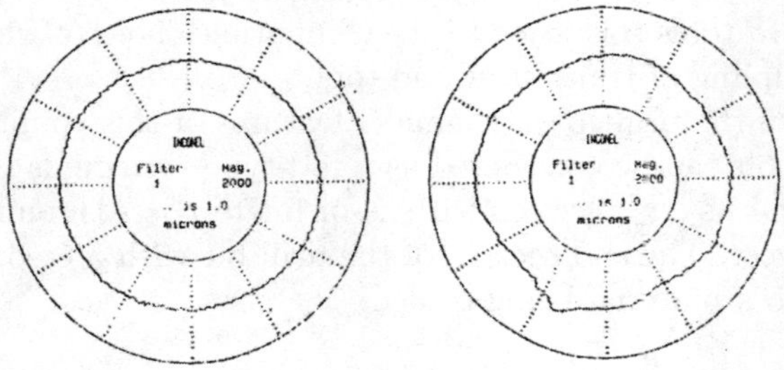

a b

Fig. 5.32.

The roundness plots for Inconel-718 are shown in Fig.5.32 for cutting speeds $17m/min$. From the plots, it can be seen that ultrasonic vibration cutting (Fig. 5.32,a) improves roundness compared to conventional cutting (Fig. 5.32,b) considerably. For the two particular materials analysed the peak-to-valley improvements are 40% for Inconel-718 and 50% for C263. Similar profiles were obtained for vibration cutting in the tangential direction by several researches previously (Han et al. (1998), Ji et al. (1991)).

Results are very similar to the case of surface roughness analysis and improvement of roundness throughout the investigated cutting speeds was observed for both materials. In this occasion also large data fluctuations were observed in comparable conditions of conventional cutting.

5.4 Ultrasonically assisted drilling

1. Ultrasonically assisted drilling takes place when ultrasonic vibration is superimposed on to the relative cutting motion between a drill bit and the workpiece being drilled. Usually this is achieved by excitation of the drill vibration either torsionally (Kumabe (1979)) or axially (Devine (1985), Neugebauer & Stoll (2004)). The latter was used in a majority of applications and will be mainly analysed below. A reduction in the cutting forces, an increase in the penetration speed, and the elimination of burrs are among the main benefits of drilling with ultrasonic assistance.

Effective vibration of the drill bit is achieved only when it is used as a wave guide to produce amplification of vibration amplitude. Only when the drill's length $l \ll \Lambda/4$ (Λ is an ultrasonic wave length) a drill can be treated as a small lamp mass attached to the transducer. With an increase in diameters of holes a successful application of ultrasonically assisted drilling needs acoustical matching of the drill bit as a continuous system with a transducer. Effectively, the drill bit has to be treated as an additional step of the wave-guide. When using the standard tools this leads to the compatible choice of the transducer and accurate coupling of transducer and tool.

Let us analyse the principal dynamical features of this coupling. We propose that a drill bit can be considered as a rod-type waveguide (see Fig.5.33) with a length l and its cross section with co-ordinate $x = s$ is connected rigidly with the transducer. The end section of the drill bit with a co-ordinate $x = l$ is interacted with a material being drilled.

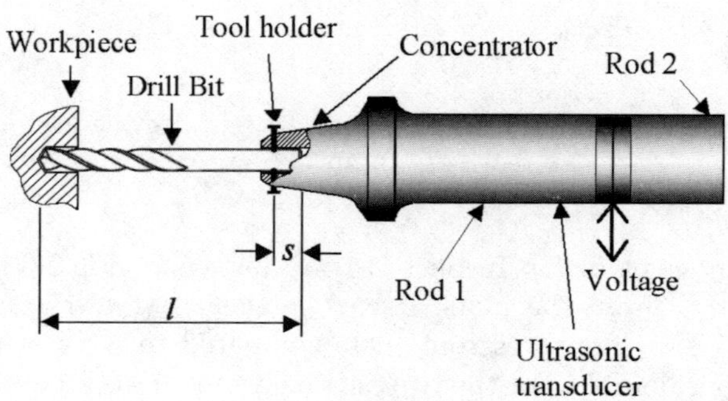

Fig. 5.33.

Vibration of arbitrary cross section with co-ordinate x of the drill can be described by a periodic function

$$u_x(t) = \tilde{a}_x \exp j\omega t = a_x \exp j(\omega t - \varphi_x) \tag{5.77}$$

where a_x, ω and φ are an amplitude, an angular frequency and an initial phase of vibration. A load of the drill bit from the drilling process we describe, as usual, by nonlinear dynamic characteristic $f(u_l, \dot{u}_l)$ of interaction, which can be transformed by harmonic linearisation as follows:

$$f_l = f(u_l, \dot{u}_l) \approx [k(a_l) + j\omega b(a_l)]\, u_l + P(a_l) \tag{5.78}$$

An interaction of the drill bit and transducer can be described by an axial force acting in the cross section $x = s$:

$$f_s(t) = \tilde{F}_s \exp j\omega t = F_s \exp j\omega t \tag{5.79}$$

We describe vibration of the drill bit as a wave guide with a help of dynamic compliances (receptances) $L_{kx}(j\omega) \quad (k = l, s)$ coupling a displacement u_x of cross section x with forces acting in cross sections l, s:

$$u_x(t) = L_{sx}(j\omega)\tilde{F}_s \exp j\omega t - L_{lx}(j\omega)\left[k(a_l) + \beta(a_l)\right] u_l(t) \tag{5.80}$$

According to (5.77), we have

$$\exp j\omega t = \frac{u_l}{a_l} \exp j\varphi_l, \quad u_x = u_l \frac{a_x}{a_l} \exp j(\varphi_l - \varphi_x) \tag{5.81}$$

Substituting (5.81) in equation (5.80) we have for $x = s$ and $x = l$ correspondingly:

$$a_s = \frac{F_s}{W_{ss}(j\omega)} \exp j\varphi_s - \frac{a_l\left[k(a_l) + j\omega b(a_l)\right]}{W_{ls}(j\omega)} \exp j(\varphi_s - \varphi_l) \tag{5.82}$$

$$a_l W(a_l, j\omega) = F_s \exp j\varphi_l \tag{5.83}$$

Here $W(a_l, j\omega) = \left[W_{ll}(j\omega) + k(a_l) + j\omega b(a_l)\right] \frac{W_{sl}(j\omega)}{W_{ll}(j\omega)}$, $W_{kx}(j\omega) = L_{kx}^{-1}(j\omega)$ is a dynamic stiffness of the drill bit.

Separating in (5.83) a real and an imaginary components we find the expressions for amplitude a_l and phase φ_l for the cutting end of the drill bit $(x = l)$:

$$a_l = \frac{F_s}{|W(a_l, j\omega)|} = \frac{F_s}{\sqrt{\left[\mathrm{Re}W(a_l, j\omega)\right]^2 + \left[\mathrm{Im}W(a_l, j\omega)\right]^2}} \tag{5.84}$$

$$\cos\varphi_l = \frac{a_l}{F_s}\mathrm{Re}W(a_l, j\omega), \quad \sin\varphi_l = \frac{a_l}{F_s}\mathrm{Im}W(a_l, j\omega) \tag{5.85}$$

Let us present the dynamic stiffness $W_{sx}(j\omega)$ in the form:

$$W_{sx}(j\omega) = U_{sx}(\omega) + jV_{sx}(\omega) \tag{5.86}$$

In the following consideration the parameters $V_{sx}(\omega)$ and $b(a_l)$, which characterise dissipation in the drill bit structure and the drilling load, are treated as the small ones and all transformations are limited by variables of the first order of magnitude.

Equation (5.84) defines amplitude-frequency characteristic of the drill bit cutting edge excited by harmonic force applied at the cross section $x = s$. Maximum amplitude is achieved under a condition $\mathrm{Re}W(a_l, j\omega) = 0$, which with account of (5.86) can be presented as follows:

$$U_{ll}(\omega) + k(a_l) = 0 \tag{5.87}$$

Under condition (5.87) any amplitude a_l is achieved with a minimal force F_s . Hence, (5.87) is a condition of the best matching of the transducer and the loaded drill bit.

Excluding F_s from equations (5.82), (5.83), and taking into account equations (5.86), (5.87), we have

$$K_s = \frac{a_l}{a_s} = \left| \frac{U_{ss}(\omega)}{U_{sl}(\omega)} \right| \tag{5.88}$$

We will name it as an amplification ratio from an exciter to a drilling section of the bit. The coefficient K_s is not influenced by the load and for each magnitude of a_s it defines an axial amplitude of the cutting edge of the drill under the best matching (5.87) with the transducer.

2. Connection of the drill bit with the transducer leads to additional loading of the latter. Suppose that unloaded transducer vibrates in its connecting cross section according to equation $u^*(t) = a^* \exp j(\omega t - \varphi)$. A connection with the loaded drill bit leads to arrival of an additional force $-f_s(t)$ applied to the transducer. As a result its vibration $u(t)$ will be modified as follows:

$$u(t) = u^*(t) - \frac{f_s(t)}{W_n(j\omega)} \tag{5.89}$$

where $W_n(j\omega) = U_n(\omega) + jV_n(\omega)$ is a dynamic stiffness of the transducer in the cross section of connection with the drill bit.

With account of (5.83), (5.86)-(5.88) we can rewrite the expression for $f_s(t)$ as follows:

$$f_s(t) = \dot{u}_s(t) K_s^2 \left[V_{ll}(\omega) + \omega b(a_l) \right] \tag{5.90}$$

Ultrasonic transducer is a resonant electro-acoustical device. Its resonant tuning corresponds to condition $U_n(\omega) = 0$. Taking into consideration that in the connecting cross section of the transducer and the drill bit $u(t) = u_s(t)$, we have from (5.89), (5.90) for the loaded transducer:

$$u(t) = u^*(t) \left[1 + \frac{V_{ll}(\omega) + \omega b(a_l)}{V_n(\omega)} K_s^2 \right]^{-1} \tag{5.91}$$

It is follows from (5.91) that connecting of the matched drill bit with the transducer does not change the resonant tuning of the latter. Its vibration amplitude is defined by relationship of damping characteristics of the transducer, the drill bit and the load as well as an axial coordinate of the excitation. From equation (5.88) with account of (5.90) we find the amplitude of vibration for the cutting end of the loaded drill bit:

$$a_l = a^* \left[\frac{1}{K_s} + \frac{V_{ll}(\omega) + \omega b(a_l)}{V_n(\omega)} K_s \right]^{-1} \tag{5.92}$$

3. Let us suppose that the drill bit can be modelled as a rod with unified cross section area S. A dynamic stiffness of such rod is as follows (see (5.48):

$$L_{sx}(j\omega) = \begin{cases} -\frac{\lambda}{\rho \cdot S\omega^2} \cdot \frac{ch\lambda x ch\lambda(l-s)}{sh\lambda l} & 0 \leq x \leq s \\ -\frac{\lambda}{\rho \cdot S\omega^2} \cdot \frac{ch\lambda(l-x) ch\lambda s}{sh\lambda l} & s \leq x \leq l \end{cases} \tag{5.93}$$

where $\lambda = \left(j + \frac{\psi}{4\pi}\right)\frac{\omega}{c}$, $c = \sqrt{E/\rho}$ is sound speed in the rod material, E, ρ are elastic modulus and density of the material, ψ is a coefficient of dissipation of a material. From (5.93) we find dynamic stiffnesses for equations (5.87), (5.88):

$$U_{ll} = -\frac{ES\zeta}{l}\frac{\sin\zeta}{\cos\zeta} \tag{5.94}$$

$$U_{ss} = -\frac{ES\zeta}{l}\frac{\sin\zeta}{\cos\zeta\frac{s}{l}\ \cos\zeta\left(1-\frac{s}{l}\right)} \tag{5.95}$$

$$U_{sl} = -\frac{ES\zeta}{l}\frac{\sin\zeta}{\cos\zeta\frac{s}{l}} \tag{5.96}$$

where $\zeta = \omega l/c$.

While running without load (idle regime) we have from (5.87), (5.94) a condition of ideal match:

$$\zeta = n\pi \quad (n = 1, 2, ...) \tag{5.97}$$

It defines the natural frequencies of a drill bit. As it is physically clear, the condition does not depend on position of the contact cross section with the transducer. For the amplification ratio (5.88) with account of (5.95)-(5.97) we have for the case:

$$K_s = \left|\cos^{-1}\pi n\left(1-\frac{s}{l}\right)\right| \tag{5.98}$$

It follows from (5.98) that the amplification ratio changes its value from $K_s = 1$, when the excitation is applied in an anti nodal cross section of standing wave ($s = lm/n$, $m = 0, 1, \ldots, n$), to $K_s = \infty$, when approaching a nodal cross section ($s = l(2m-1)/n$).

Let us analyse a proper vibration of the cutting cross section (5.92). In absence of dissipation and load the force of interaction $f_s(t) = 0$ and connection of the matched drill bit with the transducer does not influence the latter. Dissipation of energy in the drill and due to the drilling process has to be compensated by the transducer. This changes its vibration amplitude a^* and amplitude of the cutting edge a_l. For the further estimation this relationship, we find from (5.92) a dissipative component of the dynamic stiffness:

$$V_{ll} = \frac{ES}{l}\frac{\psi\zeta}{4\pi}\frac{\zeta - 0.5\sin 2\zeta}{\cos^2\zeta} \tag{5.99}$$

Substituting (5.97)-(5.99) into (5.92), we have when $b = 0$:

$$a_l = a^* \left[\left| \cos \pi n \frac{s}{l} \right| + \frac{ES}{4lV_n(\omega)} \frac{\pi \psi \zeta n^2}{\left| \cos \pi n \frac{s}{l} \right|} \right]^{-1} \tag{5.100}$$

It follows from (5.100) that when application of excitation approaches the nodal points the amplitude tends to zero, $a_l \to 0$, due to increase of a transducer dissipative loading. This means that there exists some optimal position of excitation cross section, which depends on relationship between elasto-dissipative characteristics of the transducer and the drill bit. The optimal coupling preserves the tuning of the transducer and compensates the additional energy losses in drill bit and from processing load. Additionally it produces an amplification of vibration amplitude.

4. A prototype of an ultrasonic drilling system has been designed and manufactured. The system employs an autoresonant control system described in 5.3 to stabilise the drilling process.

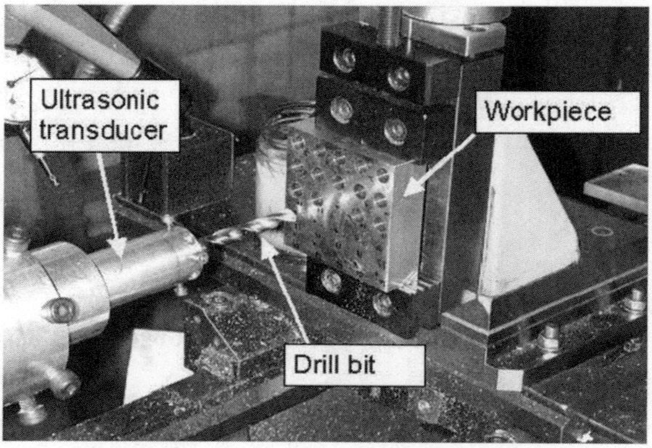

Fig. 5.34.

Experiments with ultrasonically assisted drilling were conducted on a lathe as shown in Fig.5.34. The transducer was held in the three jaw chuck of the lathe through the intermediate bush and was energised by means of the slip ring assembly fitted to the hollow shaft of the lathe at the end remote from the chuck. Once firmly clamped, the working end of concentrator was drilled in situ so that the hole, used for securing drills to the transducer, coincided perfectly with the rotational axis of the lathe. Samples to be drilled were clumped in a vice with a vertical traverse that was rigidly attached to the saddle of the lathe or held in a simple holder attached to the load cell, which could also be positioned on the saddle as an alternative. The usual tool holder assembly with its traverse having been previously removed. This arrangement allowed rows of holes to be drilled in quick succession. The saddle of the lathe can be power driven in a line parallel to the lathe axis and the tool post mounting stage can be similarly power driven in a lane at right angles to this for facing.

A wide variety of materials was drilled. It was found universally that the application of ultrasonic vibration significantly increased the penetration rate of the drills. In some cases, the penetration rate was increased by a factor of four depending on the rotational speed. Fig.5.35 shows a typical example of penetration speed when ultrasonic vibration is switched on and off (denoted as "US on" and "US off", respectively) while drilling in titanium with constant force. The drilling is conducted with a standard twist drill having diameter $3mm$ and rotation speed $125rpm$. The axial force $P = 127.5N$. The frequency of the longitudinal ultrasonic vibration was $20kHz$ and the amplitude at the cutting edge was $10\mu m$. The wear state of the drill is obviously an important parameter in determining the penetration rate. Examples of this have been recorded where, owing to drill wear, the drill has been unable to penetrate the material except when ultrasonically excited.

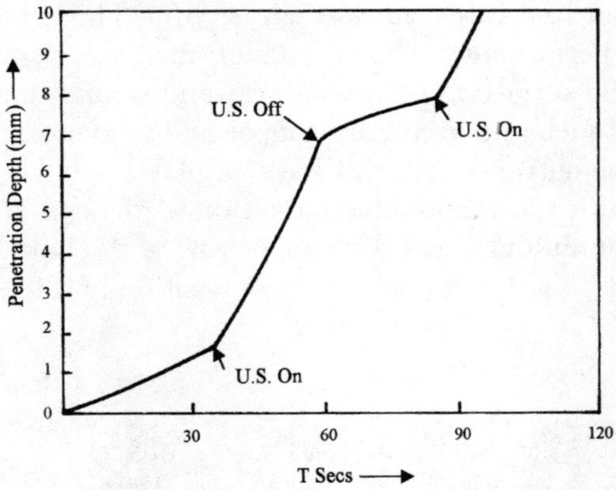

Fig. 5.35.

The axial reactive force on the drill is usually reduced significantly by the application of ultrasonic vibration. Fig.5.36 gives an example of force reduction while drilling with constant feed rate in 3mm thick plate of a carbon fibre/resin composite material using a standard twist drill mounted on a longitudinal transducer. Cutting was started with no excitation at first to establish the steady level of force and then the ultrasonic excitation was switched on. As it apparent from the Figure there is a dramatic reduction in the reactive force when the drill is excited ultrasonically demonstrating that the cutting efficiency is substantially improved.

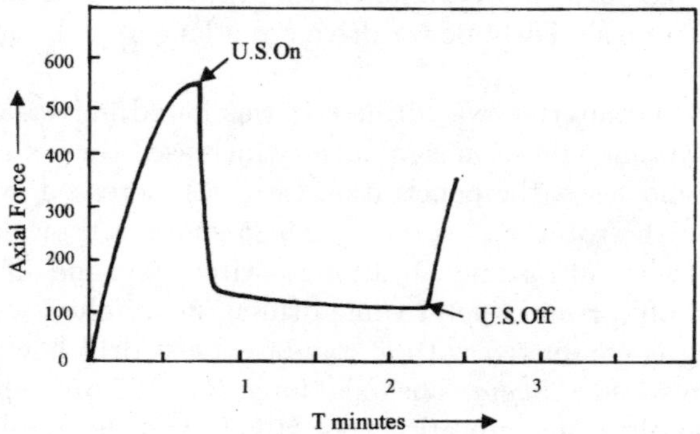

Fig. 5.36.

In some drilling experiments, the reduction in the reactive force for given rotational speed and feed rates can as high as 90%. This leads to significant benefit in drilling performance. These include increase of drill bit lifetime, improvements in the surface finish, roundness and straightness of holes and, in ductile materials such as aluminium, copper and mild steel, the elimination of burrs on both the entrance and exit faces of plates.

Fig.5.37 illustrates the remarkable reduction in the exit burring achieved when drilling an aluminium plate. The upper row of the holes corresponds to conventional drilling, the lower one - to superposition of ultrasonics.

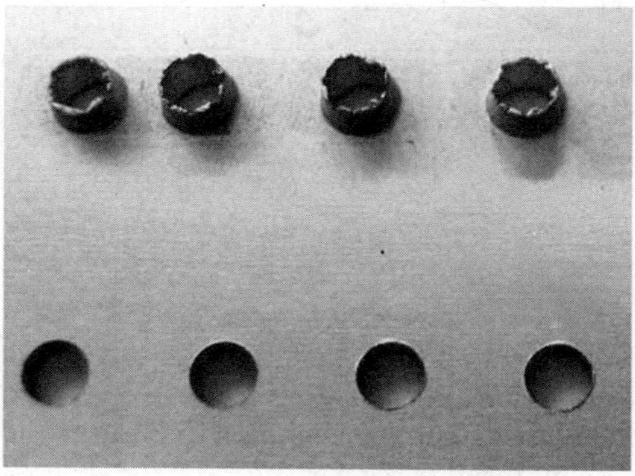

Fig. 5.37.

In the case of drilling through brittle materials, the reduction of stress with superposition of ultrasonic vibration yields large reductions in the size of cratering on the exit face.

The reduction in the reactive force experienced also causes greatly reduces deformation when drilling through thin, flexible plates. When superimposing ultrasonic vibration it was possible, for example, to drill $6mm$ holes cleanly through aluminium strips ($0.25mm$ thick) supported just at the ends (Fig.5.38,a). Attempt to drill through such strips conventionally only resulted in severe deformation of the strip and no penetration (Fig.5.38,b). This experiment demonstrates increase in the dynamic stiffness of the plate due to superimposing of the high frequency vibration.

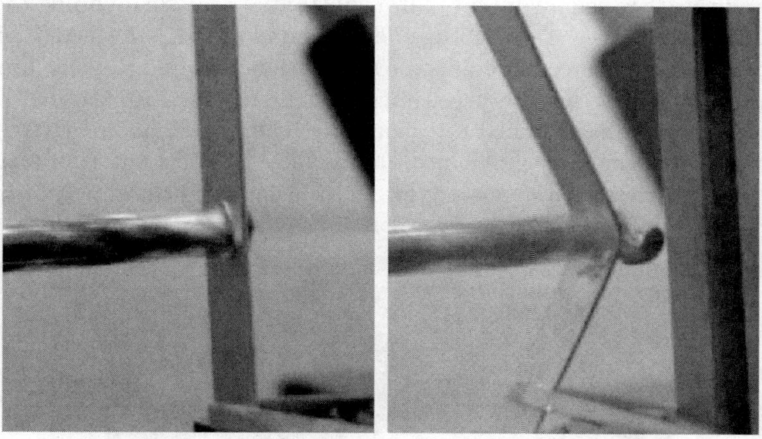

a b

Fig. 5.38.

A sample of a $3mm$ thick plate made from a special purpose composite material comprising silicon carbide particles dispersed in an aluminium matrix has been successfully drilled using an ultrasonically excited diamond impregnated cutter which was substituted for the conventional twist drills used for the previously described drilling trials in metals. The composite material under test proved to be very brittle and extremely abrasive. Attempts to drill through the sample with a new solid carbide drill produced a very ragged hole with a huge exit crater. The drill was severely blunted to such a degree that further attempts to drill completely through the plate failed.

In contrast, the material proved to be very easy to drill with the diamond impregnated cutter excited ultrasonically and rotated at only a very slow speed typically 40 to $125rpm$, and there was very little exit damage apparent. Diamond drills for penetrating such materials would usually be rotated at very high speeds in excess 20, 000*rpm* and would require the use of very expensive, special purpose machine tools.

Reduction in axial force with longitudinal excitation was not observed, however, in the case of drilling through glass. For this material a special purpose torsional transducer was developed by Dynamics Research Group at Loughborough University.

Fig.5.39 shows an experimental set up while drilling with torsional transducer. A significant reduction in the axial force was measured when drilling glass with the torsional transducer. A very beneficial effect is that, when such a drill is allowed to proceed through the thickness until it emerges on the opposite side, it is found that the area of damage surrounding the exit point of the drill is considerably reduced. In the case of conventional drilling at the same feed rate it is usually found that a substantial area surrounding the exit point of the drill is severely damaged and a large crater is formed. This occurs because, before breakthrough, the stresses surrounding the hole are very large and eventually the fracture strength of the material is exceeded before the tip of the drill has reached the geometric boundary. The greatly reduced damage surrounding the exit point of the drill is independent verification of the substantial reduction in cutting forces resulting from the superposition of ultrasonic vibration on the usual motion of the drill.

Fig. 5.39.

Fig. 5.40 demonstrates a comparison in glass milling with a conventional slot drill. The upper part of the picture shows an unsuccessful attempt of conventional milling of the slot, the lower part is an example of slot produced after superimposing of ultrasonic vibration. All the other parameters of machining regime were the same in both cases.

The beneficial effect of vibration on the cutting process in glass has also been illustrated by measurements of the axial reactive force and penetration

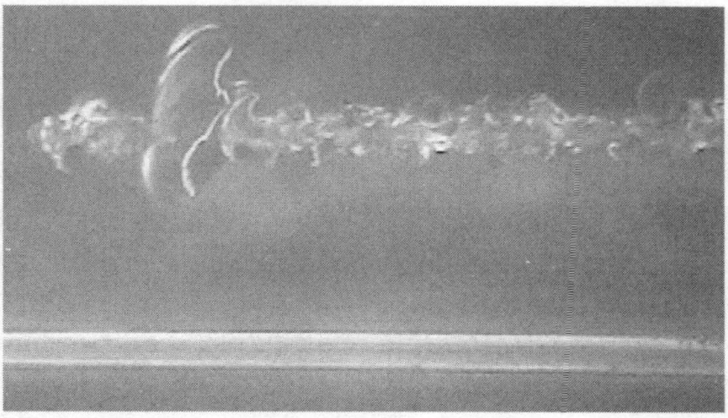

Fig. 5.40.

rates on drills that were not excited externally, which reveal large differences when drilling with and without a lubricant. It has been observed that the presence of a lubricant such as water or oil greatly increased the axial reactive force on the drill because the cutting process can be virtually halted and no penetration is then possible.

A laser vibrometer was used to measure the vibration of the drill in the two cases when it was found that a dry drill developed a high level of vibration as it was penetrating and cutting the glass but, in presence of a lubricant, the stick-slip vibration of the drill was no longer excited and cutting ceased. The higher level of friction in the dry case appears to induce vibration by a stick-slip process which, once initiated, is sustained by the natural resonances in the drill and the maintenance of physical contact by being continuously advanced into the material.

Opposite to the glass drilling, no reduction in axial force was observed when metals were drilled using torsional vibration. Recently, there were published some promising results on drilling assisted by ultrasonic elliptical vibration (Ma et al. (2005)).

5.5 Finite element modelling of tool-material interaction

1. The finite element method (FEM) is a main computational tool for simulation of the tool-workpiece interaction in metal cutting. A detailed review of FE models of cutting can be found in various monographs, e.g. (Childs et al. (2000), Trent & Wright (2000)). Up to now, FE models were used to simulate only conventional turning (CT) processes, e.g. (Ceretti et al. (1999), Anagonye & Stephenson (2002), Pantale et al. (2004)). A number of three-dimensional (3D) FE models of CT were developed and used to study three-dimensional forces in the tool; stress, strain and temperature distributions in the cutting

zone, and chip formation mechanisms (Strenkowski et al. (2002), Pantale et al. (2004)).

The majority of the suggested computational schemes employ the method of chip separation along a predefined surface, "unzipping" adjoining finite elements in the initial discretization of the area, hence reducing flexibility (and adequacy) of the analysis. Only a few schemes use other techniques, such as elements deletion based upon penetration of cutting tool tip into the elements of workpiece (Anagonye & Stephenson (2002)), adaptive remeshing of elements in the workpiece (Pantale et al. (2004)), and combination of both manual deletion and remeshing (Ceretti et al. (1999)).

A FE analysis of heat generation in machining of isotropic materials was conducted in (Ramesh et al. (1999)) in order to study the effects of the convective heat transfer. A different approach, using an orthogonal FE model coupled with an analytical 3D model of cutting, was developed in (Strenkowski et al. (2002)) to predict a chip flow angle and three-dimensional forces in the tool. Another 3D model was introduced in (Pantale et al. (2004)) that took into account dynamic effects, thermo-mechanical coupling, constitutive damage law and contact with friction in order to study the cutting forces and plastic deformation.

In order to understand mechanics of the tool-chip interaction with superimposed ultrasonic vibration, and to analyze distributions of stresses and strains in the cutting region, the process of heat transfer in the workpiece material and in the cutting tool as well as to estimate cutting forces, a 2D FE model was developed (Mitrofanov et al. (2003)). In that model, the steady state UAT process as well as a single cycle of ultrasonic vibration were numerically analyzed. This initial, purely mechanical model was further improved, resulting in a transient, thermomechanically coupled one for both UAT and CT (Mitrofanov et al. (2004)). The improved model also allowed replicating the multifold reduction in cutting forces in UAT numerically that was previously obtained only experimentally.

A detailed analysis of the thermomechanics of the UAT process was further conducted in (Mitrofanov, Babitsky & Silberschmidt (2005)). In that paper, the influence of various thermal factors, such as convective heat transfer and gap conductance, on the UAT process was analyzed. A three-dimensional FE model of UAT was developed as an extension of the 2D model and presented in (Ahmed et al. (2005), Mitrofanov, Babitsky & Silberschmidt (2005)). The use of the 3D model allowed the study of 3D chip formation predicting distributions of stresses, strains, cutting forces and temperatures in the workpiece and cutting tool.

Effects of lubrication and friction on the chip shapes and cutting forces in UAT and CT were studied (Mitrofanov, Ahmed, Babitsky & Silberschmidt (2005)) using both 2D and 3D FE models. It was demonstrated that the chip formed in a lubricated UAT process was significantly more curled than the one produced in dry cutting. The developed 3D model was further employed to analyze the material response to ultrasonic vibration loading in (Ahmed et al. (2006)). Thermally induced strains and residual stresses in the surface layer of the workpiece machined with UAT and CT were investigated. The current section presents a description of the 3D model of UAT as well as computational results obtained with this model.

2. The described FE model utilizes the MSC MARC/MENTAT FE code (*MSC.Marc User's Guide* (2005)) and is based on the updated Lagrangian analysis procedure that provides a transient analysis for an elasto-plastic material and accounts for the frictional contact interaction between the cutter and workpiece as well as material separation in front of the cutting edge. Below some general features of this computational scheme are discussed.

A 3D model for the orthogonal turning process, i.e. the one where the tool edge is normal to both the cutting and feed directions, is considered. The dimensions of the part of the workpiece modeled in simulations are $2.5mm$ in length by $0.5mm$ in height by $0.4mm$ in depth with the uncut chip thickness $0.1mm$ (Fig.5.41). The simulated cutting speed is $18.6m/min$. The relative movement of the workpiece and cutting tool in CT is simulated by the translation of the tool with the constant velocity.

Harmonic oscillation with vibration amplitude of $15\mu m$ (peak-to-valley) is then superimposed on this movement in the tangential direction (i.e. along X-axis in Fig.5.41) in order to model ultrasonic vibration of the tool. The vibration speed is several times greater than the chosen translational speed of the tool leading to the periodic separation of the tool from the newly formed chip, thus transforming the process of cutting into one with a vibro-impact interaction between the tool and chip. Various stages of a vibration cycle are described in detail in (Babitsky, Mitrofanov & Silberschmidt (2004)).

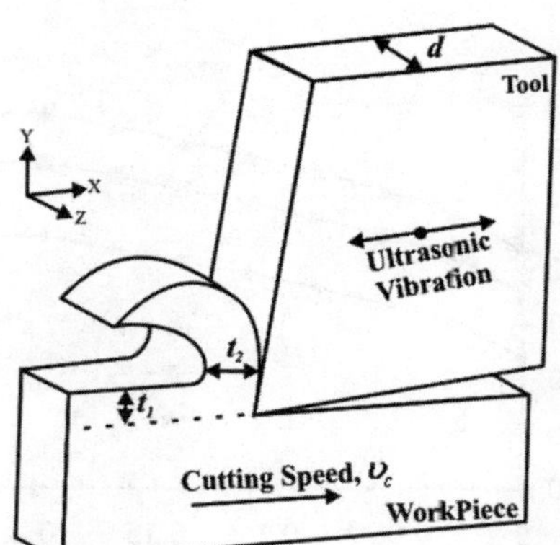

Fig. 5.41.

The developed FE model is fully thermomechanically coupled in order to properly reflect interconnection between thermal and mechanical processes in the cutting zone: excessive plastic deformation and friction at the tool-chip interface lead to high temperatures generated in the cutting region. This results in thermal stresses and volumetric expansion as well as affects material properties of the workpiece, such as thermal conductivity and specific heat. More details on thermomechanical processes in UAT in comparison to CT can be found in (Mitrofanov et al. (2004)).

The mechanical behaviour of the workpiece material (aged Inconel 718) at high strains, strain rates and elevated temperatures can be adequately described by the Johnson-Cook material model (Johnson & Cook (1985)), accounting for the strain-rate sensitivity, that is employed in simulations.

$$\sigma_{\mathrm{Y}} = \left(A + B\varepsilon_{\mathrm{p}}^{n}\right)\left(1 + C\ln\left(\frac{\dot{\varepsilon}_{\mathrm{p}}}{\dot{\varepsilon}_{0}}\right)\right)\left(1 - T^{*m}\right) \tag{5.101}$$

Here $A = 1241MPa$, $B = 622MPa$, $I = 0.0134$, $n = 0.6522$, $T^{*} = (T - T_{\mathrm{room}})/(T_{\mathrm{melt}} - T_{\mathrm{room}})$, ε_{p} and $\dot{\varepsilon}_{\mathrm{p}}$ are plastic strain and a strain rate, T_{room} and T_{melt} are the room and melting temperatures, respectively. A term T^{*m} is assumed to be negligible since thermal softening of Inconel 718 is insignificant (less then 5%) within the temperature range that is obtained in both FE simulations and infrared thermography experiments for chosen cutting parameters. This material model, utilised by various researchers (see, e.g. (Maudlin & Stout (1996), Ng et al. (2002)) as well as in the previous work on modelling of UAT (Mitrofanov et al. (2004)) has been modified to prevent unrealistically high stress values at high strains, so that maximum stress values are now limited to ultimate tensile strength of Inconel 718 at corresponding strain rates (that can reach $10^{5} 1/s$ for standard cases). Fig.5.42 shows an effect of strain rate on plastic behaviour of Inconel 718.

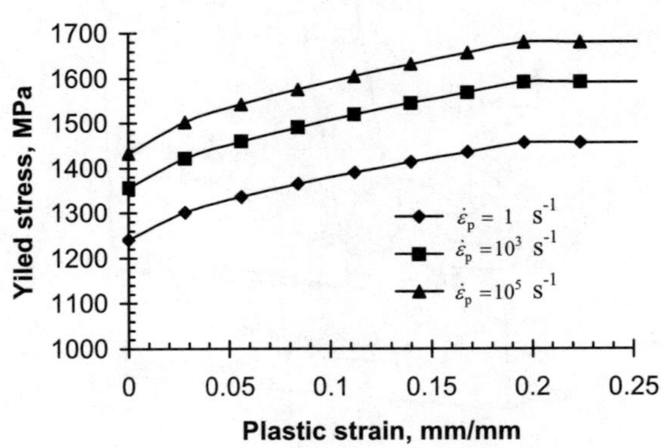

Fig. 5.42.

Fig.5.43 shows a finite-element mesh for the workpiece and cutting tool in the 3D model. In nearly all transient models developed to simulate metal cutting, an initial cut was introduced in the workpiece as the starting point for separation that leads to chip formation. The described FE model does not utilize such a cut. Hence, a more adequate study of chip formation from its very beginning to a fully formed chip is possible. This type of modelling also provides the benefit of using the same model of the workpiece for different shapes of the cutting tool thus forming a basis for optimization studies for the tool shape.

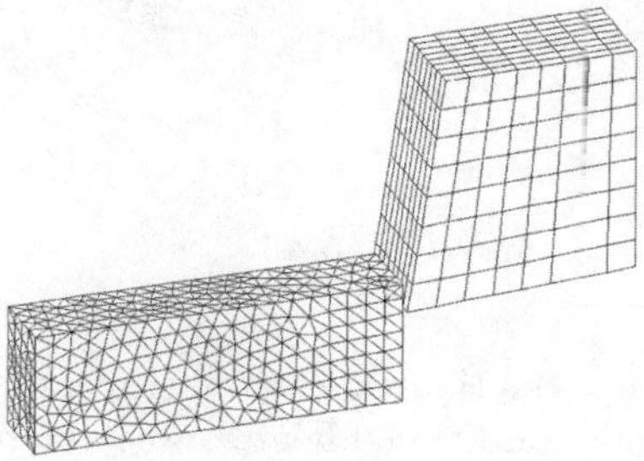

Fig. 5.43.

During the simulation, elements in the process zone can become highly distorted, and hence are no more appropriate for calculation. Automatic remeshing/rezoning is used in the workpiece and chip to replace those distorted elements with ones of better shape. Fig.5.44 shows the chip formed as a result of successful implementation of remeshing/reasoning.

High contact stresses are generated at the tool-chip interface leading to significant friction forces. The classical Coulomb model is unable to adequately reflect friction processes under these conditions resulting in unrealistically high friction force. Hence, the shear friction model was chosen for simulations, where the friction force depends on the fraction of the equivalent stress of the material and not the normal force as in the Coulomb model. Thus, friction stress is introduced in the following form:

$$\sigma_{\mathrm{fr}} \leq -\vartheta \frac{\bar{\sigma}}{\sqrt{3}} \frac{2}{\pi} \operatorname{sgn}(v_{\mathrm{r}}) \arctan(\frac{v_{\mathrm{r}}}{v_{\mathrm{cr}}}) \tag{5.102}$$

Here $\bar{\sigma}$ is the equivalent stress, v_{r} is a relative sliding velocity, v_{cr} is a critical sliding velocity below which sticking is simulated, ϑ is a friction coefficient.

Two contact conditions are considered at the tool-chip interface: (a) a frictionless contact, and (b) a contact with friction (coefficient of friction $\vartheta = 0.5$). The former case corresponds to the idealised condition where heat generation occurs only due to plastic deformation processes; it can be considered as an

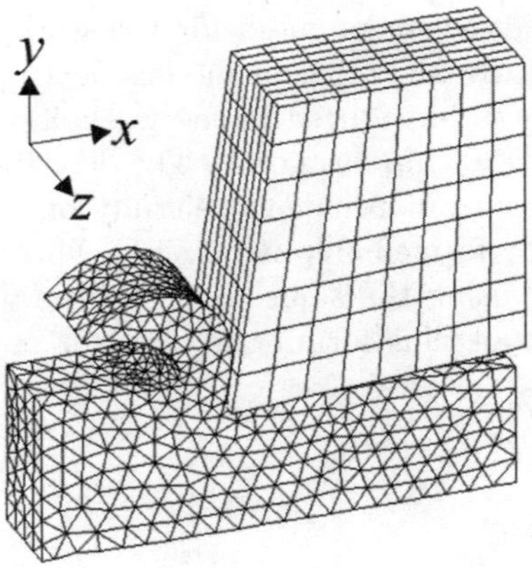

Fig. 5.44.

extreme case of friction reduction due to lubrication. Case (b) models dry cutting conditions, with an additional heat source due to friction between the tool surface and separated workpiece material.

The described model possesses a number of advantages compared to 2D models. Various 3D effects in turning, such as non-orthogonal / oblique chip formation, as well as the influence of the tool geometry on process parameters (cutting forces and stresses generated in the workpiece material) could be studied. The 3D model also permits to investigate the effect of various vibration directions of the cutting tip in UAT on the cutting process, and could serve as an optimization tool for the UAT technology. Various combinations of vibration directions can also be studied numerically, whereas their experimental implementation can be extremely laborious, as it may require new types of ultrasonic transducers and mounting systems to be designed.

Furthermore, the three-dimensional FE formulation helps to perform a direct comparison of numerical results with experimental tests for oblique cutting, thus not requiring any changes to a standard cutting setup. This is important since the FE results, e.g. cutting forces, based on the 2D model can be directly compared only with those from orthogonal turning tests. Such turning tests can be very difficult to implement for intractable materials, as they require special setup arrangements or specific workpiece shapes, e.g. thin tubes.

In addition, the 3D model does not need as many assumptions as the 2D model, for example, the workpiece thickness is introduced here explicitly as compared to its artificial introduction in 2D. The 3D model also accounts for chip expansion in the lateral dimension (along Z-axis in Fig.5.41) that was impossible in the 2D model and led to generation of excessive stresses in the cutting region. Finally, the real geometry of the cutting tool can be studied with the 3D model, thus allowing the analysis of the influence of the tool sharpness and wear on the cutting process.

3. All variants of numerical (finite element) simulations below are performed for two cutting techniques (CT and UAT) with identical parameters so that results for CT could serve as a reference for UAT. Noticeable differences are observed between chip shapes obtained in FE simulations with and without friction for both CT and UAT. The radius of curvature of the chip under the frictionless contact condition (Fig.5.45,a) at the tool-chip interface is approximately 2.5 smaller than that for the contact with friction (Fig.5.45,b) for both cutting techniques; that is supported by turning experiments with different lubricants, showing higher values of the radius of curvature for dry turning.

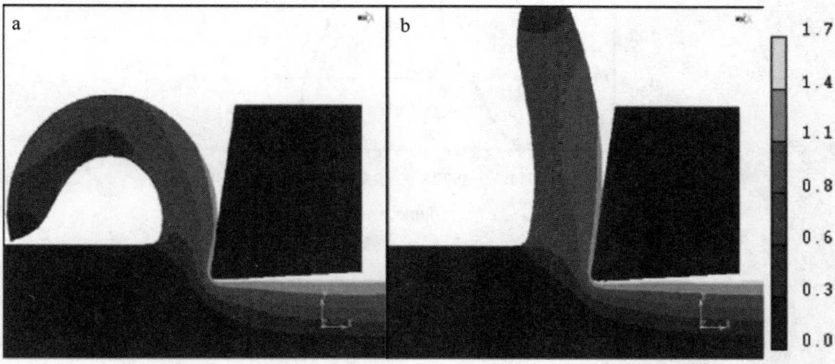

Fig. 5.45.

The chip shape in the steady-state part of the machining process depends on the type of the cutting procedure (CT vs. UAT) and especially on the amplitude of the tool vibration. As the amplitude of the tool vibration grows, the chip becomes thinner and less curved, leading to an increase in the contact length between the tool and chip when the tool is in full contact with the chip in simulations of UAT.

4. A significant difference in forces acting on the cutting tool is obtained for UAT and CT. In simulations of CT the force acting on the cutting tool remains practically unchanged, whilst in UAT simulations it changes drastically within each cycle of vibration. The cutting force grows steadily from the moment of the first contact between the vibrating tool and formed chip until it reaches the maximum level at the point of maximum penetration.

This maximum force is by 30% higher than the average force in CT. The force magnitude then starts to decline at the unloading stage until it vanishes when the cutter separates from the chip and starts to move away from it. The forces stay close to zero level until the cutter comes into contact with the chip again at the next cycle of ultrasonic vibration. The comparison of values from Fig.5.46 shows that the force in UAT (averaged over the vibration cycle) is about 40% of that in CT. Low-level fluctuations of the cutting force at the withdrawal and approach stages of the cycles are explained by the remaining contact between the cutter and freshly formed workpiece surface, as well as by the numerical error involved in FE simulations.

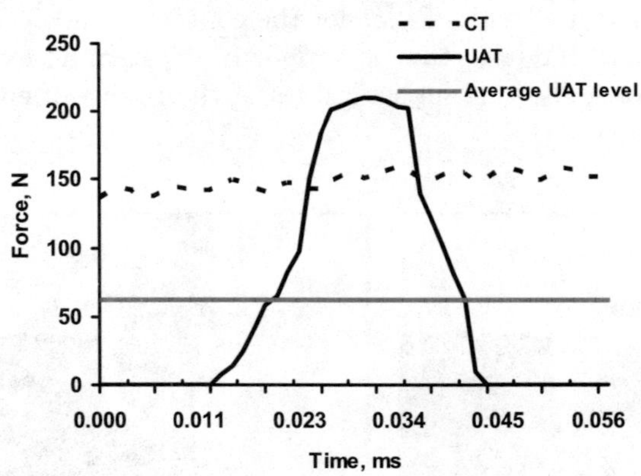

Fig. 5.46.

FE simulations were conducted to study the effect of vibration amplitude and frequency on the force acting on the cutting tool in UAT. An increase in the peak force with an increase in amplitude is observed, however, an average cutting force decreases. This is explained mainly by the increase of the relative velocity of the cutting tool resulting in a shorter period of contact with a chip for a given frequency (i.e. period) of ultrasonic vibration. A decrease of 28% in the average force is recorded for an increase in the amplitude from $7.5\mu m$ to $15\mu m$, and a further 24% decrease is observed when the amplitude increases from $15\mu m$ to $30\mu m$ (Fig.5.47).

In the analysis of the frequency's influence on the cutting force, the peak force remains nearly at the same level in all studied cases, but the average force decreases by 21% for a frequency increase from $10kHz$ to $20kHz$ (Fig.5.48). A further 26% decline in the average force is observed when vibration frequency increases from $20kHz$ to $30kHz$. This trend can be explained by the increase in the velocity of the vibrating cutting tool, which amplitude is proportional to both vibration amplitude and frequency.

5. High temperatures in the cutting region are generated due to the plastic deformation of the workpiece material and friction between the chip and workpiece and cutting tool. Maximum temperature levels in the process zone and chip are invariably higher in UAT simulations. The highest temperatures

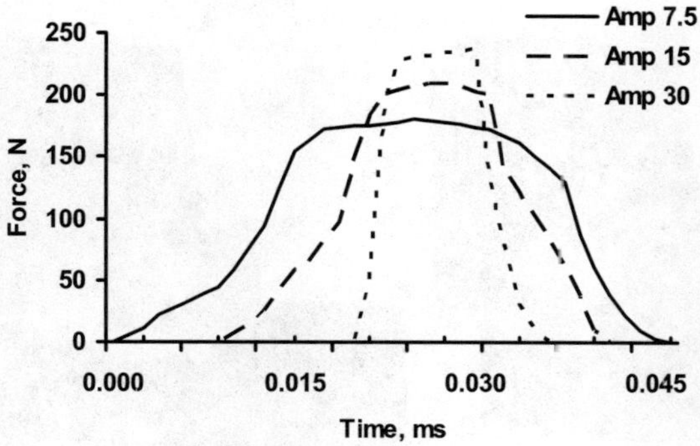

Fig. 5.47.

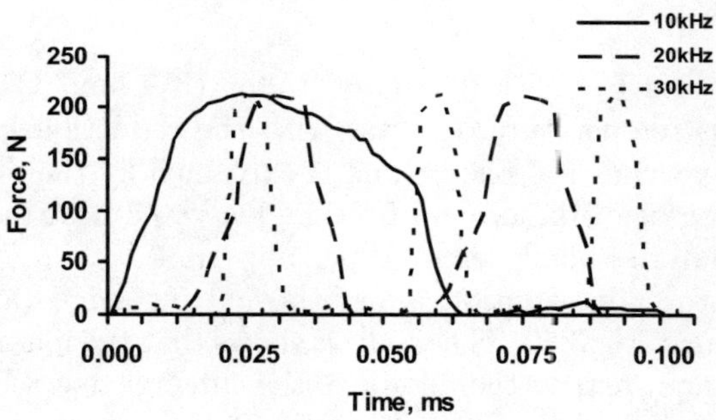

Fig. 5.48.

are registered along the contact area at the tool-chip interface in both UAT and CT models. Fig.5.49 shows the temperature distributions in the cutting regions in CT (a) and UAT (b) simulations ($\vartheta = 0$, $t_1 = 0.1mm$, $t = 6ms$). The analysis of the temperature evolution in the cutting tool shows that the tool temperature in CT grows faster than that in UAT, in spite of the final temperature being higher for UAT (largely due to the effect of the additional factor linked to dissipation of the vibration energy).

The effect of the feed rate on the cutting-tool temperatures in CT and UAT can be studied by means of FE simulations. The feed rate, that is the distance covered by the tool in the feed direction at each revolution of the workpiece, corresponds to the uncut chip thickness t_1 in the FE model. In simulations t_1 is reduced from $0.1mm$ to $0.05mm$ ($\vartheta = 0.5$). Such a reduction leads to a decrease in the maximum temperature levels in the cutting region from 440°C to 400°C and from 410°C to 375°C for UAT and CT, respectively. A drop in the cutting-tool temperature is also observed: it diminishes from

Fig. 5.49.

155°C to 125°C and from 130°C to 100°C for UAT and CT ($t = 2.5ms$), respectively. This temperature decrease with the reduction in the feed rate reproduces our experimental results and is explained by the decrease both in the work of plastic deformation and friction-generated heat due to a reduced amount of the material being removed per unit time.

The temperature distribution along the cutting edge (along Z-axis in Fig.5.41) is shown at Fig.5.50. The analysis shows that the maximum temperature is reached somewhere in the middle of the cutting edge, with insignificant drops towards the ends of the cutting length. This result can be attributed to the convective heat transfer from the tool surface into the environment. This kind of the distribution is observed throughout the simulation time, with the absolute values of the tool temperature growing with time due to the frictional heating and contact heat transfer between the chip and workpiece surfaces.

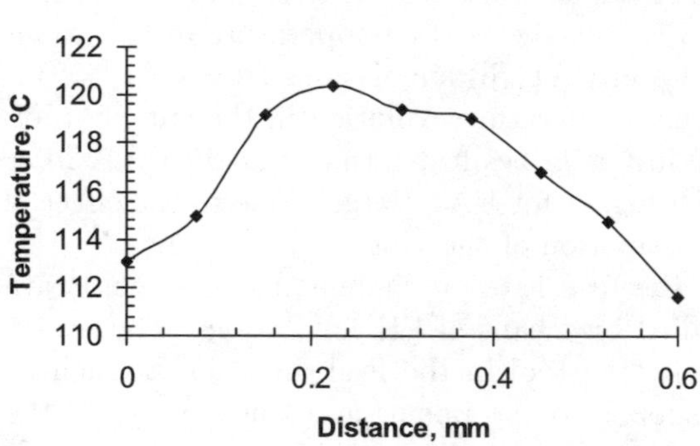

Fig. 5.50.

Bibliography

I quote from others only the better to express myself.

Michel Eyquem de Montaigne (1533-1592)

Abramov, O. (1972), *The crystallisation of metals in an ultrasonic field*, Metallurgia, Moscow. (in Russian).

Agranat, B., ed. (1974), *Ultrasonic technology*, Metallurgia, Moscow. (in Russian).

Ahmed, N., Mitrofanov, A. V., Babitsky, V. I. & Silberschmidt, V. V. (2005), Finite element modeling of ultrasonically assisted turning, *in* '8th CIRP International Workshop on Modeling of Machining Operations', Chemnitz, Germany.

Ahmed, N., Mitrofanov, A. V., Babitsky, V. I. & Silberschmidt, V. V. (2006), 'Analysis of material response to ultrasonic vibration loading in turning inconel 718', *Materials Science and Engineering* **A 424**(1-2).

Anagonye, A. & Stephenson, D. (2002), 'Modelling cutting temperatures for turning inserts with various tool geometries and materials', *Trans. ASME. Journal of manufacturing science and engineering* **124**(3).

Ananiev, I. (1946), *Handbook for calculating natural modes of elastic systems*, Gostechizdat, Moscow. (in Russian).

Andronov, A. & Maier, A. (1946), 'Simple time-delayed linear systems', *Automatica and telemechanica (Automatics and telecommunications)* **7**(2–3). (in Russian).

Andronov, A., Vitt, A. & Khaikyn, S. (1966), *Theory of oscillations*, Pergamon Press, New York. (Translation from 2nd Russian ed., Fizmatgiz, Moscow, 1959).

Andronov, V. (1967*a*), 'Investigations of a hard indenters penetration into granular media when either static or subjected to vibration', *Mechanika tverdogo tela (Mechanics of solids)* (3). (in Russian).

Andronov, V. (1967*b*), 'On the pseudo transformation of dry friction into viscous friction', *Mechanika tverdogo tela (Mechanics of solids)* (2). (in Russian).

Andronov, V. (1970), 'A method of defining friction force and the corresponding coefficient of friction', *Machinovedenie (Machine science)* (4). (in Russian).

Andronov, V. (1975), 'Forced vibration in systems with transformed dry friction', *Machinovedenie (Machine science)* (5). (in Russian).

Arkhangelsky, M. (1963), 'On the transformation of surface ultrasonic vibration into rotational and translational body motion', *Acoustical journal* **9**(3). (in Russian).

Arkhangelsky, M. (1969), 'The reduction of dynamic dry friction by ultrasonic vibration', *Mechanika tverdogo tela (Mechanics of solids)* (2). (in Russian).

Astashev, V. (1968), 'Dynamics of vibro-impact system with distributed parameters', *Mechanika machin (Machine mechanics)* (17–18). (in Russian).

Astashev, V. (1971), 'The dynamics of an oscillator impacting a limiter', *Machinovedenie (Machine science)* (2). (in Russian).

Astashev, V. (1972), 'The tuning of ultrasonic machines whilst under load', *Stanki i instrumenti (Machine-tools and tools)* (10). (in Russian).

Astashev, V. (1978), 'Matching an oscillating system with drive and nonlinear load', *Machinovedenie (Machine science)* (3). (in Russian).

Astashev, V. (1981), 'Calculation of rod concentrators with a nonlinear load', *Soviet Physics-Acoustics* **27**(6).

Astashev, V. (1983), 'The influence of ultrasonic vibration on plastic deformation processes', *Machinovedenie (Machine science)* (2). (in Russian).

Astashev, V. (1992), 'Effect of ultrasonic vibrations of a single point tool on the process of cutting', *Journal of machinery manufacturing and reliability* **5**(3).

Astashev, V. & Babitsky, V. (1972), 'Resonant oscillations of a visco-elastic rod with a limiter', *Mechanika tverdogo tela (Mechanics of solids)* (4). (in Russian).

Astashev, V. & Babitsky, V. (1974), 'Vibro-impact interactions of visco-elastic rods', *Machinovedenie (Machine science)* (5). (in Russian).

Astashev, V. & Babitsky, V. (1981), 'The efficiency of resonantly tuned ultrasonic machines', *Machinovedenie (Machine science)* (6). (in Russian).

Astashev, V. & Babitsky, V. (1982*a*), 'Methods of tuning of the resonant machines', *Machinovedenie (Machine science)* (1). (in Russian).

Astashev, V. & Babitsky, V. (1982*b*), 'Methods of ultrasonic machine efficiency increase', *Stanki i instrumenti (Machine-tools and tools)* (3). (in Russian).

Astashev, V. & Babitsky, V. (1998), 'Ultrasonic cutting as nonlinear (vibro-impact) process', *Ultrasonics* **36**(1–5).

Astashev, V., Babitsky, V. & Hertz, M. (1973), 'To the synthesis of autoresonant systems', *Vibro engineering* **20**(3). (in Russian).

Astashev, V., Babitsky, V. & Hertz, M. (1977), 'Adaptive control system for ultrasonic machines', *Pribori i sistemi upravleniya (Instruments and control systems)* (2). (in Russian).

Astashev, V., Babitsky, V. & Kolovsky, M. (2000), *Dynamics and control of machines*, Springer, Berlin.

Astashev, V. & Hertz, M. (1973), 'The self-sustained vibration of a visco-elastic rod with a limiter under the influence of a time-delayed force', *Machinovedenie (Machine science)* (5). (in Russian).

Astashev, V. & Hertz, M. (1976*a*), 'The excitation and stabilisation of resonant vibration in ultrasonic rod systems', *Acoustical journal* **22**(2). (in Russian).

Astashev, V. & Hertz, M. (1976*b*), 'A system that automatically tunes an ultrasonic machine', *Electrophysichiskie i electrochemichiskie methodi obrabotki (Electrophysical and electro-chemical methods of machining)* (6). (in Russian).

Astashev, V. & Hertz, M. (1978), 'To the theory of vibration transportation', *Mechanika tverdogo tela (Mechanics of solids)* (1). (in Russian).

Astashev, V., Hertz, M. & Tresviatsky, A. (1981), 'Usage of a focused ultrasonic for contactless measurements', *Acoustical journal* **27**(5). (in Russian).

Astashev, V. & Sakaian, A. (1967), 'Experimental investigation of ultrasonic machine dynamics', *Machinovedenie (Machine science)* (4). (in Russian).

Astashev, V. & Turich, V. (1986), The surface quality that results when ultrasonic vibration is superimposed on to a drawing process, *in* 'Increase in efficiency of drawing', Rigas politechnical institute. (in Russian).

Averianova, V., Diachenko, P. & Mizrokhy, U. (1962), Investigation of solid dispersion under the application of ultrasonics, *in* 'Friction and wear in machines, issue 15', USSR Academy of Sciences, Moscow. (in Russian).

Averianova, V. & Milovidov, A. (1964), 'An investigation of ultrasonic machining', *Advances in scientific photography* **9**. (in Russian).

Babikov, O. (1960), *Ultrasonics and its industrial applications*, Consultants Bureau.

Babitsky, V. (1971), 'Parametric oscillations of vibro-impact systems', *Machinovedenie (Machine science)* (1). (in Russian).

Babitsky, V. (1995), 'Autoresonant mechatronic systems', *Mechatronics* **5**(5).

Babitsky, V. (1998), *Theory of vibro-impact systems and applications*, Springer, Berlin. (Revised translation from Russian, Nauka, Moscow, 1978).

Babitsky, V., Astashev, V. & Kalashnikov, A. (2004), 'Autoresonant control of nonlinear mode in ultrasonic transducer for machining applications', *Ultrasonics* **42**(1–9).

Babitsky, V., Kalashnikov, A., Meadows, A. & Wijesundara, H. (2003), 'Ultrasonically assisted turning of aviation materials', *Journal of materials processing technology* **132**(1–3).

Babitsky, V., Kalashnikov, A. & Molodtsov, F. (2004), 'Autoresonant control of ultrasonically assisted cutting', *Mechatronics* **14**(1).

Babitsky, V. & Krupenin, V. (2001), *Vibration of strongly nonlinear discontinuous system*, Springer, Berlin. (Revised translation from Russian, Nauka, Moscow, 1985).

Babitsky, V., Mitrofanov, A. & Silberschmidt, V. (2004), 'Ultrasonically assisted turning of aviation materials: Simulations and experimental study', *Ultrasonics* **42**(1–9).

Babitsky, V. & Tresviatsky, A. (1976), 'Forced oscillations of viscous-elastic beam with lumped mass between limiters', *Machinovedenie (Machine science)* (2). (in Russian).

Bancevicius, R. & Ragulskis, K. (1981), *Vibromotors*, Mokslas, Vilnius. (in Russian).

Biderman, V. (1978), *Theory of mechanical oscillations*, Vishaya skola, Moscow. (in Russian).

Blank, D. (1961), 'Glasbearbeitung durch stosslappen bei ultraschallfrequenz', *Glastechn. Bear* **34**(11).

Blekhman, I. (1999), *Vibrational mechanics*, World scientific, Syngapore. (Translation from Russian, Nauka, Moscow, 1994).

Blekhman, I., ed. (1979), *Vibration in engineering, Handbook, v. 2: Nonlinear system vibration*, Machinostroenie, Moscow. (in Russian).

Blitz, J. (1971), *Ultrasonics: methods and applications*, Butterworths, London. (Translation from Russian, Nauka, Moscow, 1994).

Bogoliubov, N. & Mitropolskii, Y. (1961), *Asymptotic methods in the theory of nonlinear oscillations*, Gordon & Breach, New York. (Translation from 2nd Russian ed., Fizmatgiz, Moscow, 1958).

Bogomolov, S. & Simson, E. (1981), 'Optimal design of ultrasonic concentrators', *Acoustical journal* **27**(4). (in Russian).

Boothroyd, G. & Knight, W. (1989), *Fundamentals of machining and machine tools*, Marcel Dekker, New York.

Borst, R. (2003), 'Numerical aspects of cohesive-zone models', *Engineering fracture mechanics* **70**.

Bronshtein, I. & Semendyayev, K. (1998), *Handbook of mathematics*, Springer, Berlin.

Butkovsky, A. (1977), *Structural theory of continual systems*, Nauka, Moscow. (in Russian).

Butkovsky, A. (1979), *Characteristics of distributed parameters systems*, Nauka, Moscow. (in Russian).

Ceretti, E., Lucchi, M. & Altan, T. (1999), 'FEM simulation of orthogonal cutting: serrated chip formation', *Journal of materials processing technology* **95**.

Chen, W. (2000), 'Cutting forces and surface finish when machining medium hardness steel using cbn tools', *International journal of machine tools and manufacture* **40**.

Childs, T., Maekawa, K., Obikawa, T. & Yamane, Y. (2000), *Metal machining. Theory and applications*, Arnold, London.

Cornier, A. (1992), 'Turning assisted by non-conventional processes: ultrasonic assistance, high pressure water jet assistance. influence on productivity and workpiece quality', *Bull. cercle etud. metaux* **16**(4).

Crawford, A. (1955), *Ultrasonic engineering*, Butterworths, London.

Dam, H., Quist, P. & Schreiber, M. (1995), 'Productivity, surface quality and tolerances in ultrasonic machining of ceramics', *Journal of materials processing technology* **51**.

Daniels, H. (1965), 'Ultrasonic welding', *Ultrasonics* (3).

David, J. & Cheeke, N. (2002), *Fundamentals and applications of ultrasonic waves*, CRC Press.

Dawson, G., Winsper, C. et al. (1970*a*), 'Application of high- and low-frequency oscillations to the plastic deformation of metals. part 1: Discussion', *Metal form* **37**(8).

Dawson, G., Winsper, C. et al. (1970*b*), 'Application of high- and low-frequency oscillations to the plastic deformation of metals. part 2: Review of literature', *Metal form* **37**(9).

Deng, G., Li, S. et al. (1993), 'Ultrasonic vibration drilling of small-diameter deep holes of difficult-to-machine stainless steel: theoretical analysis and comparative experimental research', *ASME Production engineering division (publication)* **64**.

Devine, J. (1985), *Ultrasonically assisted metal removal*, American Society for Metals, Fabrication of composite materials: Source Book.

Diachenko, P., Mizrokhy, U. & Averianova, V. (1959), Some aspects of the ultrasonic machining of metals, *in* 'Application of ultrasonics in industry', Mashgiz, Moscow. (in Russian).

Donskoy, A., Keller, O. & Kratish, G. (1968), *Ultrasonic electro-technological equipment*, Energia. (in Russian).

Drews, P. (1970), 'Untersuchungen zum ultraschallpunktschweissen von metallen', *Schweissen und schneiden* **22**(2).

Eaves, A., Smith, A., Waterhouse, W. & Sansome, D. (1975), 'Review of the application of ultrasonic vibrations to deforming metals', *Ultrasonics* **13**(4).

Evan-Iwanowski, R. (1976), *Resonance vibration in mechanical systems*, Elsevier, New York.

Farrer, J. (1948), 'Method of abrading', *Patent of Great Britain* (602801).

Feodosiev, V. (1974), *Strength of materials*, Nauka, Moscow. (in Russian).

Floyd, T. (1992), *Electronic devices*, Merrill, New York.

Frederick, J. (1965), *Ultrasonic engineering*, John Wiley & Sons.

Fridman, H. & Levensque, P. (1959), 'Reduction of static friction by sonic vibrations', *Journal of applied physics* **30**(10).

Frolov, K., ed. (1995), *Vibration in engineering, Handbook, v. 6: Vibration and impact protection*, 2 edn, Machinostroenie, Moscow. (in Russian).

Ganeva, L., Goliamina, I. & Margolin, V. (1981), 'Mechanical resistance during the ultrasonic machining of brittle materials', *Acoustical journal* **27**(5). (in Russian).

Gao, G., Zhao, B., Jiao, F. & Liu, C. (2002), 'Research on the influence of the cutting conditions on the surface microstructure of ultra-thin wall parts in ultrasonic vibration cutting', *Journal of materials processing technology* **129**.

Gershgal, D. & Freedman, V. (1976), *Ultrasonic technological equipment*, Energia, Moscow. (in Russian).

Gilmore, R. (1991), 'Ultrasonic machining: a case study', *Journal of materials processing technology* **28**.

Goetze, D. (1956), 'Effect of vibration amplitude, frequency and composition of abrasive slurry on the rate of ultrasonic machining in ketoes tool steel', *Journal of Accoust. Sos. Amer.* **28**(6).

Goetze, D. (1957), 'Effect of pressure between tool tip and workpiece on the rate of ultrasonic machining in ketoes tool steel', *Journal of Accoust. Sos. Amer.* **23**(4).

Goliamina, I., ed. (1979), *Ultrasonics. Small encyclopaedia*, Soviet encyclopaedia, Moscow. (in Russian).

Goliamina, I. & Isaakovich, M. (1973), 'Frequency characteristics of mechanical harmonic oscillator when dipped into a medium', *Acoustical journal* **19**(4). (in Russian).

Gonorovsky, N. (1958), 'The theory of high frequency auto-generators with delayed feedback', *Radiotechnika (Radio engineering)* **13**(5). (in Russian).

Graf, K. (1975), 'Macrosonics in industry: 5, ultrasonic machining', *Ultrasonics* **13**.

Granovsky, G. & Granovsky, V. (1985), *Cutting of metals*, Vishaya skola, Moscow. (in Russian).

Han, L., Xu, W. & Tso, S. (1998), 'Ultrasonically assisted and piezoelectric actuators integrated cutting tool', *Japanese journal of applied physics* **37**(8).

Hansen, H. (1997), 'Optimal design of an ultrasonic transducer', *Structural optimisation* **14**(2–3).

Hartley, M. (1956), 'Ultrasonic machining of brittle materials', *Electronics* (1).

Hayashi, M., Jin, M., Thipprakmas, S., Murakawa, M., Hung, J., Tsai, Y. & Hung, C. (2003), 'Simulation of ultrasonic-vibration drawing using the finite element method (FEM)', *Journal of material processing technology* **140**(1–3).

Hertz, M. (1979), 'Autoresonant excitation of high vibration modes in mechanical systems', *Machinovedenie (Machine science)* (6). (in Russian).

Hocheng, H. & Hsu, C. (1995), 'Preliminary study of ultrasonic drilling of fiber-reinforced plastics', *Journal of Materials Processing Technology* **48**(1–4).

Hocheng, H., Kuo, K. & Lin, J. (1999), 'Machine ability of zirconia ceramics in ultrasonic drilling', *Materials and manufacturing processes* **14**(5).

Holopov, Y. (1972), *Ultrasonic welding*, Machinostroenie, Leningrad. (in Russian).

Isaev, A. & Anokhin, V. (1961), 'Application of ultrasonically vibrating tool to metal cutting', *Vestnik machinostroenia (News in machine building)* (5). (in Russian).

Ishikawa, K., Suwabe, H. et al. (1998), 'Study on combined vibration drilling by ultrasonic and low- frequency vibrations for hard and brittle materials', *Precision engineering* **22**(4).

Izumi, O., Oyama, K. & Suzuki, Y. (1966*a*), 'Effects of superimposed ultrasonic vibration during compressive deformation of metals', *Trans. of the Japan Institute of metals* **7**(3).

Izumi, O., Oyama, K. & Suzuki, Y. (1966*b*), 'On the superimposing of ultrasonic vibration on compressive deformation of metals', *Trans. of the Japan Institute of metals* **7**(3).

Ji, Z., Li-Jiang, W. & Hua, L. (1991), 'A new side-clamped ultrasonic vibration cutting system with two- stage amplification', *International journal of machine tools and manufacture* **31**(3).

Jin, M. & Murakawa, M. (2001), 'Development of a practical ultrasonic vibration cutting tool system', *Journal of materials processing technology* **113**(1–3).

Johnson, G. & Cook, W. (1985), 'Fracture characteristics of three metals subjected to various strains, strain rates temperatures and pressures', *Engineering fracture mechanics* **2**(1).

Katkovnik, V. (1968), 'Vibro-impact excitation of high-frequency elastic vibration', *Mechanika tverdogo tela (Mechanics of solids)* (5). (in Russian).

Kazantsev, V. (1963*a*), 'The dependence of material stresses on vibration amplitude and compressional forces during ultrasonic machining', *Acoustical journal* **9**(1). (in Russian).

Kazantsev, V. (1963*b*), 'The dependence of ultrasonic machining productivity on cutting regime types', *Stanki i instrumenti (Machine-tools and tools)* (3). (in Russian).

Kazantsev, V., Metchetner, B. & Rozenberg, L. (1966), 'Methods of increasing productivity and accuracy during ultrasonic machining', *Stanki i instrumenti (Machine-tools and tools)* (4). (in Russian).

Kharkevich, A. (1973), Theory of electro-acoustical transducers, *in* 'Selected works', Vol. 1, Nauka. (in Russian).

Kikuchi, Y., ed. (1969), *Ultrasonic transducers*, Corona.

Kim, J. & Choi, I. (1997), 'Micro surface phenomenon of ductile cutting in the ultrasonic vibration cutting of optical plastics', *Journal of materials processing technology* **68**.

Kim, J. & Choi, I. (1998), 'Characteristics of chip generation by ultrasonic vibration cutting with extremely low cutting velocity', *International journal of advanced manufacturing technology* **14**.

Kim, J. & Kwonb, O. (2003), 'Vibration characteristics of piezoelectric torsional transducers', *Journal of sound and vibration* **264**.

Kim, J. & Lee, E. (1996), 'A study of ultrasonic vibration cutting of carbon fibre reinforced plastics', *International journal of advanced manufacturing technology* **12**.

Klocke, F., Demmer, A. & Heselhaus, M. (2004), 'Material removal mechanisms in ultrasonic-assisted diamond turning of brittle materials', *International journal of materials and product technology* (4).

Klocke, F. & Rübenach, O. (2000), 'Ultrasonic-assisted diamond turning of glass and steel', *Industrial diamond review* **60**(586).

Klubovich, V. & Sakevich, V. (1992), 'Vibro-impact interaction of rods', *Journal of machinery manufacturing and reliability* (6).

Klubovich, V. & Vagapov, I. (1991), 'A model of the plastic deformation of a hardening material due to superimposition of ultrasonic vibration', *Proc. of Belorussian Academy of Sciences* **35**(4). (in Russian).

Koepfer, C. (2002), 'Ultrasonic technology helps machine hard materials', *Modern machine shop* **74**(12).

Kolovsky, M. (1999), *Nonlinear dynamics of active and passive systems of vibration protection*, Springer, Berlin.

Komaraiah, M., Manan, M., Reddy, P., Narashimha, P. & Victor, S. (1988), 'Investigation of surface roughness and accuracy in ultrasonic machining', *Precision engineering* **74**(10).

Komaraiah, M. & Reddy, P. (1993), 'Relative performance of tool materials in ultrasonic machining', *Wear* **161**.

Krasovsky, A. (1948), 'About vibration linearisation of some nonlinear systems', *Avtomatika and telemechanika (Automatics and telecommunications)* **9**(1). (in Russian).

Kremer, D. (1995), 'Ultrasonically assisted machining', *Mechanique industrielle et materiaux* **48**(1).

Kubota, M., Tamura, I. & Shimamura, N. (1977), 'Ultrasonic machining with diamond impregnated tool', *Bull. of Japanese society of precision engineering* (11).

Kudinov, V. (1967), *Dynamics of machine-tools*, Machinostroenie, Moscow. (in Russian).

Kumabe, J. (1979), *Vibratory cutting*, Dzikke sjuppan, Tokyo. (in Japanese).

Kumabe, J., Fuchizawa, K., Soutome, T. & Nishimoto, Y. (1989), 'Ultrasonic superposition vibration cutting of ceramics', *Precision engineering nanotechnology* **11**(2).

Kuznetsov, V. & Loschilov, V. (1972), *Ultrasonic welding of metals and plastics*, Machinostroenie, Moscow. (in Russian).

Langenecker, E. (1966), 'Effects of ultrasound on deformation characteristics of metals', *IEEE Trans. on sonics and ultrasonics* **SU-13**(1).

Lependin, L. (1978), *Acoustics*, Vishaya skola, Moscow. (in Russian).

Lin, S. & Zhang, F. (2000), 'Measurement of ultrasonic power and electro-acoustic efficiency of high power transducers', *Ultrasonics* **27**.

Littmann, W., Storck, H. & Wallaschek, J. (2001), 'Sliding friction in the presence of ultrasonic oscillations: superposition of longitudinal oscillations', *Archive of applied mechanics* **71**(8).

Liu, J., Yan, J. & Yang, S. (2001), 'A preliminary study on ultrasonic cutting process for carbon fibre prepreg', *Journal of materials science & technology (China) (USA)* **17**.

Liu, K., Li, X., Rahman, M. & Liu, X. (2004), 'Study of ductile mode cutting in grooving of tungsten carbide with and without ultrasonic vibration assistance', *Int. journal of advance manufacturing technology* **24**, 389–394.

Lucas, M., Graham, G. & Smith, A. (1996), 'Enhanced vibration control of an ultrasonic cutting process', *Ultrasonics* **16**.

Ma, C., Shamoto, E. & Moriwaki, T. (2005), 'Drilling assisted by ultrasonic elliptical vibration', *Key engineering materials* **291–292**.

Maekawa, K., Kubo, A. & Childs, T. (2001), 'A friction model for free machining steels and its applicability to machine ability analysis', *Key engineering materials* **196**.

Maezava, S. (1961), 'Steady forced vibration of unsymmetrical piecewise-linear system', *Bull. JSME* **4**(14).

Maezava, S. & Watanabe, T. (1973), Steady impact vibration of a body having hysteresis collision characteristics, *in* 'Theoretical and Applied Mechanics', Vol. 21, Tokyo University Press.

Magnus, K. (1961), *Schwingungen*, Teubner, Stuttgart.

Mandelstam, L. (1972), *Lectures in oscillation theory*, Nauka, Moscow. (in Russian).

Markov, A. (1962), *Cutting of intractable materials with sonic and ultrasonic vibration*, Machgiz, Moscow. (in Russian).

Markov, A. (1966), *Ultrasonic machining of intractable materials*, Illife Books, London.

Markov, A. (1980), *Ultrasonic machining of materials*, Machinostroenie, Moscow. (in Russian).

Mase, G. (1970), *Theory and problems of continuum mechanics*, McGraw-Hill, New York.

Mason, W. (1939), 'A dynamic measurement of the elastic, electric and piezoelectric constants of rochelle salt', *Phys. rev.* **55**.

Mason, W., ed. (1964), *Physical acoustics, v.1: Methods and devices*, Academic Press, New York.

Maudlin, P. & Stout, M. (1996), Metal cutting simulation of 4340 steel using an accurate mechanical description of material strength and fracture, *in* 'Minerals, metals and materials society'.

Meirovitch, L. (2001), *Fundamentals of vibration*, McGraw-Hill, New York.

Merkulov, L. (1957), 'Ultrasonic concentrator theory', *Acoustical journal* **3**(3). (in Russian).

Merkulov, L. & Kharitonov, A. (1959), 'Theory and calculation of sectional concentrators', *Acoustical journal* **5**(2). (in Russian).

Metal working products, Turning Tools (1998), Sandvik Coromant.

Mitrofanov, A., Babitsky, V. & Silberschmidt, V. (2003), 'Finite element simulations of ultrasonically assisted turning', *Computational material science* **28**(3–4).

Mitrofanov, A., Babitsky, V. & Silberschmidt, V. (2004), 'Finite element modelling of ultrasonically assisted turning of inconel 718', *Journal of materials processing technology* **153-154C**.

Mitrofanov, A., Babitsky, V. & Silberschmidt, V. (2005), 'Thermomechanical finite element simulations of ultrasonically assisted turning', *Computational materials science* **32**(3–4).

Mitrofanov, A. V., Ahmed, N., Babitsky, V. I. & Silberschmidt, V. V. (2005), 'Effect of lubrication and cutting parameters on ultrasonically assisted turning of inconel 718', *Journal of materials processing technology* **162–163**.

Mitzkevich, A. & Pugachev, S. (1979), *Ultrasonic welding and metal coating*, Machinistroenie, Moscow. (in Russian).

Moriwaki, T. & Shamoto, E. (1991), 'Ultra precision diamond turning of stainless steel by applying ultrasonic vibration', *Annals of the CIPR* **40**(1).

Moriwaki, T. & Shamoto, E. (1995), *Ultrasonic elliptical vibration cutting*, Hallwag Publishers.

Moriwaki, T., Shamoto, E. & Inoue, K. (1992), 'Ultra precision ductile cutting of glass by applying ultrasonic vibration', *Annals of the CIPR* **41**(1).

MSC.Marc User's Guide (2005), MSC Software Corporation.

Nashif, A., Jones, D. & Henderson, J. (1985), *Vibration damping*, Wiley, New York.

Neimark, J. (1978), *Dynamical systems and control processes*, Nauka, Moscow. (in Russian).

Neppiras, E. (1956), 'Report on ultrasonic machining', *Metalworking production* **100**(29).

Neppiras, E. (1960), 'Very high energy ultrasonics', *British journal of applied physics* **11**(4).

Neugebauer, R. & Stoll, A. (2004), 'Ultrasonic application in drilling', *Journal of materials processing technology* **149**.

Ng, E., El-Wardany, T., Dumitrescu, M. & Elbestawi, M. (2002), 'Physics-based simulation of high speed machining', *Machining science and technology* **6**(3).

Nikolaev, G. (1962), *The application of ultrasonics in welding*, Gosstroyizdat, Moscow. (in Russian).

Nikolaev, G. & Olshansky, N. (1966), *New methods of metals and plastic welding*, Machinistroenie, Moscow. (in Russian).

Nishimura, G., Inagishima, K. & Shina, T. (1959), 'Ultrasonic mechanical machining, part vii', *J. Fac. Eng. Univ. (Tokyo)* **26**(1).

Nishimura, G. & Shimakawa, S. (1959), 'Ultrasonic mechanical machining, part iv', *J. Fac. Eng. Univ. (Tokyo)* **26**(1).

Nosal, V. & Rimsha, O. (1966), 'Reduction of force and determination of parameters of tube drawing with ultrasonic vibration of a die', *Steel* (2). (in Russian).

Nozdrev, V. & Fedorischenko, N. (1978), *Molecular acoustics*, Vishaya skola, Moscow. (in Russian).

Onikura, H., Ohnishi, O. et al. (1996), 'Effects of ultrasonic vibration on machining accuracy in microdrilling', *International journal of the Japan society for precision engineering* **30**(3).

Palmov, V. (1998), *Vibration of elasto-plastic bodies*, Springer, Berlin. (Revised translation from Russian, Nauka, Moscow, 1976).

Panov, S. & Piskunov, Y. (1966), 'The influence of ultrasonic vibration on drawing processes', *Ultrasonic technology* (3). (in Russian).

Panovko, J. (1960), *Inner damping due to the vibration of elastic systems*, Fizmatgiz, Moscow. (in Russian).

Pantale, O. et al. (2004), '2d and 3d numerical models of metal cutting with damage effects', *Computational methods for applied mechanical engineering* (193).

Pei, Z. & Ferreira, P. (1998), 'Modelling of ductile-mode material removal in rotary ultrasonic machining', *Int. journal of machine tools and manufacture* **38**.

Perepeliatnik, P. (1961), 'Self-sustaining vibration in a generator which contains a delay', *Radiotechnika and electronica* **6**(10). (in Russian).

Petruha, P., Markov, A. & Ustinov, I. (1970), 'Ultrasonic drilling of deep holes', *Vestnik machinostroenia (News in machine building)* (10). (in Russian).

Poduraev, V. (1970), *Machining with vibration*, Machonostroenie, Moscow. (in Russian).

Pohlman, R. & Lehfeldt, E. (1966), 'Influence of ultrasonic vibration on metallic friction', *Ultrasonics* (4).

Popov, E. & Paltov, I. (1960), *Approximate methods of analysis for nonlinear automatic systems*, Fizmatgiz, Moscow. (in Russian).

Puskar, A. (1982), *The use of high-intensity ultrasonics*, Elsevier.

Rabotnov, Y. (1979), *Mechanics of deformable solids*, Nauka, Moscow. (in Russian).

Ramesh, M. et al. (1999), 'Finite element modelling of heat transfer analysis in machining of isotropic materials', *Int. journal of heat and mass transfer* **42**.

Rayleigh, J. (1945), *The theory of sound*, Dover publications, New York.

Reiner, M. (1958), *Reology*, Springer, Goettingen.

Rimsky-Korsakov, A. (1973), *Electro acoustics*, Sviaz, Moscow. (in Russian).

Robinson, A. (1964), 'The application of ultrasonic energy to metal wire drawing', *Wire and wire production* **39**(12).

Rosenberg, L. & Kazantsev, V. (1959*a*), 'Investigation of ultrasonic machining by high-speed photography', *Stanki i instrumenti (Machine-tools and tools)* (5). (in Russian).

Rosenberg, L. & Kazantsev, V. (1959*b*), 'On the physics of ultrasonic machining of the hard materials', *Proc. of the USSR Academy of Sciences* **124**(1). (in Russian).

Rosenberg, L., Kazantsev, V., Makarov, L. & Yahimovich, D. (1964), *Ultrasonic cutting*, Consultants Bureau.

Rozenberg, L., ed. (1969), *Sources of high intensity ultrasound*, Vol. 1, Plenum.

Rozenberg, L., ed. (1971), *High intensity ultrasonic fields*, Plenum.

Rubanik, V. (1969), *The oscillations of quasi-linear delayed systems*, Nauka, Moscow. (in Russian).

Severdenko, V., Klubovich, V. & Stepanenko, A. (1973), *Working of plastic metals with ultrasonic vibration*, Nauka and technika, Minsk. (in Russian).

Severdenko, V., Klubovich, V. & Stepanenko, A. (1976), *Ultrasonics and plasticity*, Nauka and technika, Minsk. (in Russian).

Shamoto, E., Ma, C. & Moriwaki, T. (1999), 'Ultra precision ductile cutting of glass by applying ultrasonic elliptical vibration cutting', *Precision engineering nanotechnology* **1**.

Shamoto, E., Suzuki, N., Moriwaki, T. & Naoi, Y. (2002), 'Development of ultrasonic elliptical vibration controller for elliptical vibration cutting', *Annals CIRP* **51**(1).

Shaw, M. (2005), *Metal cutting principles*, 2 edn, Oxford University Press, New York.

Shuyu, L. (1994), 'Study of the sandwiched piezoelectric ultrasonic torsional transducer', *Ultrasonics* **32**(6).

Shuyu, L. (1996), 'Study on the longitudinal-torsional composite mode exponential ultrasonic horns', *Ultrasonics* **34**(7).

Shuyu, L. (1997), 'Study on the longitudinaltorsional composite vibration of a sectional exponential horn', *The Journal of the acoustical society of America* **102**(3).

Silin, L., Balandin, G. & Kogan, M. (1962), *Ultrasonic welding*, Machgiz, Moscow. (in Russian).

Skelton, C. (1969), 'Effect of ultrasonic vibration on the turning process', *International journal of machine tool design research* **9**.

Skudrzyk, E. (1971), *The foundations of acoustics*, Springer, Wien. (in Russian).

Sokolov, I. & Babitsky, V. (2001), 'Phase control of self-sustained vibration', *Journal of sound and vibration* **248**(4).

Spur, S., Bruker, T. & Holl, S. (1997), 'Ultrasonic machining of ceramics', *Industrial ceramics* **17**(1).

Strenkowski, J., Shih, A. & Lin, J. (2002), 'An analytical finite element model for predicting three- dimentional tool force and chip flow', *Int. journal of machine tools and manufacture* **42**.

Sun, X., Masuzawa, T. & Fujino, M. (1996), 'Micro ultrasonic machining and its applications in mems', *Sensors and Actuators* **A 57**.

Takeyama, H. & Lijima, N. (1998), 'Machinability of glass fibre reinforced plastics and application of ultrasonic machining', *Annals of the CIPR* **37**(1).

Tchechina, L. & Yahimovich, D. (1974*a*), 'Ultrasonic machining by a diamond tool', *Electrophysichiskie i electrochemichiskie methodi obrabotki (Electro-physical and electro-chemical methods of machining)* (3). (in Russian).

Tchechina, L. & Yahimovich, D. (1974*b*), 'Ultrasonic machining using a diamond tool', *Electrophysichiskie i electrochemichiskie methodi obrabotki (Electro-physical and electro-chemical methods of machining)* (2). (in Russian).

Tekeyama, H. & Lijima, N. (1988), 'Machinability of glassfiber reinforced plastics and application of ultrasonic machining', *Annals of CIRP* **37**.

Teodorchik, K. (1952), *Self-sustaining oscillation systems*, Gostechizdat, Moscow. (in Russian).

Teumin, I. (1959), *Ultrasonic oscillating systems*, Machgiz, Moscow. (in Russian).

Thoe, T., Aspinwall, D. & Wise, M. (1998), 'Review on ultrasonic machining', *Int. journal of machine tools and manufacture* **38**(4).

Timoshenko, S. (1955), *Vibration problems in engineering*, D Van Nostrand Co., Toronto.

Tlusty, J. (2000), *Manufacturing processes and equipment*, Prentice-Hall, New York.

Trent, E. & Wright, P. (2000), *Metal cutting*, Butterworth-Heinemann, London.

Tsujino, J. (1985), 'Studies of ultrasonic: welding with two longitudinal systems', *Jap. journal of applied physics* **24**(1).

Tsujino, J., Ueoka, T. et al. (2000), 'One-dimensional longitudinal-torsional vibration converter with multiple diagonally slitted parts', *Ultrasonics* **38**(1).

Vagapov, I. (1987), *Nonlinear effects in ultrasonic machining*, Nauka and technika, Minsk. (in Russian).

Vagapov, I. (1994), 'Resonant characteristics of double-rod oscillating system under load', *Journal of machinery manufacturing and reliability* (3).

Wang, L. & Zhao, J. (1987), 'Influence on surface roughness in turning with ultrasonic vibration tool', *International journal of machine tools and manufacture* **27**(2).

Wang, X., Zhou, M., Gan, J. & Ngoi, B. (2002), 'Theoretical and experimental studies of ultra precision machining of brittle materials with uiltrasonic vibration', *Int. journal of advanced manufacturing technology* **20**.

Weare, N., Antonevich, J. & Monrol, R. (1960), 'Fundamental studies of ultrasonic welding', *Welding journal* **39**.

Weber, H., Herberger, J. & Piltz, R. (1984), 'Turning of machinable glass ceramics with an ultrasonically vibrated tool', *Annals of the CIPR* **33**(1).

Whitehouse, D. (1997), 'Surface metrology', *Measurement science technology* **8**.

Wiltshire, A. & Pilkington, D. (1986), 'The development of the ultrasonic cutting process for aerospace prepreg materials', *Automated composites* .

Winsper, C. & Sansome, D. (1968), 'The influence of oscillatory energy on the stresses during plastic deformation', *Journal of the Institute of metals* **96**.

Xiao, M., Sato, K., Karube, S. & Soutome, T. (2003), 'The effect of tool nose radius in ultrasonic vibration cutting of hard metal', *Int. journal of machine tools and manufacture* **43**.

Ya, G., Qin, H. et al. (2001), 'An experimental investigation on rotary ultrasonic machining', *Key engineering materials* **202–203**.

Zhang, D., Feng, X. et al. (1991), 'Study on the drill skidding motion in ultrasonic vibration microdrilling', *ASME Production engineering division (publication)* **55**.

Zhang, D., Feng, X. et al. (1994), 'Study on the drill skidding motion in ultrasonic vibration microdrilling', *International journal of machine tools and manufacture* **34**(6).

Zhao, J., Wang, L. et al. (1991), 'A new side-clamped ultrasonic vibration cutting system with two- stage amplification', *International journal of machine tools and manufacture* **31**(3).

Zharkov, I. (1986), *Vibrations due to cutting*, Machinostroenie, Leningrad. (in Russian).

Zhixin, J., Jianhua, Z. & Xing, A. (1997), 'Study on a new kind of combined machining technology of ultrasonic machining and electrical discharge machining', *Int. journal of machine tools and manufacture* **37**(2).

Zhou, M., Eow, Y., Ngoi, B. & Lim, E. (2003), 'Vibration-assisted precision machining of steel with PCD tools', *Materials and manufacturing processes* **18**.

Zhou, M., Wang, X., Ngoi, B. & Gan, J. (2002), 'Brittle-ductile transition in the diamond cutting of glasses with the aid of ultrasonic vibration', *Journal of material processing technology* **121**.

Index

Foundations of Engineering Mechanics

Series Editors: Vladimir I. Babitsky, Loughborough University, UK
Jens Wittenburg, Karlsruhe University, Germany

Further volumes of this series can be found on our homepage: springer.com

(Continued from page ii)

Perelmuter, A.V., Slivker, V.I.
Numerical Structural Analysis, 2003
ISBN 3-540-00628-1

Lurie, A.I.,
Analytical Mechanics, 2002
ISBN 3-540-42982-4

Manevitch, L.I., Andrianov, I.V., Oshmyan, V.G.
Mechanics of Periodically Heterogeneous Structures, 2002
ISBN 3-540-41630-7

Babitsky, V.I., Krupenin, V.L.
Vibration of Strongly Nonlinear Discontinuous Systems, 2001
ISBN 3-540-41447-9

Landa, P.S.
Regular and Chaotic Oscillations, 2001
ISBN 3-540-41001-5

Alfutov, N.A.
Stability of Elastic Structures, 2000
ISBN 3-540-65700-2

Astashev, V.K , Babitsky, V.I., Kolovsky, M.Z.
Dynamics and Control of Machines, 2000
ISBN 3-540-63722-2

Kolovsky, M.Z., Evgrafov, A.N., Semenov Y.A., Slousch, A.V.
Advanced Theory of Mechanisms and Machines, 2000
ISBN 3-540-67168-4

Astashev, V.K., Babitsky, V.I.
Ultrasonic Processes and Machines, 2007
ISBN 978-3-540-72060-7

Printing: Krips bv, Meppel
Binding: Stürtz, Würzburg